XIV Conference on Theoretical Nuclear Physics in Italy 2013

Journal of Physics: Conference Series Volume 527

Cortona, Italy
29 – 31 October 2013

ISBN: 978-1-63439-308-9
ISSN: 1742-6588

Printed from e-media with permission by:

Curran Associates, Inc.
57 Morehouse Lane
Red Hook, NY 12571

Some format issues inherent in the e-media version may also appear in this print version.

Printed by Curran Associates, Inc. (2014)

For permission requests, please contact the Institute of Physics
at the address below.

Institute of Physics
Dirac House, Temple Back
Bristol BS1 6BE UK

Phone: 44 1 17 929 7481
Fax: 44 1 17 920 0979

techtracking@iop.org

Additional copies of this publication are available from:

Curran Associates, Inc.
57 Morehouse Lane
Red Hook, NY 12571 USA
Phone: 845-758-0400
Fax: 845-758-2634
Email: curran@proceedings.com
Web: www.proceedings.com

TABLE OF CONTENTS

NUCLEAR ASTROPHYSICS

NUCLEAR PHYSICS WITH ELECTROWEAK PROBES

STRUCTURE OF HADRONS AND HADRONIC MATTER

SPECIAL TOPICS

Journal of Physics: Conference Series **527** (2014) 011001 doi:10.1088/1742-6596/527/1/011001

Preface

This volume contains the invited and contributed papers presented at the 14th Conference on Theoretical Nuclear Physics in Italy held in Cortona, Italy, from 29–31 October, 2013. The meeting was held at the *Palazzone*, an elegant Renaissance Villa, commissioned by the Cardinal Silvio Passerini (1469–1529), Bishop of Cortona, and presently owned by the *Scuola Normale Superiore di Pisa.*

The aim of this biennial Conference is to bring together Italian theorists working in various fields of nuclear physics to discuss their latest results and confront their points of view in a lively and informal way. This offers the opportunity to stimulate new ideas and promote collaborations between different research groups.

The Conference was attended by 46 participants, coming from 13 Italian Universities and 11 Laboratories and *Sezioni of the Istituto Nazionale di Fisica Nucleare* – INFN. The program of the conference, prepared by the Organizing Committee (Ignazio Bombaci, Aldo Covello, Laura Elisa Marcucci and Sergio Rosati) focused on the following main topics:

(i) Few-Nucleon Systems
(ii) Nuclear Structure
(iii) Nuclear Matter and Nuclear Dynamics
(iv) Relativistic Heavy Ion Collisions and Quark-Gluon Plasma
(v) Nuclear Astrophysics
(vi) Nuclear Physics with Electroweak Probes
(vii) Structure of Hadrons and Hadronic Matter.

In the last session of the Conference there were two invited review talks related to experimental activities of great current interest. Giacomo De Angelis from the Laboratori Nazionali di Legnaro spoke about the INFN SPES radioactive ion beam project. Sara Pirrone, INFN Sezione di Catania, gave a talk on the symmetry energy and isospin physics with the CHIMERA detector. Finally, Mauro Taiuti (Università di Genova), National Coordinator of the INFN-CSN3 (Nuclear Physics Experiments), reported on the present status and future challenges of experimental nuclear physics in Italy. We gratefully acknowledge the financial support of INFN who helped make the conference possible.

I Bombaci, A Covello, L E Marcucci, S Rosati

Journal of Physics: Conference Series **527** (2014) 011002 doi:10.1088/1742-6596/527/1/011002

List of Participants

Alba Paolo (Università di Torino)
Becattini Francesco (Università di Firenze)
Bombaci Ignazio (Università di Pisa)
Bonaccorso Angela (INFN Pisa)
Colonna Maria (INFN-LNS Catania)
Coraggio Luigi (INFN Napoli)
Covello Aldo (Università di Napoli)
Di Toro Massimo (Università di Catania)
De Angelis Giacomo (INFN-LNL Legnaro)
Gargano Angela (INFN Napoli)
Gattobigio Mario (INLN, Université de Nice-Sophia Antipolis, CNRS, France)
Gensini Paolo (INFN Lecce)
Giannini Mauro (Università di Genova)
Girlanda Luca (Università del Salento)
Giusti Carlotta (Università di Pavia)
Greco Vincenzo (Università di Catania)
Grossi Eduardo (Università di Firenze)
Itaco Nunzio (Università di Napoli)
Kievsky Alejandro (INFN Pisa)
Lanza Edoardo (INFN Catania)
Lavagno Andrea (Politecnico di Torino)
Logoteta Domenico (Universidade de Coimbra, Coimbra, Portugal)
Lo Iudice Nicola (Università di Napoli)
Lombardo Maria Paola (INFN-LNF Frascati)
Lo Meo Sergio (ENEA Bologna)
Mannarelli Massimo (INFN-LNGS Assergi)
Marcucci Laura Elisa (Università di Pisa)
Matera Francesco (Università di Firenze)
Orlandini Giuseppina (Università di Trento)
Pacati Franco (Università di Pavia)
Pederiva Francesco (Università di Trento)
Pirrone Sara (INFN Catania)
Puglisi Armando (Università di Catania)
Radici Marco (INFN Pavia)
Rinaldi Matteo (Università di Perugia)
Roggero Alessandro (Università di Trento)
Rolando Valentina (Università di Ferrara)
Rosati Sergio (Università di Pisa)
Ruggieri Marco (Università di Catania)
Salmè Gianni (INFN Roma)
Santopinto Elena (INFN Genova)
Scopetta Sergio (Università di Perugia)
Taiuti Mauro (Università di Genova)
Vigezzi Enrico (INFN Milano)
Viviani Michele (INFN Pisa)
Vorabbi Matteo (Università di Pavia)

Journal of Physics: Conference Series **527** (2014) 011003 doi:10.1088/1742-6596/527/1/011003

Peer review statement

All papers published in this volume of *Journal of Physics: Conference Series* have been peer reviewed through processes administered by the proceedings Editors. Reviews were conducted by expert referees to the professional and scientific standards expected of a proceedings journal published by IOP Publishing.

Journal of Physics: Conference Series **527** (2014) 012001 doi:10.1088/1742-6596/527/1/012001

Universality in few-body systems: from few-atoms to few-nucleons

A Kievsky[1], **M Gattobigio**[2], **E Garrido**[3]

[1]Istituto Nazionale di Fisica Nucleare, Largo B. Pontecorvo 3, 56127 Pisa, Italy
[2]Université de Nice-Sophia Antipolis, Institut Non-Linéaire de Nice, CNRS, 1361 route des Lucioles, 06560 Valbonne, France
[3]Instituto de Estructura de la Materia, CSIC, Serrano 123, E-28006 Madrid, Spain

E-mail: kievsky@pi.infn.it

Abstract. Some aspects of universal properties are discussed in few-atom as well as in few-nucleon systems. When the two-body scattering length a is large the three-body system presents some peculiar characteristics such as the Efimov effect. In the present contribution, the zero-range theory deduced by V. Efimov is compared to results obtained using potential models. Accordingly range corrections are introduced. These corrections appear as a shift in the variable $\kappa_* a$ with κ_* the three-body parameter. The three-boson spectrum, atom-dimer scattering and the recombination rate at threshold are analyzed using the finite-range theory proposed. In the case of nuclear systems, $n-d$ scattering is discussed using the universal formula derived by Efimov. Furthermore, the spectrum up to $N=6$ is discussed using the central Volkov potential.

1. Introduction

The scattering of two particles at very low energy E shows universal behavior encoded on the effective range function in that regime

$$k \cot\delta = -\frac{1}{a} + \frac{1}{2} r_s k^2 \ , \tag{1}$$

where $k^2 = E/(\hbar^2/m)$, δ is the phase shift and r_s is the effective range of the interaction. In fact, systems interacting differently but having equal values of the scattering length a and effective range r_s share the same low energy behavior. Two different limits can be studied that produce interesting effects in the three-body system. The first one is the scaling limit: the range, r_0, of the potential goes to zero. In this limit the two-body system presents a continuous scale invariance (CSI): the low energy observables depend on a single parameter, the two-body scattering length a. For example the s-wave phase shift reads $k\cot\delta = -1/a$ and, for positive values of a, a two-body bound state appears (called dimer) with binding energy $E_2 = \hbar^2/ma^2$. In this context a appears as a control parameter. As it has been demonstrated by Thomas in 1935 [1], the three-body system collapses in this limit (this effect is known as the Thomas collapse). Using a variational argument Thomas showed that the three-body ground state energy scales as $\hbar^2/mr_0^2$ producing a system unbound from below as $r_0 \to 0$. Remarkably, this collapse is independent of the magnitude of the two-body binding energy.

The second limit, which is subject of intense investigations, is the unitary limit. In this limit $|a| >> r_0$ and, in the case of positive and large values of a, a shallow dimer appears with binding energy $E_2 \approx \hbar^2/ma^2$. As pointed out by V. Efimov in 1970 and subsequent papers [2, 3], when $a >> r_0$ the three-boson system presents a new class of universal behavior. The spectrum of three interacting bosons results from a geometric series: $E^{n+1}/E^n \to e^{-2\pi/s_0}$ with $s_0 \approx 1.00624$ a universal constant. What is observed, introducing the three-body hyperradial variable ρ, is that a very extended hyperradial potential is generated which, as $\rho \to \infty$, goes to $\hbar^2/ma^2$. In the limit $a \to \infty$ the hyperradial potential goes to zero as $1/\rho^2$ supporting an infinite number of bound states. The short range physics can be implemented as a boundary condition at short distances (of the order of the interaction range r_0) and, as n increases, the universal spectrum results. More specifically when $a >> r_0$ a continuous scale invariance appears in the two-body system while, in the three-body system, it is broken to a discrete scale invariance (DSI).

The DSI imposes strong constrains on the observables. In addition to the geometric spectrum already mentioned in the case of three-boson bound states, the atom-dimer scattering length a_{AD} has the following functional form deduced by Efimov [4]

$$a_{AD}/a = d_1 + d_2 \tan[s_0 \ln(\kappa_* a) + d_3] \,, \tag{2}$$

where d_1, d_2, d_3 are universal constants and κ_* is the three-body parameter that fixes the energy scale at the unitary limit (see Ref. [5] for details).

For collisions below the dimer breakup threshold the DSI imposes the following universal form for the effective range function

$$ka \cot \delta_{AD} = c_1(ka) + c_2(ka) \cot[s_0 \ln(a\kappa_*) + \phi(ka)] \,, \tag{3}$$

with δ_{AD} the atom-dimer phase-shift and c_1, c_2, ϕ universal functions depending on the product ka, where $k^2 = (4/3)E/(\hbar^2/m)$, being E the center of mass energy of the process. As $k \to 0$, $ka \cot \delta_{AD} \to -1/a_{AD}$ and at $k = 0$ the constants d_1, d_2, d_3 and $c_1(0), c_2(0), \phi(0)$ are related by simple trigonometric equalities. A parametrization of the functions c_1, c_2, ϕ can be found in Ref. [5].

The universal character of the effective range function can be used to evaluate a very different system: low energy nucleon-deuteron scattering [6]. The effective range function corresponding to nucleon-deuteron scattering at low energies, in the $J = 1/2^+$ state, does not show the expected linear behavior but it presents a pole structure. First observations of this particular behavior have been done in Refs. [7, 8, 9] whereas in Ref. [10] explicit calculations of the effective range function have been done using nucleon-nucleon potentials. The pole structure has been related to a virtual state and, from the calculations, it was possible to extract the pole energy fitting the effective range formula with the form suggested by Delves [11]. Recently it was shown that the pole structure of the effective range function can be quantitatively related to the universal form given by equation (3) and, using the parametrization determined in the atomic three-helium system, that equation can be used to describe nucleon-deuteron scattering as well. In particular, using the universal function ϕ, the pole energy can be used to extract the three-body parameter κ_*. In this way, the universal behavior imposed by the DSI is analyzed in systems with natural lengths that differ of several order of magnitude.

The DSI imposes constrains also on the form of the S-matrix for energies above the dimer threshold. The following peculiar form for the recombination rate at threshold [12, 13] has been deduced:

$$K_3 = \frac{128\pi^2(4\pi - 3\sqrt{3})}{\sinh^2(\pi s_0) + \cosh^2(\pi s_0)\cot^2[s_0 \ln(\kappa_* a) + \gamma]} \frac{\hbar a^4}{m} \,, \tag{4}$$

and using the large value of the factor $e^{2\pi s_0} \approx 515$, the above equation can be approximated by

$$K_3 = \alpha\, a^4\hbar/m \approx 67.1 \sin^2[s_0 \ln(\kappa_* a) + \gamma] a^4\hbar/m \,, \tag{5}$$

where $\gamma = 1.16$ [5]. As it has been shown recently [14], this equation can be used to describe very different systems as recombination of three atoms or three nucleons.

The above discussion shows that the study of universality in few-body systems is of interest in several fields of research, ranging from cold-atoms to nuclear physics. In atomic physics, where the Efimov effect has been observed for the first time [15], discrepancies arise between the theoretical prediction and the experimental determination of the ratio between a_* and a_- [16, 17], that means between the scattering lengths at which the Efimov state disappears in the atom-dimer and in the three-atom continuum, respectively. The solution to this puzzle is probably hidden in finite-range corrections to the universal formulas (see Ref. [18] and references therein for a recent account to the problem). A recent study has discussed these corrections from a new perspective [19].

In nuclear physics these studies can be applied to describe halo nuclei where a cluster description is justified. We can mention the observation of universal aspects in the scattering of a neutron on a neutron-halo nucleus having a large scattering length, as for example the $n-^{19}$C system (see Refs. [20, 21] and Ref. [22] for a recent review). Also the light nuclei spectrum shows some universal characteristic as it has been discussed in Ref. [23].

2. The three-boson spectrum

The spectrum of the three-boson system (called trimer) having a large two-body scattering length a has been derived by Efimov in the zero-range limit [4]. The Efimov's binding energy equation is

$$E_3^n + \frac{\hbar^2}{ma^2} = \mathrm{e}^{-2(n-n^*)\pi/s_0} \exp\left[\Delta(\xi)/s_0\right] \frac{\hbar^2\kappa_*^2}{m}\,, \tag{6}$$

where $E_3^n/E_2 = \tan^2\xi$ and the function $\Delta(\xi)$ is an universal function defined in the range $-\pi \le \xi \le -\pi/4$ (a parametrization of $\Delta(\xi)$ in this range can be found in [5]). The two-body binding energy is defined in the zero-range limit as $E_2 = \hbar^2/ma^2$ and the three-body parameter κ_* is determined by the energy E_u^* of the n^* level at the unitary limit. The scattering length a appears as a control parameter, fixing its value the above equation has to be solved simultaneously with the condition $E_3^n/(\hbar^2/ma^2) = \tan^2\xi$ to find the the binding energies E_3^n at each value of the angle ξ. In this context the three-body parameter appears as a scale parameter. The universal character of the above equation can be seen using dimensionless quantities as

$$\epsilon^n = \mathrm{e}^{-2(n-n^*)\pi/s_0} \sin^2\xi \, \exp\left[\Delta(\xi)/s_0\right] \tag{7}$$

with $\epsilon_n = E_3^n/E_u^*$. This form of the equation explicitly shows DSI. At fixed values of the angle ξ the ratio between two different levels is constant and it results

$$E_3^n/E_3^{n'} = \mathrm{e}^{-2(n'-n)\pi/s_0}\,. \tag{8}$$

In the case of two consecutive levels the ratio $E_3^{n+1}/E_3^n = \mathrm{e}^{-2\pi/s_0} \approx 1/515.03$ is obtained. The behavior of the energy for two consecutive levels, $n = 0, 1$, as given in equation (7), is shown in figure 1. In order to see the two levels in a single plot the quantity $(\epsilon_n)^{1/4}$ is plotted as function of $\sqrt{x}$ with $x = \kappa_* a$. In the figure the following points are indicated: a_-^n and a_n^* are the values of the two-body scattering length at which the trimer disappears into the three-body continuum or into the atom-dimer continuum, respectively.

The Efimov equations are exact in the zero-range limit. They have been derived imposing a boundary condition in the short-range region encoded in the three-body parameter κ_*. The same problem has been studied using Effective Field Theories (EFT) [24]. In this approach the short-range physics enters through low energy constants, fitted to reproduce selected data. One particular form to apply this theory was suggested by Lepage [25]. The interaction between two

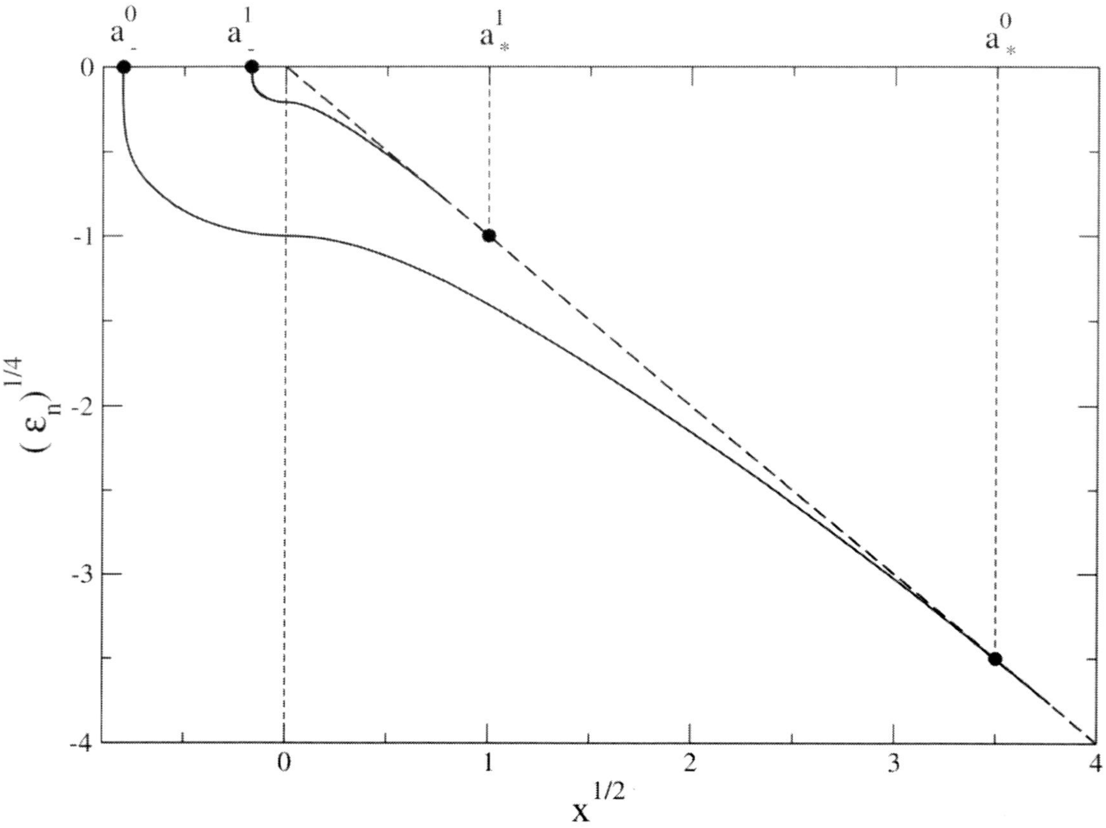

Figure 1. The dimensionless quantity $(\epsilon_n)^{1/4}$ as a function of $(x)^{1/2}$ for $n = 0, 1$. The particular values of the two-body scattering length a_-^n and a_*^n are indicated. The long-dashed line indicates the atom-dimer threshold.

particles is replaced by an effective potential parametrized in such a way that some relevant data are reproduced. For example, a two-parameter potential (as a gaussian) can be constructed to reproduce the scattering length and the effective range as given in equation (1). This effective interaction will describe the low energy dynamics of the system with good accuracy. The level of accuracy is given by the energy scale. Equation (1) is accurate at energies $E << E_0$, with $E_0 = \hbar^2/mr_0^2$, and errors in that formula appear at order of $(E/E_0)^2$. In the three-body system the leading order (LO) prescription of the EFT is that a three-body counter term has to be considered. At this order the EFT approach is based in a Hamiltonian with two coupling constants, one for the two body sector, related with the control parameter a, and one for the three-body sector, related to the scale parameter κ_*. The Efimov equations can be derived in this approach as has been shown if Ref. [26].

Following the above discussion, universal aspects of a three-boson system can be studied using a potential model. Following Refs. [6, 14, 27] the three-helium system is taken as a reference system. At the two-body level, one of the most commonly used He-He potentials, i.e., the LM2M2 interaction [28], is taken as the reference interaction. In particular, in order to explore the (a^{-1}, κ) plane ($\kappa = \text{sign}(E)[|E|/(\hbar^2/m)]^{1/2}$ and E the energy level), we modify this potential as:

$$V_\lambda(r) = \lambda V_{LM2M2}(r) \ . \tag{9}$$

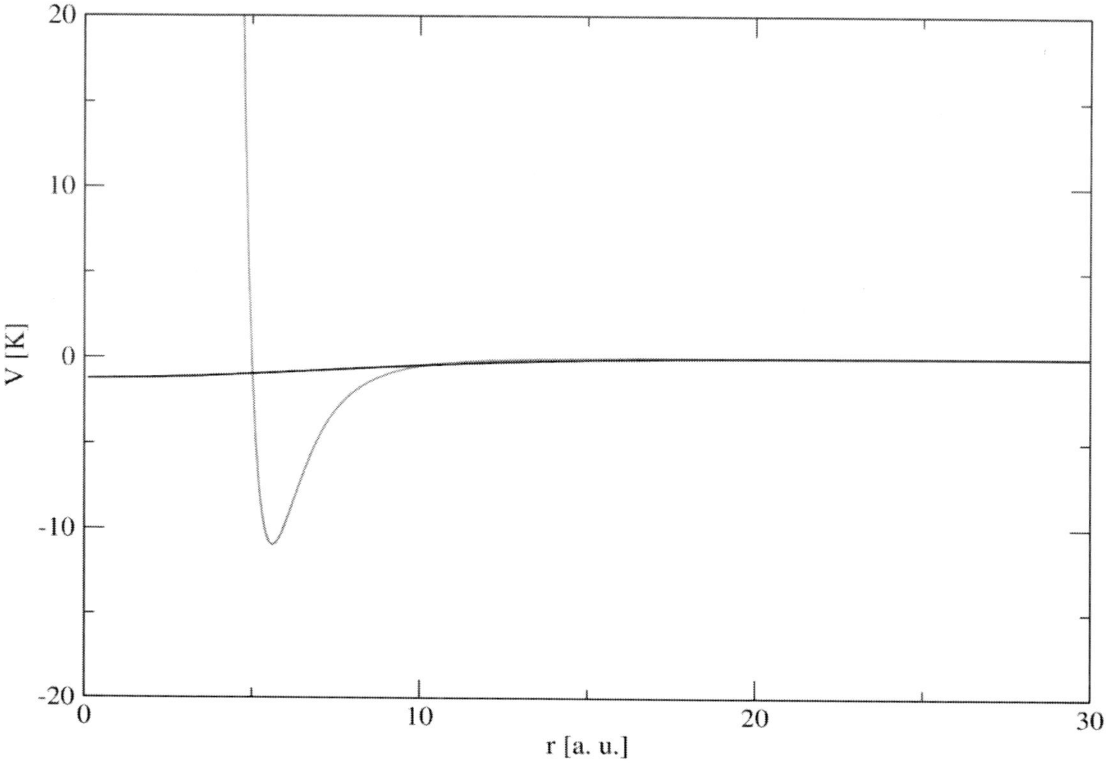

Figure 2. (Color online) The LM2M2 potential (red line) and the TBG potential (black line) as a function of the inter-particle distance.

Examples of this strategy exist in the literature [29, 30]. For $\lambda \approx 0.9743$ the interaction is close to the unitary limit ($a \to \infty$) and for $\lambda = 1$ the values predicted by the LM2M2 are recovered. The scattering length is $a = 189.41\ a_0$, the two-body energy $E_2 = -1.303$ mK and the effective range $r_s = 13.845\ a_0$, with the mass parameter $\hbar^2/m = 43.281307\,(a_0)^2$ K. Noticing the $r_0 \approx r_s$, the two-helium system has the $r_0/a << 1$ allowing a description based on the Efimov universal theory, having in mind that the range corrections have to be considered as well. As an effective potential we define an attractive two-body gaussian (TBG) potential [27, 31]

$$V(r) = V_0 \mathrm{e}^{-r^2/r_0^2} \tag{10}$$

with range $r_0 = 10\ a_0$ and strength V_0 fixed to reproduce the values of a given by $V_\lambda(r)$. For example with the strength $V_0 = -1.2343566$ K, corresponding to $\lambda = 1$, the LM2M2 low-energy data are closely reproduced, $E_2 = -1.303$ mK, $a = 189.42\ a_0$, and $r_s = 13.80\ a_0$. In figure 2 the two interactions are compared, the differences at short distanced are evident.

When the TBG potential is used in the three-atom system it produces a ground state binding energy appreciable deeper than that one calculated using the $V_\lambda(r)$ interaction. For example, at $\lambda = 1$ the LM2M2 helium-trimer ground-state binding energy is $E_3^0 = 126.4$ mK, whereas the one obtained using the TBG potential is 151.32 mK. Following the LO EFT prescription an hypercentral three-body (H3B) (repulsive) interaction can be introduced

$$W(\rho_{123}) = W_0 \mathrm{e}^{-\rho_{123}^2/\rho_0^2} \tag{11}$$

Journal of Physics: Conference Series **527** (2014) 012001 doi:10.1088/1742-6596/527/1/012001

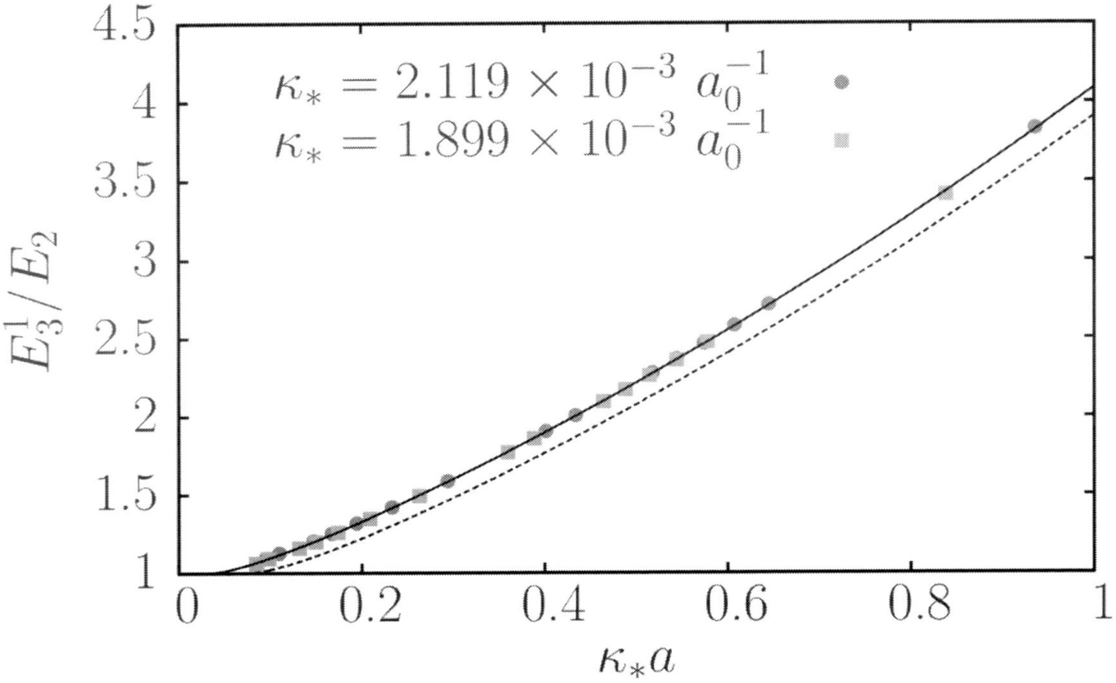

Figure 3. (Color online) The ratio between the excited state of the trimer and the dimer binding energy as a function of $\kappa_* a$ calculated for the TBG potential (red circles) and for the TBG+H3B potential (green squares). The dashed line represents the zero-range theory of equation (6) whereas the solid line has been obtained using equation (13).

with the strength W_0 tuned to reproduce the trimer E_3^0 obtained with $V_\lambda(r)$ for the different values of λ. In the above equation $\rho_{123}^2 = \frac{2}{3}(r_{12}^2 + r_{23}^2 + r_{31}^2)$ is the hyperradius of three identical particles and ρ_0 gives the range of the three-body force that can be fixed as $\rho_0 = r_0$.

Varying λ from the unitary limit to $\lambda = 1.1$ a set of values for the ground state binding energy E_3^0 and the first excited state E_3^1 using the TBG and TBG+H3B potentials can be obtained in a broad range of a. The results can be compared to the predictions given by the Efimov's binding energy equation. Fixing $n^* = 1$, the three-body parameter κ_* is determined by calculating E_3^1 at the unitary limit; we obtain $\kappa_* = 2.119 \times 10^{-3}\, a_0^{-1}$ and $\kappa_* = 1.899 \times 10^{-3}\, a_0^{-1}$ for the TBG and TBG+H3B, respectively.

As it has been shown in Refs. [6, 14, 19], in order to be in accord with the numerical results obtained solving a finite-range potential, the universal relation equation (6) must be modified in the following way

$$E_3^n/E_2 = \tan^2\xi \qquad (12)$$
$$\kappa_* a + \Gamma_n = e^{(n-n^*)\pi/s_0} \exp[-\Delta(\xi)/2s_0]/\cos\xi\,. \qquad (13)$$

The finite-range nature of the interaction has been taken into account by the substitution $\hbar^2/ma^2 \to E_2$ in the first row. The main modification however is the introduction of the shift Γ_n. In figure 3 we collect our numerical results for the ratio E_3^1/E_2 as a function of $\kappa_* a$ for the TBG potential (circles) and for the TBG+H3B potential (squares). The dashed line corresponds to the zero-range theory given in equation (6), whereas the solid line, which goes on top of the numerical results, has been obtained using equation (13) with $\Gamma_1 \approx 4 \times 10^{-2}$.

Journal of Physics: Conference Series **527** (2014) 012001 doi:10.1088/1742-6596/527/1/012001

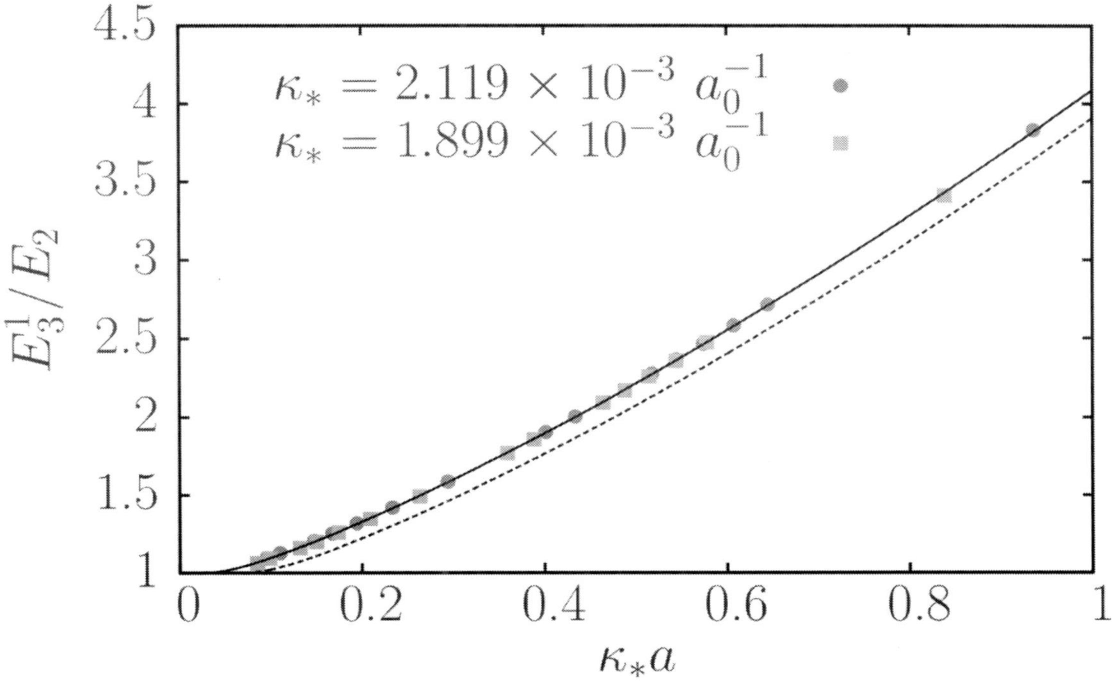

Figure 4. (Color online) The atom-dimer scattering length a_{AD} in units of a_B calculated for the TBG (red circles) and TBG+H3B (green squares) potentials. The dashed line represents the zero-range theory of equation (2) whereas the solid and dotted line are the finite-range theory of equation (14). In the latter a slightly different parametrization of the constants d_1, d_2 and d_3 is used.

3. Atom-dimer scattering

The effective potential models, TBG and TBG+H3B, described in the previous section can be used to describe atom-dimer scattering and the recombination rate K_3 at threshold. To this end the dimer-atom scattering length a_{AD} and the s-wave atom-dimer phase δ can be calculated. The calculations have been done using the hyperspherical harmonic (HH) method as described in Refs. [31, 32]. The results for the atom-dimer scattering length are collected in figure 4 where the ratio a_{AD}/a_B is given in terms of the product $\kappa_* a$ (the energy scattering length a_B is defined from the dimer binding energy as $E_2 = \hbar^2/ma_B^2$). It can be observed that the calculated points, given as full circles (TBG potential) and full squares (TBG+H3B potential), are on a curve shifted with respect to the dashed line representing the zero-range theory of equation (2) with the parametrization given in Ref. [5]. We can interpret again the shift as produced by the finite-range character of the calculations. Accordingly equation (2) can be modified to describe finite-range interactions as

$$a_{AD}/a_B = d_1 + d_2 \tan[s_0 \ln(\kappa_* a + \Gamma_*) + d_3] \,, \tag{14}$$

Using the values $\Gamma_* = \Gamma_1$, the solid line is obtained in figure 4. Furthermore, modifying slightly the values of the universal constants as $d_1 = 1.531$, $d_2 = -2.141$ and $d_3 = 1.100$, from those given in Ref. [5], a better description of the numerical results is obtained as it is shown by the dotted line in the figure.

The calculations have been extended to describe atom-dimer scattering at energies below the dimer breakup threshold for different values of a. As it has been done for the zero-range theory of the binding energies and atom-dimer scattering length, equation (3) has to be modified to consider finite-range interactions. Following Ref. [6], the effective range function can be written as

$$ka_B \cot\delta = c_1(ka) + c_2(ka)\cot[s_0\ln(\kappa_* a + \Gamma_*) + \phi(ka)] \ . \tag{15}$$

The results are collected in figure 5 (full squares) at different values of $\kappa'_* a = \kappa_* a + \Gamma_*$. In the figure different patterns can be observed. For the smallest values of $\kappa'_* a$ the behavior is almost linear in all the energy range. Starting at values $\kappa'_* a \approx 0.4$ a curvature appears close to zero energy, pointing out to an emergent pole structure that becomes evident at larger values of $\kappa'_* a$. Specifically, the pole appears when a_{AD} changes sign (see figure 4) or when the argument of the cotangent function in equation (15) becomes zero (or $n\pi$). The shadow plot in the first row of figure 5 corresponds to the case $\lambda = 1$ and describes ^{4}He-^{4}He$_2$ scattering (full triangles). The shadow plot in the second row corresponds to nucleon-deuteron scattering as discussed below. The solid curves are obtained using the finite-range theory of equation (15) with the parametrization of Ref. [5]. We can observe a noticeable agreement along the whole range of values.

The effective range function in the case of neutron-deuteron scattering has a peculiar form [10]. Moreover Efimov claimed that this process can be described with the universal formula of equation (3) [4]. In order to verify this property, equation (15) can be applied to quantitatively describe $n-d$ scattering at low energies. To this end the results of Ref. [10], in which $n-d$ scattering has been described using a spin dependent central potential, can be used. In that reference the value of $a_{nd} = 0.71$ fm has been obtained for the $n-d$ scattering length, and the effective range function has been parametrized as

$$k\cot\delta = \frac{-1/a_{nd} + r_s k^2/2}{1 + E_{c.m.}/E_p} \tag{16}$$

with $E_p = -160$ keV and $r_s \approx -127$ fm. It should be noticed that this particular parametrization of the effective range function can be simple related to equation (15) in the low energy limit. From the values of E_p and a_{nd} it is possible to determine the values of a and κ'_* using the universal function ϕ and equation (14). The values $a = 4.075$ fm and $\kappa'_* a = 0.5779$ are obtained. The negative energy E_p of the pole implies that it appears in the negative region similar to what happens in figure 5 at intermediate $\kappa'_* a$ values. With these numerical values the shadow panel of the second row in figure 5 shows a comparison between equation (15) and the $n-d$ scattering results of Ref. [10] (full circles). A noticeable agreement is observed. Is should be stressed that the universal functions used in equation (15) are those obtained through the analysis of the atomic system.

Finally the application of the finite-range theory is discussed in the case of the recombination rate at threshold corresponding, in the case of three-helium atoms, to the process ^{4}He+^{4}He+^{4}He $\rightarrow$ ^{4}He+^{4}He$_2$. The coefficient can be defined as $K_3 = \alpha(\hbar a^4/m)$. The DSI imposes α to be a log-periodic function, as given in equation (4), whose simplified version is given in equation (5). In order to describe the numerical results obtained using finite-range potentials, the following modification in the definition of K_3 has to be introduced

$$K_3 = \frac{128\pi^2(4\pi - 3\sqrt{3})}{\sinh^2(\pi s_0) + \cosh^2(\pi s_0)\cot^2[s_0\ln(\kappa_* a + \Gamma_+) + \gamma]}\frac{\hbar a_B^4}{m} \,, \tag{17}$$

with the simplified form

$$K_3 = \alpha\, a^4\hbar/m \approx 67.1\sin^2[s_0\ln(\kappa_* a + \Gamma_+) + \gamma]a_B^4\hbar/m \,, \tag{18}$$

where, as in the effective range function, a has been replaced by a_B and the shift Γ_+ has been introduced. The results obtained in Ref. [14] are shown in figure 6 as circles (TBG potential)

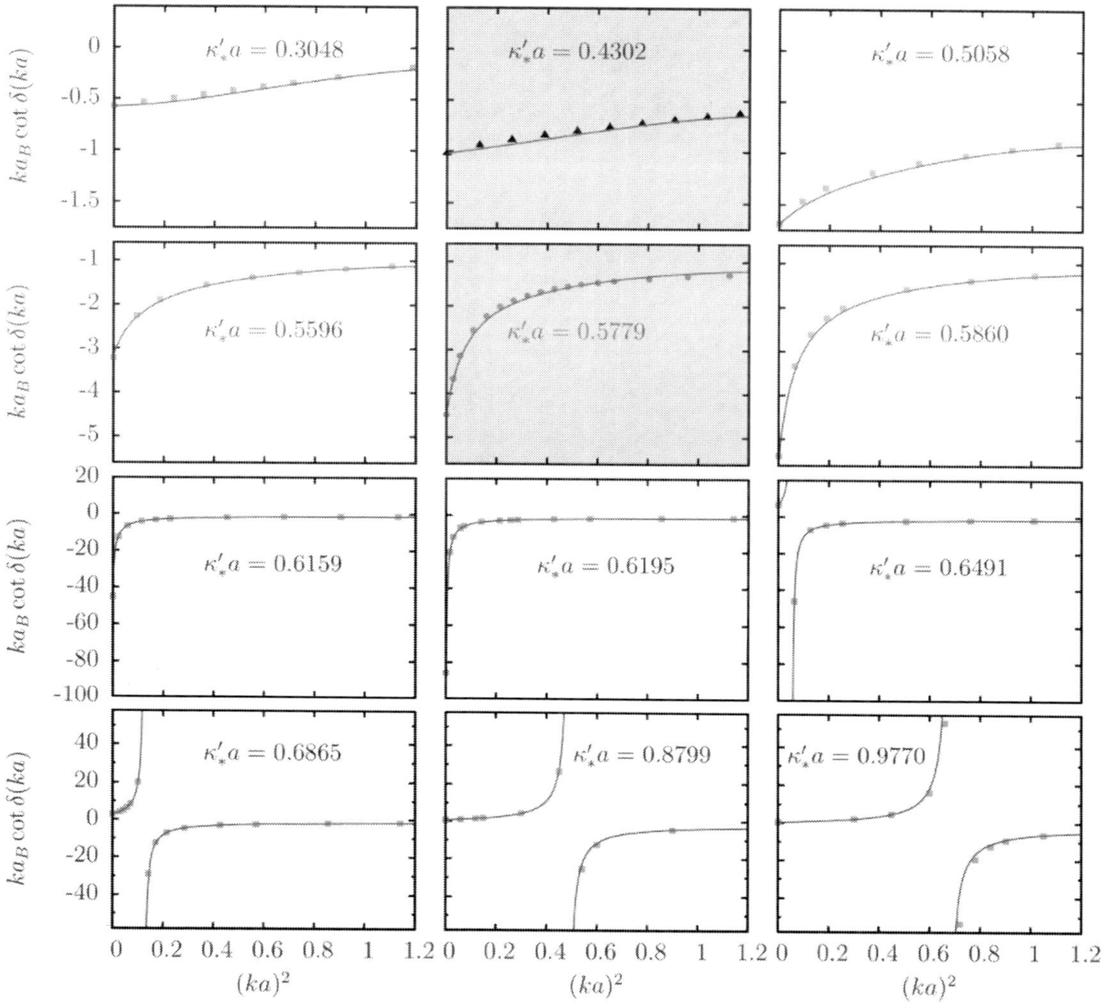

Figure 5. (Color online) The effective range function (squares) at different values of $\kappa'_* a$ as a function of $(ka)^2$. The triangle points are the calculations for real ^{4}He-^{4}He$_2$ scattering. The full circles, for $\kappa'_* a = 0.5779$, correspond to neutron-deuteron scattering in the doublet channel. The solid curves have been calculated using equation (15) for different values of $\kappa'_* a$.

and squares (TBG+H3B potential). The dashed line represents equation (4) using the value $\gamma = 1.16$ from Ref. [5]. From the figure it is evident that the calculated points organize in a curve shifted with respect to the universal curve. The solid line represents equation (17), with the same value of γ and $\Gamma_+ \approx 6 \times 10^{-2}$. Again the finite-range theory obtained by modifying the zero-range equations with a shift in the variable $\kappa_* a$ describes the calculated points with good accuracy. Moreover the values of the shifts Γ_1, Γ_* and Γ_+ are of the same order.

4. Light nuclei spectrum

In this section the spectrum obtained for $N \leq 6$ fermions is discussed using a central potential. The intention here is to see if, in the case of fermions, some of the universal properties discussed

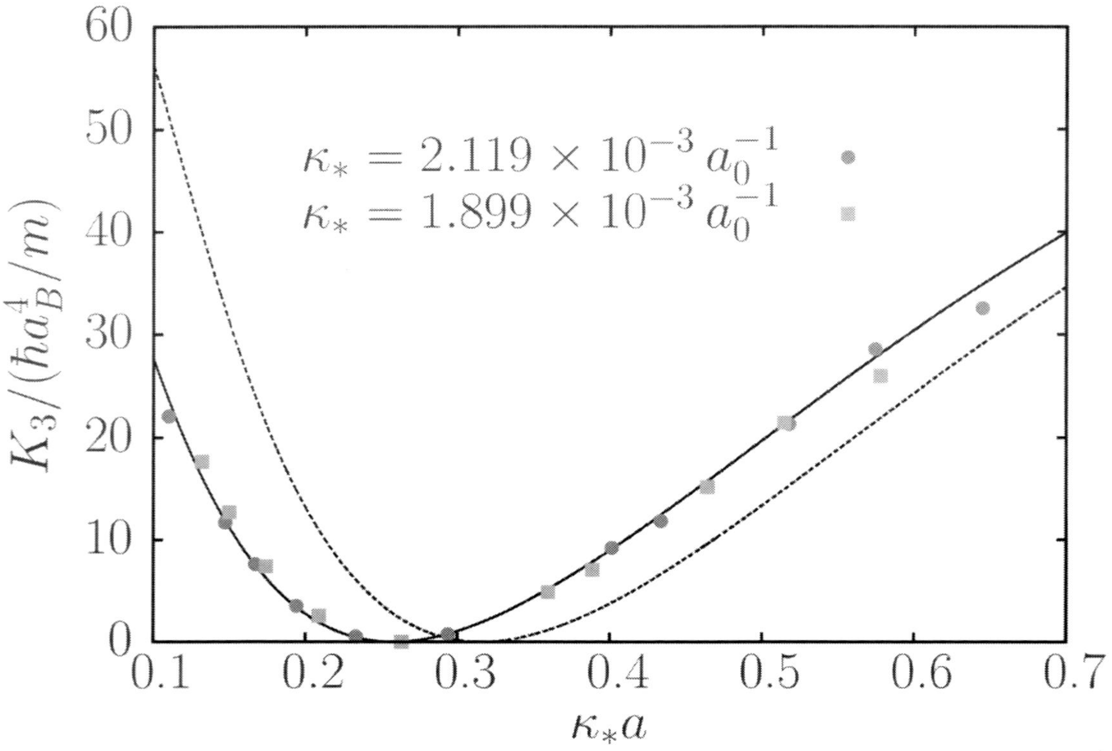

Figure 6. (Color online) The recombination rate K_3 at threshold for different values of $\kappa_* a$ calculated for the TBG (red circles) and TBG+H3B (green squares) potentials. The dotted line is the zero-range theory of equation (4) whereas the solid line is the finite-range theory of equation (17).

before can be identified in the spectrum of light nuclei. To this end the following central potential, the Volkov potential,

$$V(r) = V_R e^{-r^2/R_1^2} + V_A e^{-r^2/R_2^2} \tag{19}$$

is used with $V_R = 144.86\,\text{MeV}$, $R_1 = 0.82\,\text{fm}$, $V_A = -83.34\,\text{MeV}$, $R_2 = 1.60\,\text{fm}$. The fermions are considered to have the same mass, chosen to be equal to the reference mass m, and corresponding to $\hbar^2/m = 41.47\,\text{MeV fm}^2$ (nuclear case). With this parametrization of the potential, the two-nucleon system has binding energy $E_2 = 0.54592$ MeV and scattering length $a = 10.082\,\text{fm}$. For $N > 2$ the use of the s-wave version of the potential produces a spectrum much closer to the experimental one. This is a direct consequence of the weakness of the nuclear interaction in p-waves. The results are obtained after diagonalizing the Hamiltonian matrix constructed using the HH basis without symmetrization as discussed in Ref. [23].

With the parametrization of the potential given above, the ^{3}H and ^{3}He binding energies result to be 8.431 MeV and 7.725 MeV, very close to the experimental values. In fact, this parametrization was chosen to correctly describe the three-nucleon binding energies and, for this reason, the deuteron binding energy results artificially low. As a consequence ^{3}H has an excited state close to the nucleon-deuteron threshold. Using a more realistic interaction this state should move above the threshold becoming the virtual state discussed in the previous section. In despite of its simplicity this model predicts the ^{4}He binding energy with an error

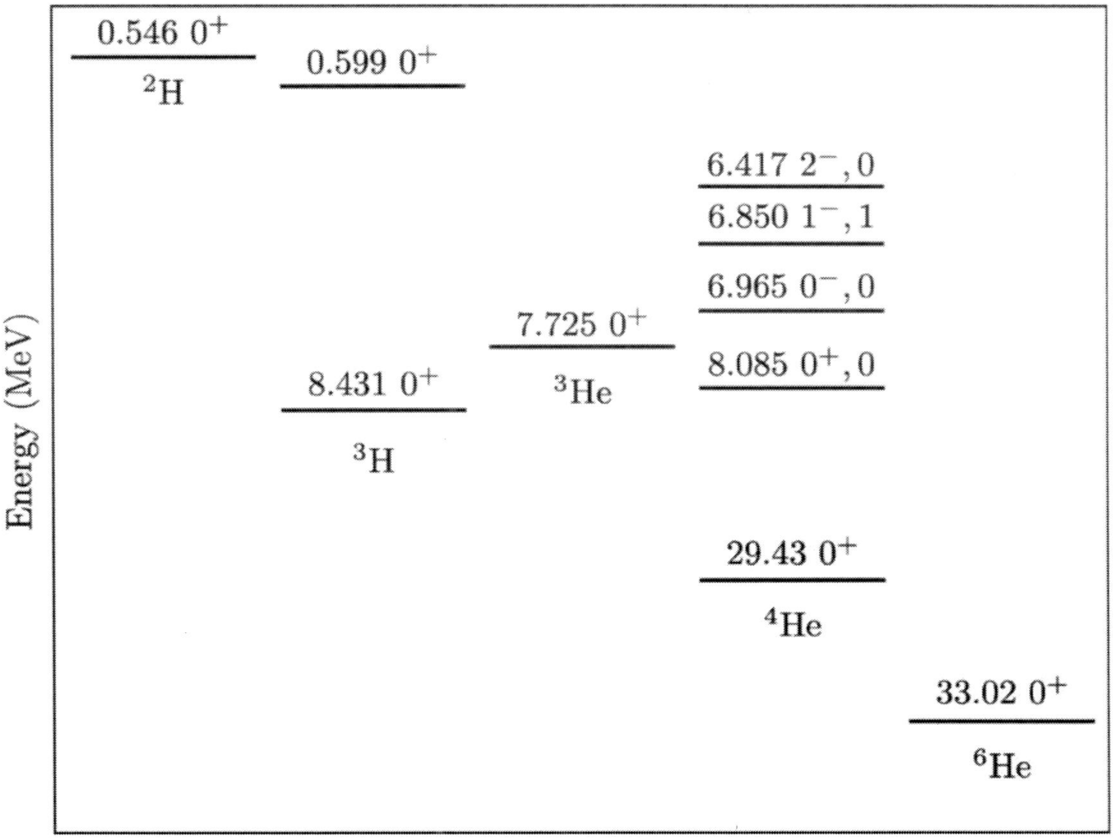

Figure 7. Calculated levels for $A = 2, 3, 4$ and 6, using the Volkov potential with the inclusion of the Coulomb interaction for the He isotopes.

below 5% and, more interesting, the almost correct position of the 0^+ excited state, as is shown in figure 7. Without considering the spin-dependence of the force the ^{6}He ground state energy results too bound.

5. Conclusions

The zero-range theory for few-body systems having a large two-body scattering length a has been studied and discussed using potential models. Following the prescription of Lepage [25] an effective interaction has been constructed that reproduces the same low energy data: the two-body scattering a, acting as a control parameter, and the energy wave number κ_* of one selected level of the trimer at the unitary limit, acting as a scale parameter. Using this model different observables have been calculated solving the Schrödinger equation: the three-boson energy spectrum, the atom-dimer scattering length, the atom-dimer effective range function and the recombination rate at threshold. The analysis of the numerical results allowed to judge the capability of the zero-range theory to describe finite-range systems. The conclusion was that a finite-range scale parameter Γ_n has to be introduced to correct the original zero-range theory. This parameter appears as a finite-range scaling parameter.

Using the finite-range theory it was possible to describe very different systems as atom-dimer scattering and neutron-deuteron scattering using the same parametrization of the universal effective range function, equation (15), obtained studying atomic systems. As a final study, the spectrum of light nuclear systems has been analyzed using a central potential, the Volkov potential. Fixing only the ^{3}H binding energy, this simple model describes the binding energies of the ground and excited states in ^{4}He with an error below 5%. The ^{6}He results slightly over bind. Studies including the spin dependence of the interaction are at present underway.

References

[1] Thomas L H 1935 *Phys. Rev.* **47** 903
[2] Efimov V 1970 *Phys. Lett. B* **33** 563
[3] Efimov V 1971 *Sov. J. Nucl.* **12** 589
[4] Efimov V 1979 *Sov. J. Nucl.* **29** 546
[5] Braaten E and Hammer H W 2006 *Phys. Rep.* **428** 259
[6] Kievsky A and Gattobigio M 2013 *Phys. Rev. A* **87** 052719
[7] Barton G and Phillips A 1969 *Nucl. Phys. A* **132** 97
[8] Whiting J S and Fuda M G 1976 *Phys. Rev. C* **14** 18
[9] Girard B A and Fuda M G 1979 *Phys. Rev. C* **19** 579
[10] Chen C R, Payne G L, Friar J L, and Gibson B F 1989 *Phys. Rev. C* **39** 1261
[11] Delves L M 1960 *Phys. Rev.* **118** 1318
[12] Nielse E and Macek J H 1999 *Phys. Rev. Lett.* **83** 1566
[13] Esry B D, Greene C H, and Burke J P 1999 *Phys. Rev. Lett.* **83** 1751
[14] Garrido E, Gattobigio M, and Kievsky A 2013 *Phys. Rev. A* **88** 032701
[15] Kraemer T *et al.* 2006 *Nature* **440** 315
[16] Zaccanti M *et al.* 2009 *Nat. Phys.* **5** 586
[17] Machtey O, Shotan Z, Gross N, and Khaykovich L 2012 *Phys. Rev. Lett.* **108** 210406
[18] Dyke P, Pollack S E, and Hulet R G 2013 *Phys. Rev. A* **88** 023625
[19] Gattobigio M and Kievsky A 2014 arXiv:1309.1927
[20] Mazumdar I, Rau A R P, and Bhasin V S 2006 *Phys. Rev. Lett.* **97** 062503
[21] Yamashita M, Frederico T, and Tomio L 2008 *Phys. Lett. B* **670** 49
[22] Frederico T, Delfino A, Tomio L, and Yamashita M 2012 *Prog. Part. Nucl. Phys.* **67** 939
[23] Gattobigio M, Kievsky A, and Viviani M 2011 *Phys. Rev. C* **83** 024001
[24] Bedaque P F, Hammer H W, and van Kolck U 1999 *Phys. Rev. Lett.* **82** 463
[25] Lepage G P 1999 *Particles and fields* eds Barata J C A, Malbouisson A P C, and Novaes S F (Singapore: World Scientific)
[26] Bedaque P F, Hammer H W, and van Kolck U 1999 *Nucl Phys. A* **646** 444
[27] Gattobigio M, Kievsky A, and Viviani M 2012 *Phys. Rev. A* **86** 042513
[28] Aziz R A and Slaman M J 1991 *J. Chem. Phys.* **94** 8047
[29] Esry B D, Lin C D, and Greene C H 1996 *Phys. Rev. A* **54** 394
[30] Barletta P and Kievsky A 2001 *Phys. Rev. A* **64** 042514
[31] Kievsky A, Garrido E, Romero-Redondo C, and Barletta P 2011 *Few-Body Syst.* **51** 259
[32] Kievsky A 1997 *Nucl. Phys. A* **624** 125

Journal of Physics: Conference Series **527** (2014) 012002 doi:10.1088/1742-6596/527/1/012002

Some aspects of universality in Efimov physics

M Gattobigio[1] **and A Kievsky**[2]

[1]Université de Nice-Sophia Antipolis, Institut Non-Linéaire de Nice, CNRS, 1361 route des Lucioles, 06560 Valbonne, France

[2]Istituto Nazionale di Fisica Nucleare, Largo B. Pontecorvo 3, 56127 Pisa, Italy

E-mail: mario.gattobigio@inln.cnrs.fr

Abstract. The Efimov effect is the appearance of an infinite tower of bound states, E_3^n, that accumulate in a geometrical way at zero energy in systems of three-identical bosons having an infinite large two-body scattering length a. In the zero-range limit the geometrical factor is $E_3^{n+1}/E_3^n = \mathrm{e}^{-2\pi/s_0}$ with $s_0 \approx 1.00624$ an universal number. The Efimov effect is the signature of the discrete scale invariance (DSI) present in these systems close to the unitary limit ($1/a = 0$). The DSI constrains also the spectrum in $N > 3$ systems. In this contribution the two-level tree structure observed in bosons systems with $N \geq 4$ will be discussed using potential models.

1. Introduction

The theoretical prediction of the Efimov effect [1, 2] and its experimental verification [3, 4, 5, 6] can be indicated as the starting points of an intense program of research in the sector of few-boson systems having large two-body scattering length a. This new field of research is today called Efimov physics; it refers to the investigation of shallow states in which some universal aspects have been discovered. At the two-body level, a large scattering length implies the appearance of a shallow dimer (a bound-state of two particles) with energy $E_2 \approx \hbar^2/ma^2$. The natural value of the binding energy of two particles should be around $\hbar^2/mr_0^2$ with r_0 the range of the interaction; however, at the unitary limit, $a \to \infty$, the two-body low energy observables are only governed by the scattering length a. Besides the dimer energy, we can also mention the mean-square radius is $\langle r^2 \rangle \approx a^2/2$ and the cross section is $\sigma \approx 8\pi a^2/(1 + a^2k^2)$, with k the modulus of the wave vector. These relations are approximate for $a \gg r_0$ but they are exact in the zero-range limit. In particular, in this limit, the two-body system presents a continuous scale invariance (CSI). When the two-body scattering length is scaled by a positive real number $a \to ua$, the mentioned observables scale with the corresponding powers: $E_2 \to u^{-2}E_2$, $\langle r^2 \rangle \to u^2\langle r^2 \rangle$ and $\sigma \to u^2\sigma$. At the level of three particles the CSI is broken and, in the three-body $L = 0$ state, the residual symmetry is the DSI. The physics is invariant under the rescaling $r \to \Lambda^n r$, with the scaling constant usually written as $\Lambda = \mathrm{e}^{\pi/s_0}$.

One question that naturally arises is in which way the N-boson spectrum, governed by the DSI, evolves as the number of bosons increases. The case of four bosons has been investigated with great detail in Refs. [7, 8, 9, 10]. The main conclusions of those studies have been that the four-boson spectrum has a tree structure of two levels, $E_4^{n,0}$ and $E_4^{n,1}$ attached to each E_3^n level. The first one is deep and the second one is shallow, very close to the three-boson threshold given by E_3^n. Moreover, the following ratios have universal character, $E_4^{n,0}/E_3^n \approx 4.611$ and $E_4^{n,1}/E_3^n \approx 1.002$. However, it should be noticed that only the $E_4^{0,0}$ and $E_4^{0,1}$ states are true bound states: all the other levels appear as resonances embedded in the three-boson continuum. Few studies exist for $N > 4$, as discussed in Refs. [11, 12]; using potential models and solving the Schrödinger equation, in Ref. [11] universal ratios have been estimated at the unitary limit up to $N = 13$, whereas in Ref. [12] the spectrum of atomic helium clusters, up to $N = 6$, has been studied in the $(1/a, \kappa)$ plane with $\kappa = \mathrm{sign}(E)[|E|/(\hbar^2/m)]^{1/2}$ and E the energy of the system. Based on these findings, in the present contribution we propose a general formula to describe the tree structure of a boson systems having a large two-body scattering length.

2. The N-boson system with large scattering length

In a sequence of papers, V. Efimov derived the three-boson spectrum in the zero-range limit. It can be expressed in a parametric form as follows

$$\begin{aligned} E_3^n/(\hbar^2/ma^2) &= \tan^2\xi \\ \kappa_* a &= \mathrm{e}^{(n-n^*)\pi/s_0}\frac{\mathrm{e}^{-\Delta(\xi)/2s_0}}{\cos\xi}\,, \end{aligned} \tag{1}$$

with κ_* the wave number corresponding to the energy of the n^*-level at the unitary limit; it is referred to as the three-body parameter. The function $\Delta(\xi)$ is an universal function and its parametrization can be found for example in [13]. According to the above equation, the knowledge of the scattering length a and of the three-body parameter κ_* completely determines the spectrum of the three-boson system. In a given system, a is related to the specific interaction between constituents and to this respect can be thought as fixed by nature. However, with the progress in manipulating trapped-ultracold atoms, the interaction between atomic species can be varied, for instance by means of magnetic fields (Feshbach resonances). Accordingly, the value of a can be varied and eventually it can take values around the unitary limit. In this way the Efimov physics can be studied.

Zero-range interactions do not describe real systems. In order to match the result of the zero-range theory with experiments and/or numerical calculations, one must take into account finite-range corrections. To this respect equation (1) has to be modified in order to consider this kind of corrections. In Refs. [15, 16] this problem was studied using potential models and, from those studies, the following modification to equation (1) has been proposed:

$$\begin{aligned} E_3^n/E_2 &= \tan^2\xi \\ \kappa_3^n a + \Gamma_n &= \frac{\mathrm{e}^{-\Delta(\xi)/2s_0}}{\cos\xi}\,. \end{aligned} \tag{2}$$

The differences between the above formula and equation (1) are the following. In the first row of equation (2) it appears the two-body binding energy whereas in equation (1) we have its zero-range limit. In the second row each level has been identified with the wave number κ_3^n corresponding to the energy at the unitary limit. The DSI is to some extent disrupted by range corrections and the ratio between the wave numbers of two consecutive levels $\kappa_3^{n+1}/\kappa_3^n \to \mathrm{e}^{-\pi/s_0}$ only for large values of n. For low n-values this relation is only approximately verified and, therefore, it is convenient to introduce a three-body parameter for each energy level. However, the main modification in the above equation is the introduction of the finite-size scaling parameter Γ_n in which the finite-range character of the interaction is absorbed.

Studies of Efimov physics has been pursued in Ref. [12] using potential models. The atomic two-helium system is taken as a reference system. To this end an attractive two-body gaussian (TBG) potential has been defined

$$V(r) = V_0\mathrm{e}^{-r^2/r_0^2} \tag{3}$$

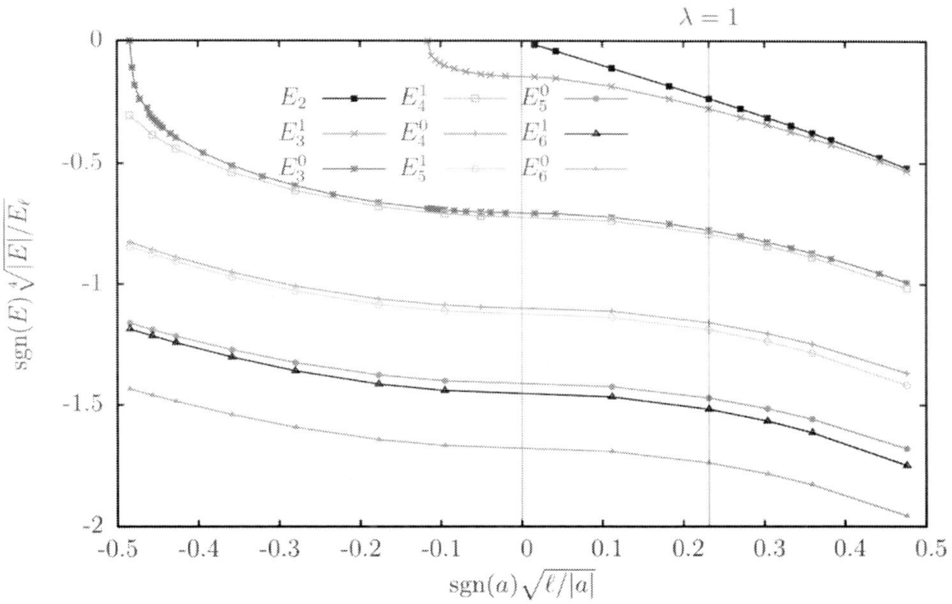

Figure 1. (Color online) Energy levels of N-cluster of bosons as a function of the scattering length. The point $\lambda = 1$ corresponds to the case of Helium, and ℓ is the van der Waals length of Helium.

with range $r_0 = 10\,a_0$ (with a_0 the atomic length) and strength V_0 fixed to reproduce the values of a given by one of the widely used He-He potentials, the LM2M2 interaction [17]. With the strength $V_0 = -1.2343566\,\mathrm{K}$, the LM2M2 low-energy data are closely reproduced, $E_2 = -1.303\,\mathrm{mK}$, $a = 189.42\,a_0$, and the effective range $r_s = 13.80\,a_0$. To explore the (a^{-1}, κ) plane, the strength V_0 of the potential has been varied to cover an extended range of a values. The results up to $N = 6$ bosons are shown in figure 1. The tree structure of the levels is evident. Starting with $N = 4$, the systems present a two-level structure, one deep state with energy E_N^0 and one shallow state with energy E_N^1. This level structure can be described with the following extension of equation (2)

$$E_N^m / E_2 = \tan^2 \xi$$
$$\kappa_N^m a + \Gamma_N^m = \frac{\mathrm{e}^{-\Delta(\xi)/2s_0}}{\cos \xi}\,, \tag{4}$$

with $m = 0, 1$. Another form of equation (4) is the following

$$y(\xi) = \kappa_N^m a + \Gamma_N^m\,, \tag{5}$$

which puts in evidence the linear relation between a and $y(\xi) = \mathrm{e}^{-\Delta(\xi)/2s_0}/\cos\xi$ for fixed values of κ_N^m and Γ_N^m. These values can be extracted from the calculations by making a fit: at each value of a, using the corresponding strength of the TBG potential, the N-boson energy is computed. Then the angle ξ is determined from first row of equation (4) and, using the universal function $\Delta(\xi)$ we obtain the value of $y(\xi)$; at the end we can verify the linear relation between a and y.

The values of κ_N^m and Γ_N^m are given in table 1 for $N = 4, 5, 6$. The corresponding straight lines for both cases, $m = 0$ (left panel) and $m = 1$ (right panel) are given in figure 2. The solid lines

Table 1. Values of κ_N^m and Γ_N^m as a function of N obtained from a fit of the calculated values using equation (4).

N	κ_N^0 $[a_0^{-1}]$	Γ_N^0	κ_N^1 $[a_0^{-1}]$	Γ_N^1
4	0.1185	0.642	0.0511	0.555
5	0.1955	0.740	0.1240	0.772
6	0.2770	0.817	0.2071	0.969

represent equation (5) whereas the circles are the calculations. The linear behavior is evident. An interesting aspect of the figure is the tendency of the lines to cross in a single point, one for $m = 0$ and one for $m = 1$. This is a consequence of the dependence of the parameters κ_N^m and Γ_N^m. In fact these parameters follows a linear relation with N as it is discussed in Ref. [18].

3. Conclusions

In the present contribution different aspects of universal behavior in N-boson systems having large two-body scattering lengths have been discussed. In first place the extension of the zero-range energy equations has been discussed. These equations have to be modified due to the finite-range character of the interaction forces [18]. In fact experiments and calculations using potential models cannot be describe directly by the zero-range theory. It has been shown that the main modification is the introduction of a shift in the parametric form of the energy equations.

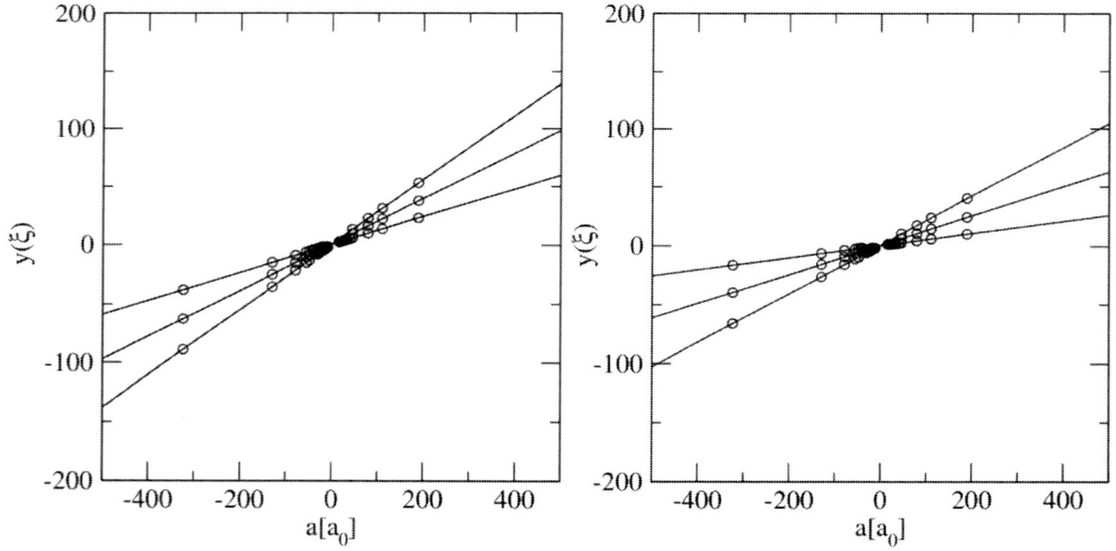

Figure 2. Same data set as figure 1 represented using equation (5). The left panel corresponds to the ground states, $m = 0$, while the right panel corresponds to the excited states, $m = 1$.

The shift however maintains the linear behavior between the scattering length a and the universal function $y(\xi)$ defined in equation (5). A further remarkable characteristic of the finite-range theory proposed in equation (4) is that the same universal function $\Delta(\xi)$, defined in the three-body system, equation (1), governs the N-boson spectrum.

Moreover , we have observed the tree structure of two levels starting from $N = 4$. Here we have discussed the lowest branch ($n = 0$) of the structure which corresponds to true bound states. States corresponding to branches with higher values of n appear embedded in the continuum of the $(N-1)$-boson system and they are difficult to study using potential models.

References

[1] Efimov V 1970 *Phys. Lett. B* **33** 563
[2] Efimov V 1971 *Sov. J. Nucl.* **12** 589
[3] Kraemer T *et al.* 2006 *Nature* **440** 315
[4] Zaccanti M *et al.* 2009 *Nat. Phys.* **5** 586
[5] Machtey O, Shotan Z, Gross N, and Khaykovich L 2012 *Phys. Rev. Lett.* **108** 210406
[6] Dyke P, Pollack S E, and Hulet R G 2013 *Phys. Rev. A* **88** 023625
[7] Hammer H W and Platter L 2007 *Eur. Phys. J. A* **32** 113
[8] von Stecher J, D'Incao J P, and Greene C H 2009 *Nat. Phys.* **5** 417
[9] Deltuva A, Lazauskas R, and Platter L 2011 *Few-Body Syst.* **51** 235
[10] Frederico T, Tomio L, Delfino A, Hadizadeh M R, and Yamashita M T 2011 *Few-Body Syst.* **51** 87
[11] von Stecher J 2010 *J. Phys. B: At. Mol. Opt. Phys.* **43** 101002
[12] Gattobigio M, Kievsky A, and Viviani M 2012 *Phys. Rev. A* **86** 042513
[13] Braaten E and Hammer H W 2006 *Phys. Rep.* **428** 259
[14] Ferlaino F *et al.* 2011 *Few-Body Syst.* **51** 113
[15] Kievsky A and Gattobigio M 2013 *Phys. Rev. A* **87** 052719
[16] Garrido E, Gattobigio M, and Kievsky A 2013 *Phys. Rev. A* **88** 032701
[17] Aziz R A and Slaman M J 1991 *J. Chem. Phys.* **94** 8047
[18] Gattobigio M and Kievsky A 2014 `arXiv:1309.1927`

Journal of Physics: Conference Series **527** (2014) 012003 doi:10.1088/1742-6596/527/1/012003

Quantum Monte Carlo with non–local chiral interactions

A Roggero[1,2], **A Mukherjee** [3], **F Pederiva**[1,2]

[1] Physics Department, University of Trento, via Sommarive 14, I-38123 Trento, Italy
[2] INFN-TIFPA, Trento Institute for Fundamental Physics and Applications
[3] ECT*, Villa Tambosi, I-38123 Villazzano (Trento), Italy

E-mail: `roggero@science.unitn.it,mukherjee@ectstar.eu,pederiva@science.unitn.it`

Abstract. Quantum Monte Carlo methods have been used very successfully to calculate the ground state properties of a variety of nuclear and other strongly interacting many body systems. An exciting recent development in low energy nuclear physics is the formulation of nuclear Hamiltonians based on the symmetries of quantum chromodynamics via chiral effective field theory. However, the usual coordinate space based quantum Monte Carlo implementation is unsuitable for these Hamiltonians due to the presence of non-localites. Here we report on our recent work in developing a new quantum Monte Carlo algorithm suitable for these non-local Hamiltonians, and present some results for pure neutron matter.

1. Introduction

Predicting the properties of nuclei from the underlying theory of strong interactions viz. quantum chromodynamics (QCD) is one of the long standing goals of nuclear physics. Chiral effective field theory (χ-EFT) serves as a bridge between QCD and low energy nuclear physics by providing a systematic expansion for the nuclear forces based on the symmetries and on the symmetry breakings of QCD [1, 2, 3]. Chiral interactions have already been employed in calculations of nuclear structure and reactions of light and medium-mass nuclei [4, 5, 6, 7, 8, 9, 10, 11, 12, 13, 14, 15, 17, 16, 18, 19], and infinite matter [20, 21, 22, 23, 24, 25, 26].

Amongst the available non-perturbative methods for strongly interacting systems, quantum Monte Carlo (QMC) methods have proven to be extremely accurate for a variety of systems [27, 28]. In fact, some of the most accurate calculations for light nuclei and neutron matter were performed using continuum diffusion based QMC methods, i.e., nuclear Green's function Monte Carlo (GFMC) and auxiliary field diffusion Monte Carlo (AFDMC) [29, 30, 31, 32], in conjunction with the semi phenomenological *local* Argonne-Urbana family of nuclear forces [33, 34]. On the other hand, the interactions obtained from chiral EFT are, in general, *non-local*, and hence are difficult to incorporate in standard continuum QMC methods [1].

In this work, we report our calculations with a non–local chiral interaction in which a proper QMC algorithm is used. The interaction we use is the recently developed chiral $\mathrm{NNLO_{opt}}$ interaction [35], which fits the experimantal phase–shifts at $\chi^2 \sim 1$ for laboratory energies less

[1] See however [25] for a recent attempt to incorporate chiral interactions in continuum QMC by removing non-localities up to next-to-next-to-leading order (NNLO).

Journal of Physics: Conference Series **527** (2014) 012003 doi:10.1088/1742-6596/527/1/012003

than 125 MeV. However, the contribution from the three nucleon forces is smaller with this parametrization than with the previous ones.

In the next section we present the QMC method and then we will present an application to neutron matter.

2. Method

Quantum Monte Carlo methods solve the Schroedinger equation for the ground state by looking at the long τ solution of the imaginary–time equation:

$$-\frac{\partial}{\partial\tau}|\Psi(\tau)\rangle = H|\Psi(\tau)\rangle = (K+V)|\Psi(\tau)\rangle \tag{1}$$

where K and V are respectively kinetic and potential parts of the total Hamiltonian H. equation (1) can be recast into the integral equation:

$$\langle\mu|\Psi(\tau+\Delta\tau)\rangle = \int d\nu\langle\mu|G_{\Delta\tau}|\nu\rangle\langle\nu|\Psi(\tau)\rangle \tag{2}$$

where we employed a (continuous) complete sets of states $|\nu\rangle$ and $|\mu\rangle$ for convenience and G is the Green's function of equation (1), the so–called imaginary-time propagator. By repeatedly integrating (2), starting from some initial ansatz for the wave-function, we will project to the eigenstate of H with the lowest eigenvalue.

Conventional QMC methods formulated in coordinate space ($\nu, \mu \equiv \mathbf{R}$) rely on the fact that for small imaginary–time step $\Delta\tau$ we have the factorization

$$\langle\mathbf{R}'|G_{\Delta\tau}|\mathbf{R}\rangle = G_{\Delta\tau}(\mathbf{R}',\mathbf{R}) \approx G^d_{\Delta\tau}(\mathbf{R}',\mathbf{R})G^b_{\Delta\tau}(\mathbf{R}') + O(\Delta\tau^2) \tag{3}$$

where $G^d_{\Delta\tau}(\mathbf{R}',\mathbf{R})$ is the free-particle Green's function coming from the kinetic energy term K and $G^b_{\Delta\tau}(\mathbf{R}')$ takes into account the effect of interactions V . Now if $G^d_{\Delta\tau}(\mathbf{R}',\mathbf{R})$ is positive–definite we can interpret it as a probability and use stochastic methods to integrate (2). This scheme however cannot be followed in the general case where the interaction is non–local in coordinate space: $\langle\mathbf{R}'|V|\mathbf{R}\rangle \neq V(\mathbf{R}')\delta(\mathbf{R}'-\mathbf{R})$. In this case in fact transitions to new states is controlled also by the potential and finding efficient ways to perform the sampling becomes a challenging task (cf. [52]). Unfortunately in general interactions coming from χ–EFT belong to this class [1, 2, 3].

These problems associated with non–local potentials are however essentially connected to fact that we are working in a continuous single–particle basis. Let us then take a finite set $\mathcal{S}$ of sp states of size $\mathcal{N}_s$ and consider a general second–quantized fermionic Hamiltonian which includes only two–body interactions

$$H = \sum_{i\in\mathcal{S}}\epsilon_i a^\dagger_i a_i + \sum_{abij\in\mathcal{S}} V^{ab}_{ij} a^\dagger_a a^\dagger_b a_i a_j\ , \tag{4}$$

where $a^\dagger_i$ creates a particle in the single-particle (sp) state labeled by i (i is a collective label for all sp quantum numbers) and the V^{ab}_{ij} are general two-body interaction matrix elements.

The resulting many–body Hilbert space would be spanned by the full set of N–particle Slater determinants that can be generated using the sp orbitals $i \in \mathcal{S}$. At this point we want to solve for the ground–state of the system by repeatedly applying a projection operator like

$$\begin{aligned}\langle\mathbf{m}|P|\mathbf{n}\rangle \equiv \langle\mathbf{m}|G_{\Delta\tau}|\mathbf{n}\rangle &= 1-\Delta\tau\,(H-E_T)|\mathbf{n}\rangle \\ &= \delta_{\mathbf{m},\mathbf{n}} - \Delta\tau\langle\mathbf{m}|H-E_T|\mathbf{n}\rangle\end{aligned} \tag{5}$$

where E_T is just an energy shift used to preserve the norm of the solution.

Since our final purpose is to interpret the off–diagonal terms in (5) as probabilities, we need the matrix elements $\langle \mathbf{m}|P|\mathbf{n}\rangle \geq 0$ (or equivalently $\langle \mathbf{m}|H|\mathbf{n}\rangle \leq 0$). In general however P will have negative–matrix elements which will cause problems, this is once again a manifestation of the sign–problem.

In principle, one can still produce a stochastic evolution absorbing the signs into the renormalization factor and sampling off-diagonal moves using $|\langle \mathbf{m}|P|\mathbf{n}\rangle|$, but this is accompanied by an exponential decay of the signal to noise ratio as a function of the projection time $\tau = M\Delta\tau$. Recently it was shown that by employing an annihilation step in the evolution this problem can be substantially alleviated [36, 37, 38]. At the end however these algorithms have still an exponential scaling with system–size, though with a reduced exponent.

In CIMC we instead deal with the sign–problem in a way which is somewhat similar to standard coordinate–space QMC: we will use an initial ansatz Φ_T for the ground–state wave–function and use that to constrain the random walk in a region of the many–body Hilbert space where $\langle \mathbf{m}|P|\mathbf{n}\rangle \geq 0$ is satisfied. In order to this scheme to be practical one needs a systematic way for reducing the bias coming from this approximation, e.g. we want the bias to go to zero as the ansatz Φ_T goes towards the ground–state Ψ_0. That's exactly what is done in coordinte-space fixed–node QMC simulations.

Following the work in [39, 40] we define a new Hamiltonian $\mathcal{H}$, whose off–diagonal matrix elements are given by

$$\langle \mathbf{m}|\mathcal{H}|\mathbf{n}\rangle = \begin{cases} 0 & \mathfrak{s}(\mathbf{m},\mathbf{n}) > 0 \\ \langle \mathbf{m}|H|\mathbf{n}\rangle & \text{otherwise} \end{cases} , \tag{6}$$

while the diagonal terms are

$$\langle \mathbf{n}|\mathcal{H}|\mathbf{n}\rangle = \langle \mathbf{n}|H|\mathbf{n}\rangle + \sum_{\substack{\mathbf{m}\neq\mathbf{n} \\ \mathfrak{s}(\mathbf{m},\mathbf{n})>0}} \mathfrak{s}(\mathbf{m},\mathbf{n}) \,. \tag{7}$$

where $\mathfrak{s}(\mathbf{m},\mathbf{n}) = \Phi_T(\mathbf{m})\langle \mathbf{m}|H|\mathbf{n}\rangle/\Phi_T(\mathbf{n})$, and a new propagator

$$\langle \mathbf{m}|\mathcal{P}|\mathbf{n}\rangle = \delta_{\mathbf{m},\mathbf{n}} - \Delta\tau\Phi_T(\mathbf{m})\langle \mathbf{m}|\mathcal{H} - E_T|\mathbf{n}\rangle/\Phi_T(\mathbf{n}) \,. \tag{8}$$

Now the propagator $\mathcal{P}$ is, by construction, free from the sign–problem and performing the corresponding random–walk allows us to filter the state $\Phi_T(\mathbf{n})\phi_0(\mathbf{n})$, where now $\phi_0(\mathbf{n})$ is the ground–state of the hamiltonian $\mathcal{H}$. The ground–state energy ϵ obtained following this procedure is a strict upper bound for the true ground–state energy E_{gs} of the true hamiltonian H and, moreover, this upper bound is tighter than the variational upper–bound provided by $\langle \Phi_T|H|\Phi_T\rangle$ [39, 40, 41].

Up to now we have just discussed real–symmetric hamiltonian matrices but in the case of χ-EFT interactions in momentum–space we have to deal with a complex–hermitian matrix. The extension to this more general case of equations (6)-(8) can be actually performed while preserving all the favorable properties of the method discussed so far: we replace $\mathfrak{s}(\mathbf{m},\mathbf{n})$ in equation (6) and equation (7) with

$$\mathfrak{s}'(\mathbf{m},\mathbf{n}) = \frac{\mathcal{R}e\left[\Phi_T^*(\mathbf{m})\langle \mathbf{m}|H|\mathbf{n}\rangle\Phi_T(\mathbf{n})\right]}{|\Phi_T(\mathbf{n})|^2} , \tag{9}$$

where $\mathcal{R}e$ selects the real part, and define the propagator

$$\langle \mathbf{m}|\mathcal{P}'|\mathbf{n}\rangle = \delta_{\mathbf{m},\mathbf{n}} - \Delta\tau\frac{\mathcal{R}e\left[\Phi_T^*(\mathbf{m})\langle \mathbf{m}|\mathcal{H} - E_T|\mathbf{n}\rangle\Phi_T(\mathbf{n})\right]}{|\Phi_T(\mathbf{n})|^2} . \tag{10}$$

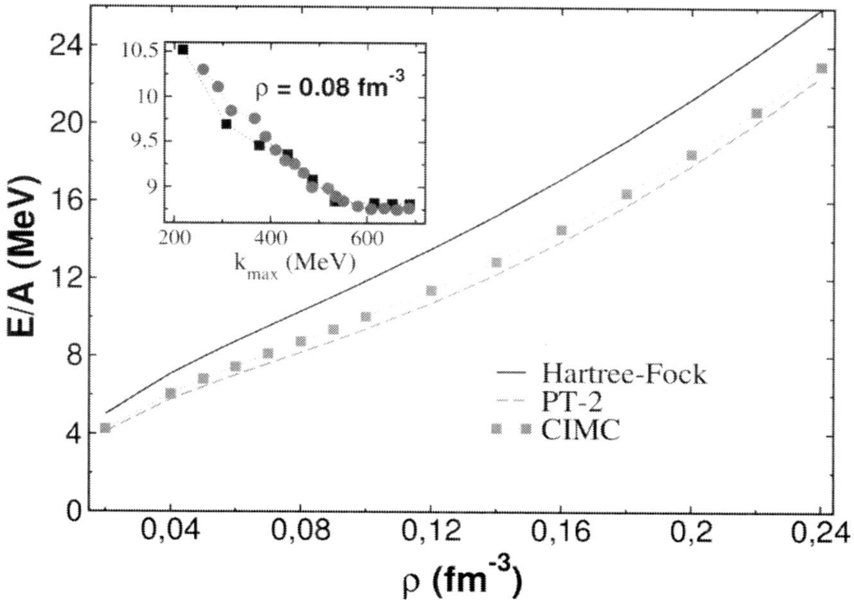

Figure 1. (Color online) The EoS of pure neutron matter (66 neutrons) using the same χ-EFT interaction with different methods: red squares - our Monte Carlo results, black solid line Hartree-Fock, green dashed lines PT-2. The inset shows the convergence of our energies as a function of $k_{\rm max}$ at $\rho = 0.08$ fm^{-3} for 14 (black squares) and 66 (blue circles) particles. The dotted lines are a guide to the eye.

The proofs of upper bound properties is analogus to the one in the real case and will be reported elsewhere [42].

Finally, to assure the success of the proposed method a good choice for the importance function $|\Phi_T\rangle$ is crucial, we need a wave–function flexible enough to account for the relevant correlations in the system and that at the same time can be evaluated sufficiently quickly on a computer. In many strongly–interacting systems coupled–cluster theory provide an ansatz that fulfills the first criterion. In [51] we showed how to evaluate the overlaps $\langle \mathbf{n}|\Phi_T\rangle$ in a computationally efficient way in the so–called coupled cluster doubles (CCD) approximation. Extensions to include singles (CCSD) and triples (CCSDT) corrections, useful for treating inhomogeneous systems or many–body interactions, are straightforward.

3. Neutron matter in momentum–space

In this work we choose for the sp basis states the eigen–states of momentum and the z components of spin and iso–spin. Calculations are performed at fixed density ρ by putting A nucleons in a cubic box of size $L^3 = A/\rho$ with periodic boundary conditions. The finiteness of the box requires the sp states to lie on a cubic lattice in momentum space with lattice constant $l = 2\pi/L$. This scale imposes thus an infra–red (IR) cut off in our scheme, and in order to obtain a finite basis we impose an additional ultra–violet (UV) cut off $k_{\rm max}$ so that only sp states with $k_i^2 \leq k_{\rm max}^2$ are included. A sequence of calculations with increasing $k_{\rm max}$ and decreasing l need to be performed till convergence is reached.

In the inset of figure 1 we plot the energy per particle as a function of the UV cut off $k_{\rm max}$ at $\rho = 0.08$ fm^{-3} for for both 14 and 66 neutrons, at all densities considered in this work we found the same smooth behaviour. The CIMC calculations are considered to have converged

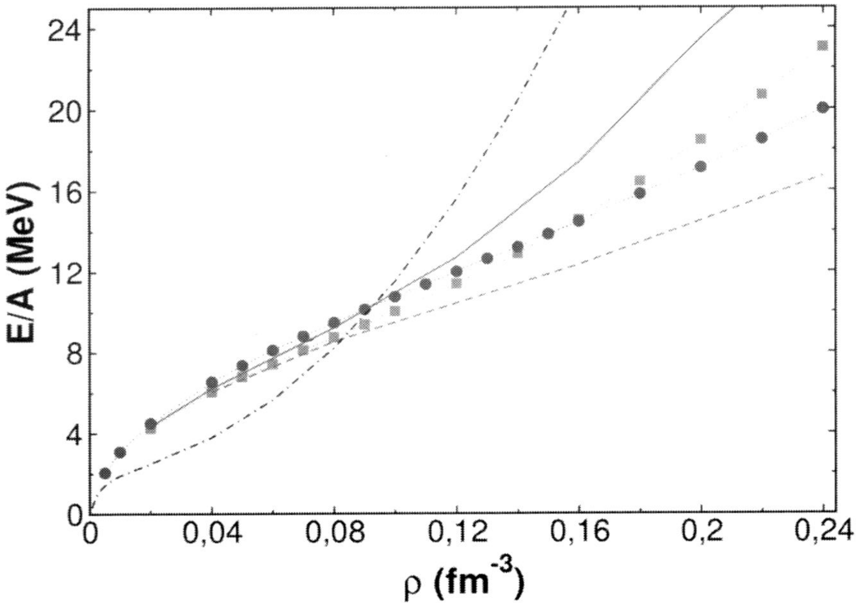

Figure 2. (Color online) The EoS of pure neutron matter: red squares - our results (66 neutrons), brown circles - AFDMC EoS with the 2b AV8' [44], blue dashed line - APR EoS with the 2b AV18 [43], blue solid line - APR EoS with the 2b AV18 + 3b UIX [43], black dashed dotted line - NL3 EoS [46].

when the difference in energy between two successive values of $k_{\max}$ is less than the statistical error ($\approx 10 - 25$ KeV at convergence).

Interestingly UV convergence is obtained for values $k_{\max} - k_f$ 450 MeV (where k_f is the Fermi momentum) which is comparable to the momentum–cutoff at 500 MeV in the N2LO interaction. We find this to be the case for all densities considered.

In figures 1 and 2 we show our results for pure neutron matter. In figure 1 we compare them with Hartree-Fock (HF) and 2nd order perturbation theory (PT-2) calculations employing the same interaction. As is apparent neutron matter with this chiral interaction is pretty much perturbative, with PT-2 numbers consistently overestimating the binding energy but still quite close to the non–perturbative answer given by CIMC .

In figure 2 we include the variational APR EoSs (two body - AV18 and two plus three body - AV18+UIX interactions) [43] and the AFDMC EoS (two body - AV8$'$ interaction) [44] obtained with realistic interactions fitting the NN phase–shifts data. Most computer simulations of supernovae use phenomenological EoSs based typically on the liquid drop model, the most popular being the Lattimer-Swesty EoS [45], or on relativistic mean field theory [46, 47, 48, 49, 50]. As a prototype of such an EoS we have included the results from the NL3 EoS [46].

As can be seen the calculations based on microscopic Hamiltonians are consistent (within about 10%) at low densities ($\lesssim 0.1$ fm^{-3}) and other many–body calculations with realistic interactions [24, 25, 22, 26, 21, 20] are also consistent with the ones shown in the figure in this density range.

Phenomenological approaches as the NL3 model on the other hand show a completely different behaviour even at very low densities. This strong deviation of predictions obtained from these popular phenomenological models from the results obtained from microscopic calculations was also pointed out recently in [26] in the context of χ-EFT interactions.

4. Conclusion

In conclusion, we reported the first quantum Monte Carlo calculations with non-local chiral interactions. In this preliminary applications we find, unsurprisingly, that the equation of state of neutron matter at low densities is reasonably model independent as long as the interaction used is fit to the low energy scattering phase shifts.

The QMC method described in this paper is quite general and can be used for nuclear matter and finite nuclei, and with three body forces. This work paves the way to a systematic non-perturbative study of chiral-forces in all this regimes. Work in these directions is in progress.

Acknowledgments

We are grateful to G Hagen and T Papenbrock for sharing their CCD code and benckmarks with us, and to A Ekström, G Hagen and T Papenbrock for sharing the numerical program for the chiral interaction. A Mukherjee would like to thank W Weise for numerous fruitful discussions. Computations were performed at the Open Facilities at the Lawrence Livermore National Laboratory.

References

[1] Epelbaum E, Hammer H W, and Meißner U G 2009 *Rev. Mod. Phys.* **81** 1773
[2] Machleidt R and Entem D R 2011 *Phys. Rep.* **503** 1
[3] Hammer H W, Nogga A, and Schwenk 2013 *Rev. Mod. Phys.* **85** 197
[4] Kalantar-Nayestanaki N, Epelbaum E, Messchendorp J G, and Nogga A 2012 *Rep. Prog. Phys.* **75** 016301
[5] Barrett B R, Navrátil P, and Vary J P 2013 *Prog. Part. Nucl. Phys.* **69** 131
[6] Roth R, Langhammer J, Calci A, Binder S, and Navrátil P 2011 *Phys. Rev. Lett.* **107** 072501
[7] Roth R, Binder S, Vobig K, Calci A, Langhammer J, and Navrátil P 2012 *Phys. Rev. Lett.* **109** 052501
[8] Hergert H, Bogner S K, Binder S, Calci A, Langhammer J, Roth R, and Schwenk A 2013 *Phys. Rev. C* **87** 034307
[9] Epelbaum E, Krebs H, Lee D, and Meißner U G 2009 *Eur. Phys. J. A* **40** 199
[10] Epelbaum E, Krebs H, Lee D, and Meißner U G 2010 *Phys. Rev. Lett.* **104** 142501
[11] Epelbaum E, Krebs H, Lee D, and Meißner U G 2011 *Phys. Rev. Lett.* **106** 192501
[12] Epelbaum E, Krebs H, Lähde T A, Lee D, and Meißner U G 2013 *Phys. Rev. Lett.* **110** 112502
[13] Hagen G, Hjorth-Jensen M, Jansen G R, Machleidt R, and Papenbrock T 2012 *Phys. Rev. Lett.* **109** 032502
[14] Hagen G, Hjorth-Jensen M, Jansen G R, Machleidt R, and Papenbrock T 2012 *Phys. Rev. Lett.* **108** 242501
[15] Otsuka T, Suzuki T, Holt J D, Schwenk A, and Akaishi Y 2010 *Phys. Rev. Lett.* **105** 032501
[16] Holt J D, Menéndez J, and Schwenk A 2013 *J. Phys. G* **40** 075105
[17] Holt J D, Otsuka T, Schwenk A, and Suzuki T 2012 *J. Phys. G* **39** 085111
[18] Somà V, Barbieri C, and Duguet T 2013 *Phys. Rev. C* **87** 011303
[19] Wienholtz F *et al.* 2013 *Nature* **498** 7454
[20] Kaiser N, Fritsch S, and Weise W 2002 *Nucl. Phys. A* **697** 255
[21] Holt J W, Kaiser N, and Weise W 2013 *Phys. Rev. C* **87** 014338
[22] Tews I, Krüger T, Hebeler K, and Schwenk A 2013 *Phys. Rev. Lett.* **110** 032504
[23] Baardsen G, Ekström A, Hagen G, and Hjorth-Jensen M 2013 *Phys. Rev. C* **88** 054312
[24] Hagen G, Papenbrock T, Ekström A, Wendt K A, Baardsen G, Gandolfi S, Hjorth-Jensen M, and Horowitz C J 2014 *Phys. Rev. C* **89** 014319
[25] Gezerlis A, Tews I, Epelbaum E, Gandolfi S, Hebeler K, Nogga A, and Schwenk A 2013 *Phys. Rev. Lett.* **111** 032501
[26] Krüger T, Tews I, Hebeler K, and Schwenk A 2013 *Phys. Rev. C* **88** 025802
[27] Kalos M H and Whitlock P A 2009 *Monte Carlo methods* (Wiley-VCH)
[28] Foulkes W M C, Mitas L, Needs R J, and Rajagopal G 2001 *Rev. Mod. Phys.* **73** 33
[29] Pudliner B S, Pandharipande V R, Carlson J, Pieper S C, and Wiringa R B 1997 *Phys. Rev. C* **56** 1720
[30] Pieper S C 2008 *Riv. Nuovo Cimento* **31** 2008
[31] Gandolfi S, Illarionov A Y, Schmidt K E, Pederiva F, and Fantoni S 2009 *Phys. Rev. C* **79** 054005
[32] Gezerlis A and Carlson J 2010 *Phys. Rev. C* **81** 025803
[33] Wiringa R B, Stoks V G J, and Schiavilla R 1995 *Phys. Rev. C* **51** 38

[34] Pieper S C, Pandharipande V R, Wiringa R B, and Carlson J 2001 *Phys. Rev. C* **21** 014001
[35] Ekström A, Baardsen G, Forssén C, Hagen G, Hjorth-Jensen M, Jansen G R, Machleidt R, Nazarewicz W, Papenbrock T, Sarich J, and Wild S M 2013 *Phys. Rev. Lett.* **110** 192502
[36] Booth G H, Thom A J W, and Alavi A 2009 *J. Chem. Phys.* **131** 054106
[37] Booth G H, Grüneis A, Kresse G, and Alavi A 2013 *Nature* **493** 365
[38] Petruzielo F R, Holmes A A, Changlani H J, Nightingale M P, and Umrigar C J 2012 *Phys. Rev. Lett.* **109** 230201
[39] ten Haaf D F B, van Bemmel H J M, van Leeuwen J M J, van Saarloos W, and Ceperley D M 1995 *Phys. Rev. B* **51** 13039
[40] Sorella S and Capriotti L 2000 *Phys. Rev. B* **61** 2599
[41] Mukherjee A and Alhassid Y 2013 *Phys. Rev. A* **88** 053622
[42] Roggero A, Mukherjee A, and Pederiva F in preparation
[43] Akmal A, Pandharipande V R, and Ravenhall D G 1998 *Phys. Rev. C* **58** 1804
[44] Gandolfi S, Carlson J, Reddy S, Steiner A W, and Wiringa R B 2013 arXiv:1307.5815
[45] Lattimer J M and Swesty D F 1991 *Nucl. Phys. A* **535** 331
[46] Shen G, Horowitz C J, and Teige S 2011 *Phys. Rev. C* **83** 035802
[47] Shen G, Horowitz C J, and O'Connor E 2011 *Phys. Rev. C* **83** 065808
[48] Steiner A W, Hempel M, and Fischer T 2013 *Astrophys. J.* **774** 17
[49] Shen H, Toki H, Oyamatsu K, and Sumiyoshi K 2011 *Astrophys. J. Suppl.* **197** 20
[50] Typel S, Röpke G, Klähn T, Blaschke D, and Wolter H H 2010 *Phys. Rev. C* **81** 015803
[51] Roggero A, Mukherjee A, and Pederiva F 2013 *Phys. Rev. B* **88** 115138
[52] Lynn J E and Schmidt K E 2012 *Phys. Rev. C* **86** 014324

Journal of Physics: Conference Series **527** (2014) 012004 doi:10.1088/1742-6596/527/1/012004

Realistic shell-model calculations and exotic nuclei

A Gargano[1], L Coraggio[1], A Covello[1,2], N Itaco[1,2]

[a]Istituto Nazionale di Fisica Nucleare,
Complesso Universitario di Monte S. Angelo, I-80126 Napoli
[b]Dipartimento di Fisica, Università di Napoli Federico II,
Complesso Universitario di Monte S. Angelo, I-80126 Napoli

E-mail: gargano@na.infn.it

Abstract. This paper presents a short overview of the shell-model approach with realistic effective interactions to the study of exotic nuclei. We first give a sketch of the current state of the art of the theoretical framework of this approach, focusing on the main ingredients and most relevant recent advances. Then, we present some selected results for neutron-rich nuclei in various mass regions, namely oxygen isotopes, $N = 40$ isotones, and nuclei around ^{132}Sn, to show the merit as well as the limits of these calculations.

1. Introduction

The shell model has long proved to be a main key to the understanding of nuclear structure. It provides the theoretical framework for a microscopic description of nuclear properties which is essentially based on the use of effective interactions. In fact, as is well known, within the shell-model approach only the particles outside a core made up of filled shells (valence particles) are considered to be active, and calculations are performed in a truncated Hilbert space, the so-called model space. The shell-model Hamiltonian acting only between the valence particles should, therefore, account for the neglected degrees of freedom, namely those of the core particles as well as of the excitations of valence particles above the chosen model space.

To this end, one can resort to empirical interactions, *i.e.* interactions containing adjustable parameters or obtained by treating the matrix elements themselves as free parameters. In both cases fitting procedures to reproduce the experimental data are required. Empirical interactions have been used in a number of shell-model calculations, providing in most cases a successful description of a variety of nuclear phenomena.

Clearly, a more fundamental approach to the shell model consists in starting from the interaction between free nucleons and constructing the Hamiltonian by means of many-body techniques, which leads to what is called "realistic effective interaction". This alternative way, based on a microscopic derivation of the shell-model effective interaction, has the great advantage that no adjustable parameter is needed and establishes a bridge between effective shell-model interactions and underlying nuclear forces. Significant progresses have been made along this line in the last two decades, and realistic shell-model calculations have been shown to provide an accurate description of nuclear structure properties for nuclei in various mass region both close to and far from the stability valley.

In this context, it should be mentioned that in many shell-model calculations, in particular those aiming at interpreting the new experimental data obtained at Radioactive Ion Beam (RIB)

Journal of Physics: Conference Series **527** (2014) 012004 doi:10.1088/1742-6596/527/1/012004

facilities, use has also been made of semi-empirical interactions, consisting in modified versions of realistic effective interactions. Here, however, we focus only on shell-model calculations employing genuine realistic effective interactions and give a brief description of the current status of this approach. We firstly review the main steps and the relevant recent developments involved in the perturbative technique used to derive the shell-model interaction from the bare nuclear potential. Then, we report some results we have obtained for various neutron-rich nuclei, as oxygen isotopes [1, 2], $N = 40$ isotones [3], and nuclei around ^{132}Sn [4], to show the practical value of this approach as well as its limits.

The basic ingredient of realistic shell-model calculations is the bare nuclear potential, for which, as is well known, there are various reliable models. However, we shall not discuss this point here. In the calculations concerned with the present contribution we have used two modern nucleon-nucleon (NN) potentials, the CD-Bonn [5] and N^{3}LOW potential [1], which fit equally well the NN scattering data. In particular, the calculations for $N = 40$ isotones and nuclei around ^{132}Sn have been performed with the former while results for oxygen isotopes have been obtained with the latter. It is worth recalling that, owing to its strong short-range repulsive behavior, the CD-Bonn potential cannot be used directly in deriving the effective interaction within the framework of a perturbative approach. In other words, it must be first renormalized, which is done by constructing a low-momentum potential $V_{\rm low-k}$ defined within a cutoff momentum Λ [6]. This is a smooth potential which preserves exactly the onshell properties of the original one. As concerns the N^{3}LOW potential, this is derived from the chiral perturbation theory with a sharp momentum cutoff at 2.1 fm^{-1} . It is conceived therefore as a low-momentum potential and no renormalization procedure is needed.

Finally, we would like to point out that our effective interactions are all based only on bare NN potentials, namely three-body forces are not explicitly considered. The study of the role of three body forces in nuclear structure has recently attracted great theoretical interest. Their effects have been evidenced within the framework of *ab initio* approaches for few nucleon systems. Until now, however, no shell-model calculation has been performed with an effective Hamiltonian derived by treating on equal footing both NN and $3N$ forces. This, which would imply the appearance of further core-polarization effects on the one- and two-body components as well as of an effective three-body term, is quite a complex project. Actually, only first-order contributions of the normal-ordered one- and two-body parts of $3N$ forces have been taken explicitly into account. In connection with the role of 3N forces, however, it should be mentioned that in a recent paper [7] an optimized NN interaction derived from the chiral effective field theory has been constructed, which seems to account for many aspects of nuclear structure without explicitly including $3N$ forces.

2. Theoretical framework

Let us start with the Schrödinger equation for a system of A nucleons interacting via two-body forces

$$H\Psi_\alpha = E_\alpha \Psi_\alpha, \tag{1}$$

where

$$H = T + V_{\rm NN}, \tag{2}$$

T being the kinetic energy and $V_{\rm NN}$ a low-momentum two-body potential, obtained through the $V_{\rm low-k}$ procedure [6] or purposely constructed to have a smooth perturbative behavior, with the addition of the Coulomb force for protons.

Now, by introducing an auxiliary one-body potential U the Hamiltonian (2) can be written as

Journal of Physics: Conference Series **527** (2014) 012004 doi:10.1088/1742-6596/527/1/012004

$$H = (T + U) + (V_{NN} - U) = H_0 + H_1, \tag{3}$$

namely as a one-body component H_0, which describes the independent motion of the nucleons, and a residual interaction H_1.

The effective Hamiltonian, $H_{\rm eff}$, is defined through the model-space eigenvalue problem

$$H_{\rm eff} P|\Psi_\alpha\rangle = H_0 P|\Psi_\alpha\rangle + H_1^{\rm eff} P|\Psi_\alpha\rangle = E_\alpha P\Psi_\alpha, \tag{4}$$

where the E_α and the corresponding Ψ_α are a subset of the eigenvalues and eigenfunctions of the original Hamiltonian. Clearly, $H_{\rm eff}$ acts only on the model space defined in terms of the eigenvalues of H_0 through the projection operator P.

A well-established approach to the determination of the effective Hamiltonian is given by the $\hat{Q}$-box folded-diagram expansion. A detailed description of this approach can be found in Refs. [8, 9], so we will not touch upon it here. We would like, however, to highlight the main points involved in the derivation of $H_{\rm eff}$, so as to make clear the present stage of development.

Within a degenerate model space, $PH_0P = \epsilon_0$, iterative techniques [10] can be used to construct the effective Hamiltonian. These, as the Krenciglowa-Kuo (KK) and the Lee-Suzuki (LS) ones, are based on an expansion of $H_1^{\rm eff}$ in terms of the $\hat{Q}$-box and its derivatives, the $\hat{Q}$-box being defined as

$$\hat{Q}(\epsilon) = PH_1P + PH_1Q\frac{1}{\epsilon - QHQ}QH_1P, \tag{5}$$

where the operator Q is the complement of P.

Once the $\hat{Q}$-box is calculated, we derive $H_1^{\rm eff}$ by means of the LS technique, which yields converged results after a small number of iterations. The calculation of the $\hat{Q}$-box, however, is the most critical step of our procedure. This is performed by writing the term $1/(\epsilon - QHQ)$ in equation (5) as a power series, which leads to a perturbative calculation to be performed under some approximations. A diagrammatic representation of the $\hat{Q}$-box, including one- and two-body diagrams up to third order in the interaction, is given in Ref. [9]. Clearly, only diagrams up to a finite order can be included and the state-of-the-art calculations do not go beyond the third order. One should also consider that the evaluation of the diagrams composing the $\hat{Q}$-box requires, in principle, a summation over all the states of the Q space. The truncation of this infinite space is, therefore, another source of approximation. Both the order-by-order and the intermediate-state convergence of the effective interaction expansion are briefly reviewed in [8, 11] where references to previous works are given, while in [9] they are discussed in detail focusing on p-shell nuclei. In section 3, results of shell-model calculations with effective interactions derived by including diagrams up to second as well as third order are reported and compared with experimental data.

In concluding this outline of the theoretical framework, it is worth mentioning two other points entering our procedure. Our effective Hamiltonian is derived for a two-valence-particle nucleus but is then used for systems with a larger number of valence particles. This means that we only include one- and two-body forces neglecting higher-body terms, which arise as an effect of the nuclear medium even if an NN potential is used. The one-body force gives the theoretical single-particle energies as resulting from the sum of the eigenvalues of H_0 and the one-body contributions of $H_1^{\rm eff}$. In most realistic shell-model calculations, however, these energies are replaced by values taken from experiment. Furthermore, we note that the perturbation expansion is performed for $H_1 = V_{\rm NN} - U$, which gives rise to $(V - U)$-insertion diagrams (see [9]), which are in general neglected with the exception of the first-order ones. As a matter of fact, these are exactly zero only when taking for U a self-consistent Hartree-Fock potential, whereas an harmonic oscillator potential is generally used. In the next section we shall present results obtained by using both theoretical and experimental single-particle energies as well as by including or not $(V-U)$-insertion diagrams beyond the first order.

3. Results for various mass regions

We discuss and compare with experiment results for O isotopes, $N=40$ isotones, and ^{132}Sn neighboring nuclei, which have been obtained within the shell-model framework using realistic effective interactions. Most of the results presented in this section have already been given in previous works [2, 3, 4], but our aim here is to give a panoramic view of them in order to show the ability of these effective interactions to describe neutron-rich nuclei in different mass regions.

3.1. Oxygen isotopes

One of the challenging problems in modern nuclear structure studies concerns the location of the neutron drip line, namely the limit of existence of neutron-rich systems. In this context, the oxygen isotopes play an important role. The drip line for $Z=8$ is quite close to the stability valley, in contrast with the situation that occurs for other isotopes in the same mass region. As a matter of fact, the limit for oxygen isotopes is established at $N=16$, the last stable one being ^{18}O. Several calculations [12, 13, 14] have been recently performed suggesting that the explanation for this limit resides in $3N$ forces.

The effective Hamiltonian for oxygen isotopes is derived for the *sd* space with ^{16}O as inert core starting from the N^3LOW potential with a sharp cutoff at 2.1 fm^{-1}, which is a low-momentum realistic NN interaction derived from the chiral perturbation theory. All diagrams up to third order are taken into account in the calculation of the $\hat{Q}$-box including the $(V-U)$-insertion diagrams. As regards the energies of the three single-particle orbitals of the *sd* space, we employ the theoretical values, as arising from the one-body contributions of $H_{\rm eff}$.

In figure 1, the calculated excitation energies of the yrast 2^+ states are compared with the experimental values [15]. We see that our predictions are in very good agreement with the observed energies. In particular, we reproduce the rise from $A=20$ to 22 as well as that from $A=22$ to 24 which are related to the $N=14$ and 16 subshell closure, respectively.

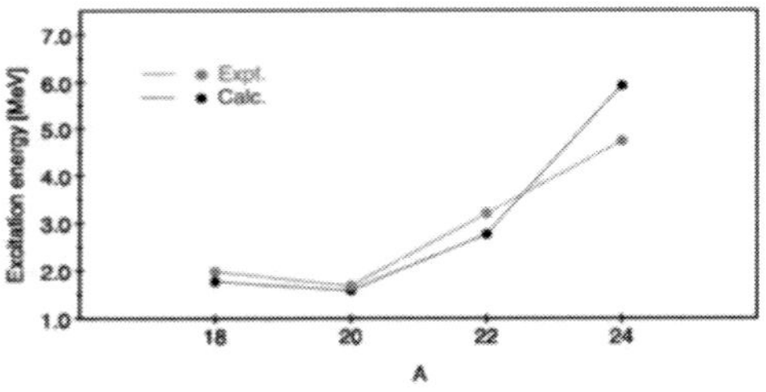

Figure 1. (Color online) Excitation energies of yrast 2^+ states in O isotopes from $A=18$ to 24 .

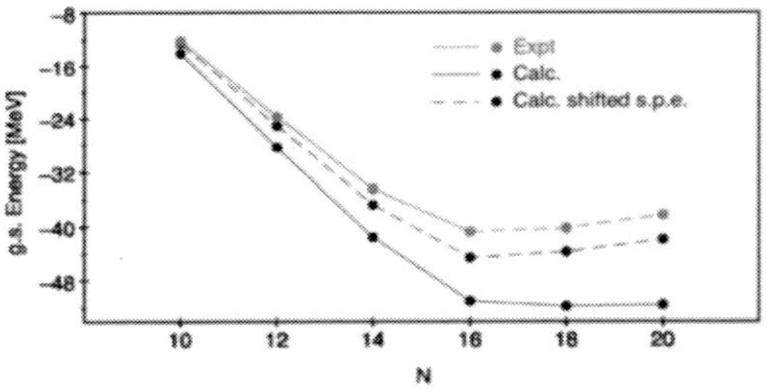

Figure 2. (Color online) Ground-state energies for O isotopes from $N=10$ to 20 (see text for details).

To test the ability of our interaction to reproduce the location of the drip line for oxygen isotopes, we have calculated the ground-state (g.s.) energies, which are compared in figure 2 with the experimental values [16]. The theoretical curve lies below the experimental one, with a discrepancy increasing with the number of neutrons. Actually, our calculations overestimate the experimental g.s. energies and fail to predict ^{26}O and ^{28}O as unbound nuclei. In the same figure, however, we show the results that are obtained by an upshift (427 keV) of the calculated single-particle spectrum so as to reproduce the experimental g.s. energy of ^{17}O relative to ^{16}O. The new curve moves up coming closer to the experimental one and shows the right slope from $N = 16$ on. This highlight some inaccuracy in our one-body effective Hamiltonian which may be traced to the lack of three-body forces.

3.2. $N = 40$ isotones

We now present and discuss our results for the $N = 40$ isotones Ca, Ti, Cr, Fe, and Ni. Before doing so, however, a few comments are in order. The experimental behavior of the excitation energy of the 2^+ yrast state in Ni isotopes, and in particular the sizable increase in ^{68}Ni with respect to the two neighboring even isotopes, evidences a subshell closure at N=40. For a decreasing number of protons this closure disappears and the onset of a collective behavior is observed. This issue is currently the subject of great experimental and theoretical interest (see [17, 18, 19, 20] and references therein), and the appearance of the collective behavior has been traced to the correlations between the quadrupole-partner neutron orbitals $0g_{9/2}$ and $1d_{5/2}$ [17].

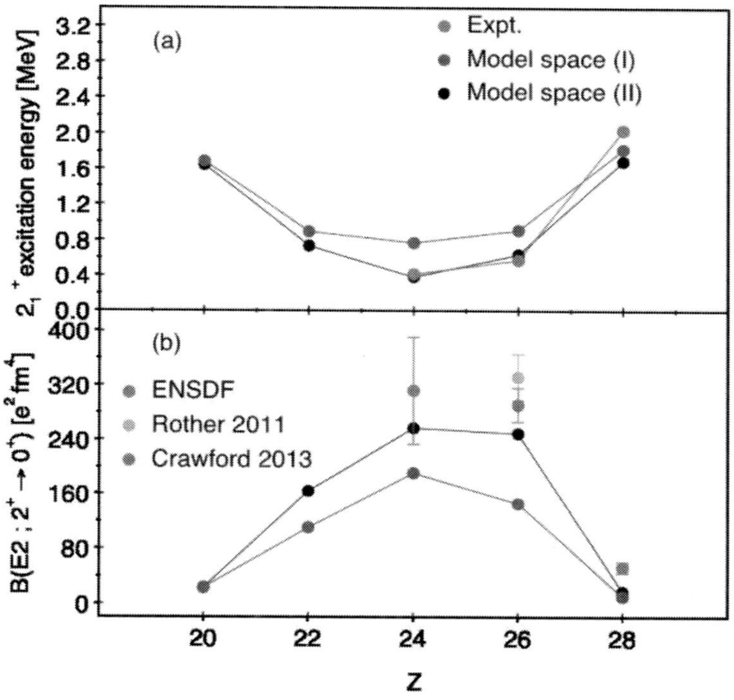

Figure 3. (Color online) (a) Experimental (ENSDF [15], Rother [18], Crawford [20]) and calculated excitation energies of the yrast 2^+ states and (b) $B(E2; 2_1^+ \rightarrow 0_1^+)$ for the $N = 40$ isotones.

With the aim of directly testing the role played by the neutron $1d_{5/2}$ orbital, we have performed realistic shell-model calculations taking ^{48}Ca as inert core and considering two different model spaces. The first one, model space (I), is spanned by the proton $0f_{7/2}$ and $1p_{3/2}$ orbitals and by the neutron $1p_{3/2}$, $1p_{1/2}$, $0f_{5/2}$, $0g_{9/2}$ orbitals while the second one, model space (II), includes the same orbitals with the addition of the neutron $1d_{5/2}$ one. We calculate the effective interaction starting from the CD-Bonn NN potential renormalized by way of the $V_{\text{low-k}}$ approach with a cutoff momentum $\Lambda = 2.6$ fm^{-1}. As for the O isotopes, all diagrams up to third order are taken into account in the calculation of the $\hat{Q}$-box, but at variance with the previous case the single-particle energies are determined using experimental data. More details on our effective Hamiltonian, including the values of the single-particle energies and a list of the two-body matrix elements in the model spaces (I) and (II) are given in Ref. [3].

In figure 3, we report the excitation energies of the yrast 2^+ states and the $B(E2; 2_1^+ \rightarrow 0_1^+)$ transition rates for the $N = 40$ isotones as a function of Z. As regards the energies, we see that both calculated curves reproduce well the observed behavior, but the inclusion of the neutron $1d_{5/2}$ orbital is essential for a quantitative agreement with the experimental data. The role of this orbital appears even more relevant when looking at the $B(E2)$ transition rates. Only the $B(E2)$'s for ^{64}Cr, ^{66}Fe, and ^{68}Ni are experimentally known, and these values for the former two nuclei are significantly underestimated by our calculations with model space (I), while this is not the case when model space (II) is used. Our results, therefore, confirm the connection between the $1d_{5/2}$ orbital and the appearance of collectivity in Ti, Cr, and Fe, as evidenced by the lowering of the 2_1^+ state and the increase in the corresponding $B(E2)$. To better understand this point, we show in table 1 the occupation numbers of the proton $\pi f_{7/2}$ and neutron $\nu g_{9/2}$ and $\nu d_{5/2}$ orbitals for the ground state of Ti, Cr, Fe, and Ni.

Table 1. Occupation numbers of the proton $\pi f_{7/2}$ and neutron $\nu g_{9/2}$ and $\nu d_{5/2}$ orbitals for the ground state of the $N = 40$ isotones (see text for details).

		Ti	Cr	Fe	Ni
$\pi f_{7/2}$	(I)	1.87	3.64	5.61	7.89
	(II)	1.69	3.31	5.31	7.85
$\nu g_{9/2}$	(I)	3.23	2.88	1.75	0.42
	(II)	3.72	3.73	2.81	0.47
$\nu d_{5/2}$	(II)	0.36	0.57	0.33	0.05

We see that there are no substantial differences between the occupation numbers obtained with model spaces (I) and (II) for Ni. In both cases, a very low occupancy of protons (0.11-0.15) is found above the $0f_{7/2}$ orbital as well as of neutrons (0.42-0.52) above the $0f_{5/2}$ orbital, the latter result implying a clear manifestation of subshell closure at $N = 40$. Note that, with model space (II), only 0.05 neutrons occupy the $1d_{5/2}$ orbital which gives reason for its minor role in the description of ^{68}Ni. However, when decreasing the number of protons the two model spaces lead to significantly different results. In fact, when going from model space (I) to model space (II) we find an increase in the occupation number of the neutron $0g_{9/2}$ orbital as well as a depletion of the proton $0f_{7/2}$ orbital. From table 1, we also see an increase in the occupation number of the neutron $1d_{5/2}$ orbital for Ti, Cr, and Fe with respect to Ni. In other words, neutron excitations above the fp orbitals are favored for protons numbers below the 28 closed shell. This may be traced to a reduction of the neutron $0g_{9/2} - 0f_{5/2}$ gap as well as to the quadrupole-quadrupole component of the effective interaction acting between the $0g_{9/2}$ and $1d_{5/2}$ orbitals. The former argument is clearly evidenced in figure 4, where we report the behavior of the effective single-neutron energies as a function of Z.

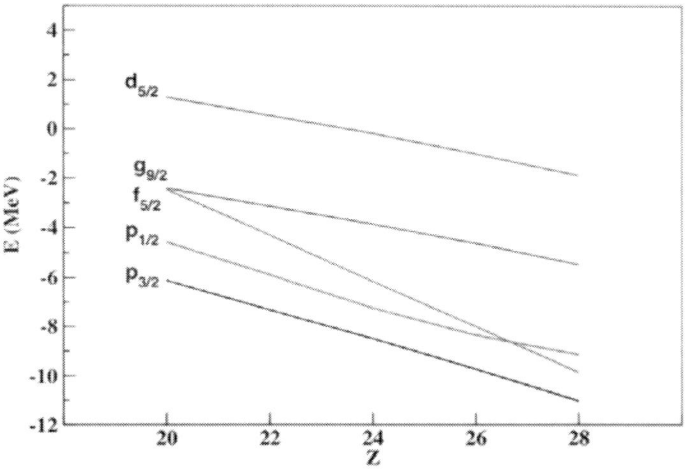

Figure 4. (Color online) Calculated effective single-neutron energies for the $N = 40$ isotones.

3.3. ^{132}*Sn neighboring*

In the last ten years or so, nuclei in the mass region around ^{132}Sn have become accessible to experimental studies thanks to new RIB facilities and the development of sophisticate detection techniques. A large amount of experimental information has been acquired, but data for nuclei with $Z \sim 50$ and $N > 82$ still remain scarce. A further step in this direction will be certainly made with the next generation of RIB facilities. The available data for $N > 82$ nuclei have not evidenced changes in the shell structure, as it is was instead the case for the lighter nuclei discussed above. However, some anomalies have been observed. A notable one, which have posed interesting questions is, for instance, the asymmetric behavior with respect to $N = 82$ of the excitation energy of the yrast 2^+ state in both Sn and Te isotopes.

We have conducted several studies [21, 22, 23, 24] on nuclei of this region, in all of them taking ^{132}Sn as closed core and assuming that proton particles and neutron holes occupy the five orbitals of the 50-82 shell, while neutron particles are in the six orbitals of the 82-126 shell. The single-particle and single-hole energies have been determined from experiment and the two-body effective interaction has been derived from the CD-Bonn NN potential renormalized by means of the $V_{\rm low-k}$ potential with $\Lambda = 2.2$ fm^{-1}. As for the calculation of the $\hat{Q}$-box, all diagrams up to second order have been taken into account, except the $(V-U)$-insertion diagrams which are limited to first order. All these studies, focused on energy spectra and electromagnetic properties, and more recently on spectroscopic factors and atomic masses, have led to results in very good agreement with experiment.

A main outcome of our work is that the pairing force plays a key role in determining the various properties of ^{132}Sn neighbors, as for instance the decrease in energy of the 2^+ state in ^{134}Sn and ^{136}Te. Our effective interaction, in fact, generates a pairing force between two neutrons in the 82-126 shell which is significantly weaker than that between two neutron holes or two proton particles in the 50-82 shell. Let us take, for instance, the $(\nu f_{7/2})^2$ configuration. The corresponding $J = 0^+$ matrix element is about -0.6 MeV to be compared to the value of -1 MeV or less for the $(\nu h_{11/2})^{-2}$ and $(\pi g_{7/2})^2$ configurations. This issue is discussed in [25], where we have investigated the origin of the pairing force within our microscopic framework and found that the above differences result from a large reduction of the core-polarization contributions to the neutron effective interaction for $N > 82$. It should be mentioned that the role of the pairing force in this mass region was also recognized in Refs.[26, 27].

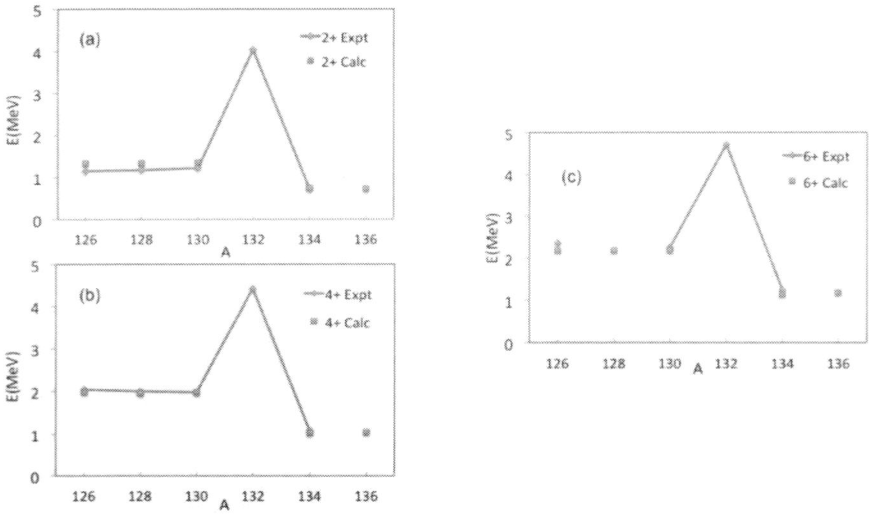

Figure 5. (Color online) Calculated and experimental [15] excitation energies of the yrast (a) 2^+, (b) 4^+, and (c) 6^+ states in tin isotopes from $A = 126$ to 136.

To illustrate the quality of our results, we focus on the low-energy spectra and $B(E2)$ transition rates in Sn and Te isotopes. Then, as a final example, we discuss the odd-even staggering (OES) of binding energies for $N = 81$ and 83 isotones to show the predictive power of our approach in the calculation of binding energies.

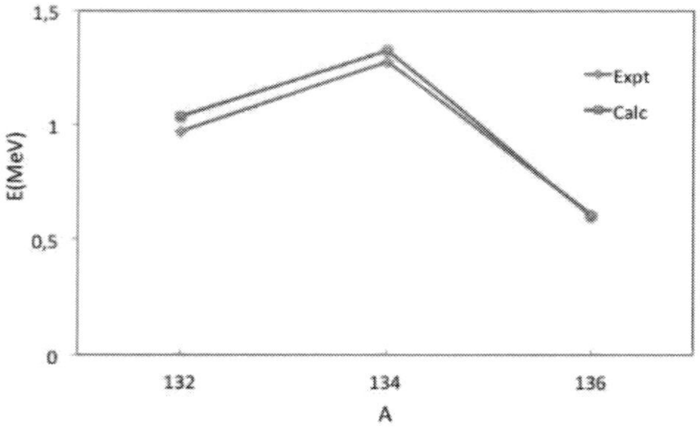

Figure 6. (Color online) Calculated and experimental [15] excitation energies of the yrast 2^+ states in tellurium isotopes with $A = 132$, 134, and 136.

The theoretical and experimental excitation energies of the yrast 2^+, 4^+, and 6^+ states in Sn isotopes from $A = 126$ to 136 are compared in figure 5, while for Te isotopes we consider only the yrast 2^+ states for $A = 132$, 134, and 136, as shown in figure 6. We see that the available experimental data are very well reproduced by our calculation, including the marked decrease in the energies at $N = 84$. Note that for ^{136}Sn, for which no spectroscopic information is available, we predict a low energy spectrum quite similar to that of ^{134}Sn.

In table 2 the calculated $B(E2; 2^+_1 \to 0^+_1)$ transition rates, obtained with an effective proton and neutron charge of 1.55 and 0.7e, respectively, are compared with the experimental values. We see that in all cases, except for ^{136}Te, the agreement between theory and experiment is very good. More precisely, the discrepancy ranges between 12 and 22%, becoming larger than 50% in ^{136}Te. It is worth mentioning, however, that a new higher precision measurement of the $B(E2)$ in ^{136}Te is certainly needed, the present experimental value being determined from a reanalysis [23] of the experiment of Ref. [28]. At the same time, the predicted overestimation may be seen as an indication that our effective interaction lacks some accuracy. Actually, we pin down the main feature of the 2^+ state wave function, which turns out to be dominated by neutron excitations, but find that a 35% remaining weight is fragmented over various components, which are likely to lead to an enhancement of the B(E2).

Table 2. Experimental [23, 29] and calculated $B(E2; 2^+_1 \to 0^+_1)$ (in W.u.) for Sn and Te isotopes .

	Expt	Calc
^{130}Sn	1.2(3)	1.4
^{134}Sn	1.4(2)	1.6
^{132}Te	10 (1)	7.8
^{134}Te	5.6 (6)	4.9
^{136}Te	5.9 (9)	9.9

As mentioned above, we conclude this section by discussing the neutron OES as defined by three-point formula

$$\Delta^{(3)}(N, Z) = \frac{1}{2}[B(N+1, Z) + B(N-1, Z) - 2B(N, Z)]. \tag{6}$$

By using our binding energies for ^{130,134}Sn, $^{132-136}$Te, and $^{134-138}$Xe, we have calculated the neutron OES for the $N = 81$ isotones ^{131}Sn, ^{133}Te, ^{135}Xe and for the $N = 83$ isotones ^{133}Sn, ^{135}Te, ^{137}Xe. They are compared with the experimental values in figure 7. We see that the agreement between theory and experiment is very good. In particular, our calculations give a quantitative description of the gap between the $N = 81$ and 83 lines at $Z = 50$ as well as of its decrease when adding two and four protons. The drop of about 0.5 MeV in the observed OES for Sn when crossing N= 82 is a consequence of the different pairing properties for neutron particles and holes with respect to the N= 82 closed shell. When going to Te and Xe, the $N = 81$ and 83 lines come closer to each other as a result of the proton-neutron effective interaction. The two lines would be indeed parallel should one ignore this interaction. From figure 7, we see that the pn interaction has an opposite effect on the $N = 81$ and $N = 83$ isotones, which is clearly related to its repulsive and attractive nature in the particle-hole and particle-particle channel, respectively. On the other hand, this effect is not very large either in ^{133,135}Te or in ^{135,137}Xe, since it results essentially from the difference between the contributions of the pn interaction to the energies of the odd and neighboring even isotopes.

Journal of Physics: Conference Series **527** (2014) 012004 doi:10.1088/1742-6596/527/1/012004

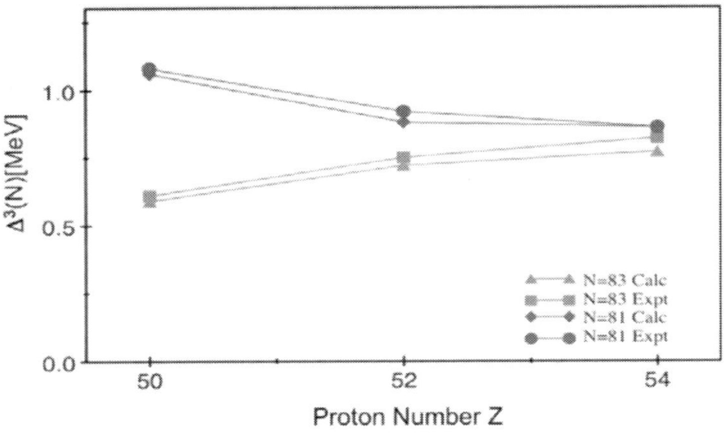

Figure 7. (Color online) Calculated and experimental [30] odd-even staggering for the $N = 81$ and 83 isotones.

4. Summary and conclusions

In this paper, we have given a short overview of the shell-model approach with realistic effective interactions to the study of exotic nuclei. We have reviewed the main steps and the relevant recent developments involved in the derivation of these interactions and reported some selected results for neutron-rich nuclei in various mass regions, as oxygen isotopes, $N = 40$ isotones, and nuclei around ^{132}Sn. Actually, we have tried to give a panoramic view of our results, most of them already given in previous works, in order to illustrate the practical value of this approch as well as some of its limits.

The quality of our results has been highlighted in section 3 by comparison with the available experimental data. This shows that, with a few exceptions, observed spectroscopic properties as well as binding energies are well reproduced by the theory, proving the predictive power of realistic shell-model calculations. We have seen, however, that, when using theoretical single-particle energies, these calculations fail to reproduce the binding energies of the O isotopes. This may be traced to the lack of a three-body force, or more precisely to the contributions of this force to the one-body term of the effective Hamiltonian, while our results do not seem to evidence the need of an explicit three-body term or of renormalization of the two-body interaction.

Based on our calculations, we can conclude that nuclear structure results do not depend substantially on the choice of the free NN potential one starts with, when employing phase-shift equivalent low-momentum potentials. In this connection, we note that the effective interactions for the $N = 40$ isotones and ^{132}Sn neighbors are derived from the $V_{\rm low-k}$ of the CD-Bonn potentials with different cutoffs. We have also verified [31] that moderate variations of the cutoff do not change significantly the shell-model results. However, further study is certainly needed on this point which has been examined essentialy for interactions derived at second order in the $\hat{Q}$-box.

The calculation of the $\hat{Q}$-box, as mentioned above, is one of the most critical point in our approach. Its convergenge properties are investigated in [9]. Here we point out that the quality of our results for nuclei around ^{132}Sn is very good when using a second-order $\hat{Q}$ box. One should consider, however, that calculations in this region are limited to nuclei near shell closures and cannot exclude that effects related to third order may appear for more valence neutrons (see comments in [11]). To conclude, we may say that realistic-shell model calculations represent by now a very effective tool to study nuclear structure. This makes it challenging to try to clarify some remaining open questions.

References

[1] Coraggio L, Covello A, Gargano A, Itaco N, Entem D R, Kuo T T S, and Machleidt R 2007 *Phys. Rev. C* **75** 024311
[2] Coraggio L, Covello A, Gargano A, Itaco N, and Kuo T T S 2011 *J. Phys. Conf. Ser.* **312** 092021
[3] Coraggio L, Covello A, Gargano A, and Itaco N 2014 submitted to *Phys. Rev. C*
[4] Coraggio L, Covello A, Gargano A, and Itaco N 2013 *Phys. Rev. C* **88** 041304(R) and references therein
[5] Machleidt R 2001 *Phys. Rev. C* **63** 024001
[6] Bogner S, Kuo T T S, Coraggio L, Covello A, and Itaco N 2002 *Phys. Rev. C* **65** 051301(R)
[7] Ekström A, Baarsden G, Forssén, Hagen G, Hjorth-Jensen M, Jansen G R, Machleidt R, Nazarewicz W, Papenbrock T, Sarich J, and Wild S M 2013 *Phys. Rev. Lett.* **110** 192502
[8] Coraggio L, Covello A, Gargano A, Itaco N, and Kuo T T S 2009 *Prog. Part. Nucl. Phys.* **62** 135
[9] Coraggio L, Covello A, Gargano A, Itaco N, and Kuo T T S 2012 *Ann. Phys.* **327** 2125
[10] Suzuki K and Lee S Y 1989 *Prog. Theor. Phys.* **64** 2091
[11] Covello A and Gargano A 2010 *J. Phys. G* **37** 064044
[12] Otsuka T, Suzuki T, Holt J D, Schwenk A, and Akaishi Y 2010 *Phys. Rev. Lett.* **105** 032501
[13] Hagen G, Hjorth-Jensen M, Jansen G R, Machleidt R, and Papenbrock T 2012 *Phys. Rev. Lett.* **108** 242501
[14] Hergert H, Binder S, Calci A, Langhammer J, and Roth R 2013 *Phys. Rev. Lett.* **110** 242501
[15] Data extracted using the NNDC On-line Data Service from the ENSDF database, file revised as of January 8, 2014
[16] Wang M, Audi G, Wapstra A H, Kondev F G, MacCormick M, Xu X, and Pfeiffer B 2012 *Chinese Phys. C* **36** 1603
[17] Lenzi S M, Nowacki F, Poves A, and Sieja K 2010 *Phys. Rev. C* **82** 054301
[18] Rother W *et al.* 2011 *Phys. Rev. Lett.* **106** 022502
[19] Baugher T *et al.* 2012 *Phys. Rev. C* **86** 011305
[20] Crawford H L *et al.* 2013 *Phys. Rev. Lett.* **110** 242701
[21] Coraggio, L, Covello A, Gargano A, and Itaco N 2009 *Phys. Rev. C* **80** 021305(R) and references therein
[22] Covello A, Coraggio L, Gargano A, and Itaco N 2011 *J. Phys. Conf. Ser.* **267** 01201 and references therein
[23] Danchev M *et al.* 2011 *Phys. Rev. C* **84** 061306(R)
[24] Coraggio L, Covello A, Gargano A, and Itaco N 2013 *Phys. Rev. C* **87** 034309
[25] Covello A, Gargano A, and Kuo T T S 2013 *Fifty Years of Nuclear BCS* ed Broglia R A and Zelevinsky V (Singapore: World Scientific) pp. 169-178
[26] Terasaki J, Engel J, Nazarewicz W, and Stoitsov M 2002 *Phys. Rev. C* **66** 054313
[27] Shimizu N, Otsuka T, Mizusaki T, and Honma M 2004 *Phys. Rev. C* **70** 054313
[28] Radford D C *et al.* 2002 *Phys. Rev. Lett.* **88** 222501
[29] Radford D C *et al.* 2005 *Nucl. Phys. A* **752** 264c
[30] Hakala J *et al.* 2012 *Phys. Rev. Lett.* **109** 032501
[31] Covello A, Coraggio L, Gargano A, and Itaco N 2005 *J. Phys. Conf. Ser.* **20** 137

Journal of Physics: Conference Series **527** (2014) 012005 doi:10.1088/1742-6596/527/1/012005

Core polarization and neutron halos

F Barranco[1], **R A Broglia**[2,3,4], **A Idini**[5], **G Potel**[6,7,8], **E Vigezzi**[3]

[1] Departamento de Física Aplicada III, Escuela Superior de Ingenieros, Universidad de Sevilla, 41092 Sevilla, Spain
[2] Dipartimento di Fisica, Università degli Studi di Milano, 20123 Milano, Italy
[3] INFN Sezione di Milano, 20123 Milano Italy
[4] The Niels Bohr Institute, University of Copenhagen, Denmark
[5] Institut für Kernphysik, Technische Universität Darmstadt, 64289 Darmstadt, Germany
[6] Lawrence Livermore National Laboratory, Livermore, CA 94551, USA
[7] National Superconducting Cyclotron Laboratory, Michigan State University, East Lansing, MI 48824, USA
[8] Department of Physics and Astronomy, Michigan State University, East Lansing, MI 48824-1321, USA

E-mail: vigezzi@mi.infn.it

Abstract. The effects of core polarization on the structure of Li and Be halos and their consequences on one-and two-neutron transfer reactions are discussed.

1. Introduction

There exists experimental evidence pointing to the fact that core polarization plays an important role in the structure of the one- and two-neutron halo nuclei ^{11}Li and ^{11}Be, ^{12}Be, as well as of the unbound nucleus ^{10}Li. Particularly relevant in this regard are the parity inversion of the ground state of ^{10}Li and ^{11}Be with respect to mean field predictions, and the cross sections associated with knockout reactions and with one- and two-neutron transfer reactions leading to excited states of the ^{10}Be and ^{9}Li cores, like ^{11}Be(p,d)^{10}Be or ^{11}Li(p,t)^{9}Li (see [1] for a recent review). While the interweaving of single-particle motion and zero-point fluctuations renormalizes the ground state mean field properties in all nuclei, its effects are particularly relevant in the weakly bound, easily polarizable halo nuclei [2]. This paper summarizes results obtained [3, 4, 5, 6] by treating explicitly the basic processes involving the collective vibrations of the core, which are associated with the diagrams shown in figure 1. The polarization diagram (figure 1(a)) describes the emission and reabsorption of a phonon by a valence neutron which then increases its binding energy. The correlation diagram (figure 1(b)) accounts for the Pauli blocking of ground state correlations: due to the particle-vibration coupling, there are processes contributing to the ground state energy of the core, in which a neutron, initially in a hole state below the Fermi energy, virtually emits a phonon and jumps into an unoccupied state above the Fermi energy, finally returning to its initial state by reabsorbing the phonon. These processes become partially forbidden when a valence neutron is added in the same state to form the halo. This loss of correlation energy associated with core fluctuations must be subtracted in calculating the binding energy of the valence neutron. This effect depends in a specific way on the angular momentum and parity of the single-particle states and of the phonons. We note

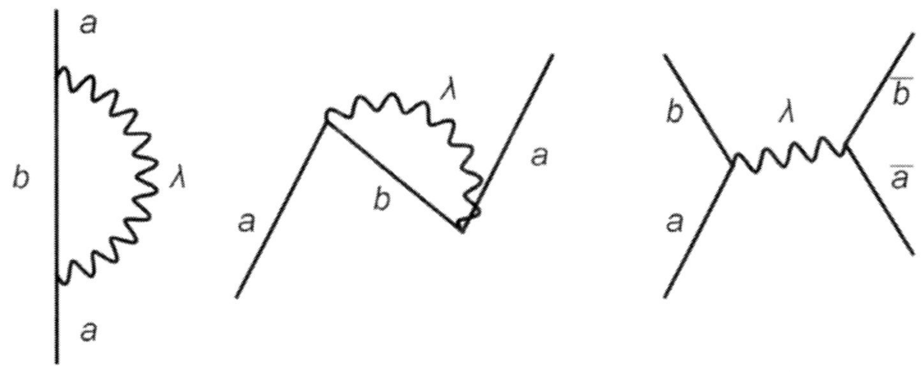

Figure 1. Processes coupling quasiparticles with collective vibrations, which renormalize the normal and abnormal densities obtained in mean field calculations: (a) polarization, (b) correlation, (c) induced pairing interaction, processes.

that a similar effect is operative in the pairing channel, associated with the blocking of collective pairing fluctuations. The latter effect, which tends to act equally on all the valence orbitals, will not be taken into account in the following. Finally, diagram (c) represents the exchange of collective vibrations between pairs of neutrons, which must be added to the bare N-N interaction and represents an essential contribution to the binding of two-neutron halos.

The importance of diagram (a) (associated with quadrupole vibrations, see below) for the phenomenon of parity inversion in halo nuclei has been recognized since a long time [7, 8]. Its effects on the structure of ^{11}Be and ^{12}Be have been studied treating the core as a deformed rotor [9, 10, 11]. It is well known that the couplings with core lead to important modifications of the radial form factors entering DWBA calculations for transfer reactions [12, 13]. They have been considered in the calculation of one-neutron transfer reactions involving ^{11}Be, both within DWBA [10, 14], and, more recently, within the Faddeev scheme [15], as well as in the calculation of breakup [16]. However these calculations assume that the hole states are completely filled, and diagram (b) has not been considered. The effect of Pauli principle associated with pairing fluctuations was instead included in the calculations of Myo *et al.* (see [17] and [18] and references therein), who, on the other hand, have not included the coupling to quadrupole surface vibrations. An early recognition of the importance of diagram (c) for the binding for two-neutron halos can be found in [19].

We notice that the dynamical processes depicted in figure 1 are taken into account to some extent in calculations of the halo structure which do not include the internal core degrees of freedom, but use parity-dependent neutron-core potentials to fit the observed position of single-particle levels in one-neutron halos, and density-dependent interactions to reproduce the experimental binding energies in two-neutron halos (see for example [20, 21, 22]). Within such an approach, however, it is not straightforward to recognize the common origin of a variety of effects observed in one- and in two-neutron halos. We also remark that reproducing the structure of halo nuclei still represents a challenge for ab initio calculations [23, 24, 25].

2. Theory and results

The one-neutron halo system will be described by the Hamiltonian [26]

$$\hat{H} = \hat{H}_F + \hat{H}_B + \delta\hat{U}, \tag{1}$$

where $\hat{H}_F$ is associated with the mean field, H_B is associated with the dynamics of the core and $\delta\hat{U}$ couples single-particle motion with the collective vibrations. We assume that the mean field potential is a spherical Saxon-Woods potential $U(r)$ parametrized as in [27] (equations (2-181, 2-182)), except for a slight adjustment of the depth. We include states of angular momentum quantum numbers $s_{1/2}, p_{1/2}$ and $d_{5/2}$, obtained by solving the Schrödinger equation taking into account the (discretized) continuum, within a box of radius R up to an energy E_{cut}.

We include collective vibrations of multipolarity $\lambda^\pi = 1^-, 2^+$ and 3^-. They are calculated within QRPA using a separable force, with a coupling constant adjusted to reproduce the energy and the transition probability of the lowest states. The coupling term is given by

$$\delta\hat{U} = \sum_{\lambda\mu} \frac{\beta_\lambda}{\sqrt{2\lambda+1}} (i)^{-\lambda} \left\{ \left[\hat{\Gamma}^\dagger_{\lambda\mu} + (-)^{\lambda+\mu} \hat{\Gamma}_{\lambda-\mu} \right] \times \sum_{kk'} \langle k| - R_0 \frac{dU}{dr} Y^*_{\lambda\mu} |k'\rangle a^\dagger_k a_{k'} \right\} \tag{2}$$

where the operators $\hat{\Gamma}^\dagger_{\lambda\mu}$ and $\Gamma_{\lambda\mu}$ create and annihilate the quantum vibrations of the core, characterized by their deformation parameter β_λ. In the case of 1^- dipole transitions, we adjust the coupling constants, so that the strength associated with the spurious translational mode is correctly shifted at zero energy. We then treat the interweaving of single-particle motion and of collective degrees of freedom arising within the framework of the Nuclear Field Theory, making use of Bloch-Horowitz perturbation theory [28, 29, 30].

Let us first consider the case of one-neutron halos. The eigenvalues of the dressed single-particle states associated with given angular momenta l, j ($s_{1/2}, p_{1/2}$ and $d_{5/2}$) were obtained by diagonalizing energy dependent matrices, schematically shown in the left part of figure 2. The matrix elements connect the basis of unperturbed single-particle states corresponding to the quantum numbers n, l, j. It is important to notice that once the mean field and the properties of the collective vibrations have been fixed, there are no free parameters in the calculations.

The effect of the renormalization processes on the parity inversion in the case of the valence neutron in ^{10}Li is shown in the right panel of figure 3. The $p_{1/2}$ single-particle state is bound by about 1.2 MeV in the initial mean field potential, while there is no $s_{1/2}$ strength close to threshold. The attractive polarization diagram (figure 2(b), left panel) couples the $s_{1/2}$ states with intermediate states consisting of one particle plus a vibrational state of the type $d_{5/2} \otimes 2^+$, leading to a correlated s state, lying close to threshold. On the other hand, the repulsive Pauli correction diagram (figure 2(c), left panel), also associated with quadrupole vibrations but involving the $p_{3/2}$ hole orbital, has a strong effect on the $p_{1/2}$ state, raising its energy from -1.2 MeV to about +0.5 MeV, and leading to parity inversion. A detailed comparison with experiment in the case of ^{10}Li requires a consideration of the odd proton, which has been considered as a spectator. A more direct comparison is possible in the case of ^{11}Be, where one finds $E_{s_{1/2}} = -0.48$ MeV and $E_{p_{1/2}} =$ -0.27 MeV, to be compared with the experimental values ($E_{s_{1/2}} = -0.50$ MeV and $E_{p_{1/2}} =$ -0.18 MeV), and with the unperturbed single-particle energies ($E_{s_{1/2}} \approx 0.14$ MeV and $E_{p_{1/2}} =$ - 3.12 MeV). We find a $d_{5/2}$ state close to threshold. A resonant state is experimentally observed at $E_{d_{5/2}} = 1.28$ MeV. The $d_{5/2}$ state lies at higher energy in the case of ^{10}Li.

We now address the case of the two-neutron halos in ^{11}Li and ^{12}Be. The associated energy-dependent matrix is schematically represented in the right part of figure 2. In this case the matrix elements connect pairs of particles lying in the $s_{1/2}, p_{1/2}$ and $d_{5/2}$ orbitals coupled to angular momentum and parity 0^+. Although the single-particle states lie in the continuum, coupling states with different number of nodes can lead to a localized wavefunction. We include the matrix elements of the Argonne bare pairing interaction in the $^1S_0$ channel (indicated by a dashed line in diagram (a)), which can scatter pairs up to high energy. This requires a high value of E_{cut}, $E_{cut} \approx 500$ MeV. We also include the attractive induced interaction arising from the exchange of collective vibrations (see diagram (d)). The exchange of 1^- vibrations is particularly

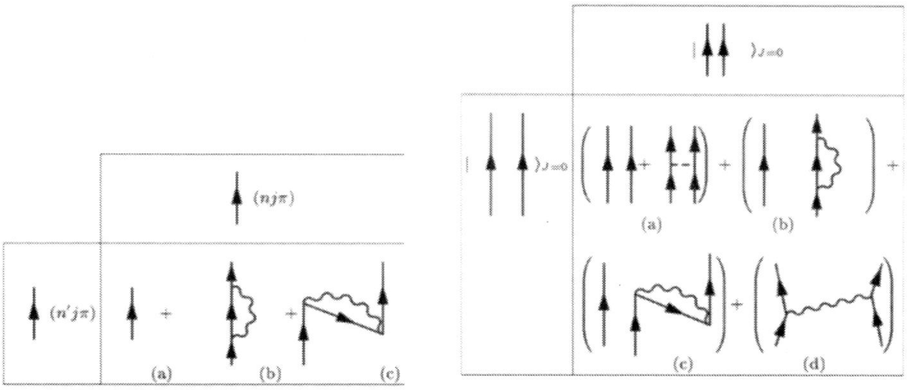

Figure 2. Energy-dependent matrices to be diagonalized self-consistently, to obtain the correlated wavefunction for one- (left panel) and two-neutron halo nuclei (right panel) (see Ref. [4]).

important in the case of ^{11}Li, where substantial dipole strength is found at low energy (pigmy dipole). Most of this low-energy wavefunction is built out of excitations which occupy, to some extent, the same particle states occupied by the loosely bound neutrons we are studying, and the corresponding particle-vibration matrix elements have been corrected from Pauli violating contributions following the nuclear field theory rules (see e.g. [26] and [31]). However, the present RPA calculation of the soft dipole resonance takes explicitly into account its interplay with the giant dipole resonance of the core, and the pigmy dipole wavefunction also includes forwards- and backwardsgoing amplitudes associated with particle-hole transitions involving core orbitals.

The resulting two-neutron separation energies for ^{11}Li (0.33 MeV) and ^{12}Be (3.58 MeV) are in overall agreement with the experimental values (0.37 MeV and 3.67 MeV). The effect of the pairing interaction leads from the uncorrelated wavefunction $|s^2_{1/2}(0) >$ to the correlated one, $|\tilde{0} >$ (see figure 3 (II) in the case of ^{11}Li). The main contribution to the binding energy arises from the induced interaction, while that associated with the bare nucleon-nucleon Argonne potential is quite small (of the order of 100 keV in the case of ^{11}Li). The bare interaction is not efficient in coupling low-angular momentum, extended wavefunctions. It is difficult to compare this result with three-body calculations which start from single-particle levels fitted to the experimental value or use density dependent interactions. It is known that the binding energy depends in a delicate way on the assumed position of the resonance energies [21]. We also calculate the energy of the excited 0^+ states; in particular, the first excited state in ^{12}Be is calculated to lie at 2.04 MeV, to be compared with the experimental value of 2.24 MeV.

The diagonalization of the energy-dependent matrix provides detailed information about the "hidden" (virtual) components of the halo wavefunction namely, of the components of the wavefunction involving core vibrations. This, coupled with reaction softwares, leads to definite predictions for the spectroscopic amplitudes and the absolute cross sections associated with one- and two-neutron transfer reactions. We remark that the coupling with core vibrations can affect the radial dependence of transfer form factors in an important way [10, 12, 13, 14, 15], so that a comparison of theory with experiment should be carried out by comparing absolute cross sections. A comparison with spectroscopic factors derived from the experimental cross sections can be at most indicative. With these limitations, the calculated admixture of the $d_{5/2} \otimes 2^+$ component in the wavefunction of the ^{11}Be ground state (15%) is to be compared to the value found in the analysis of the ^{11}Be(p,d)^{10}Be transfer reaction (17%) [14]. It is also of

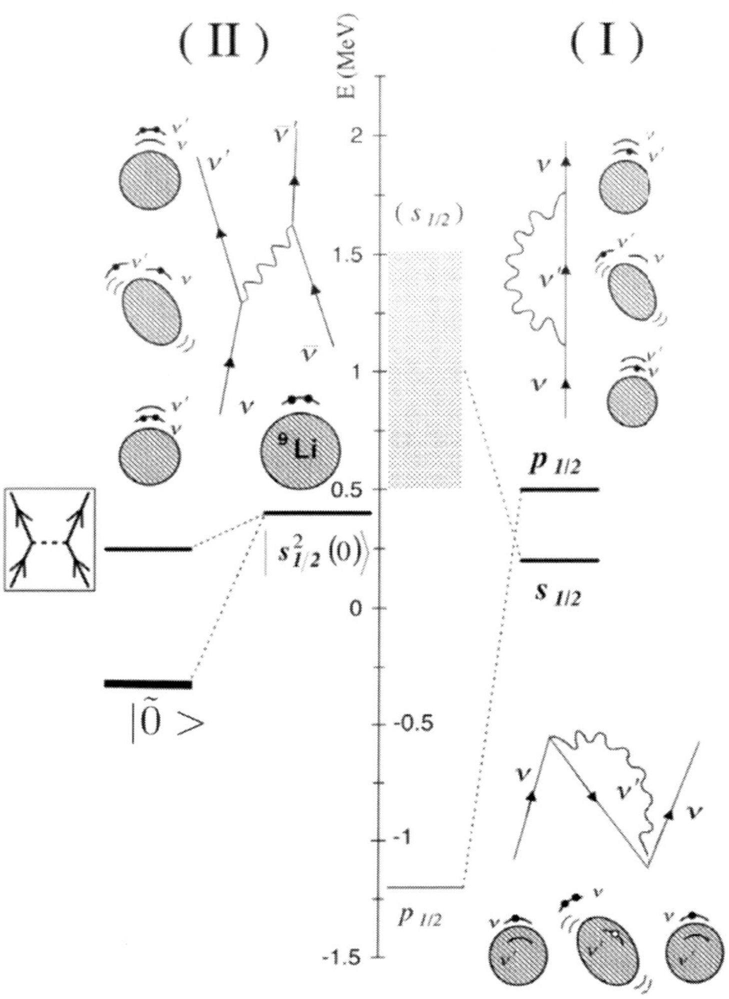

Figure 3. Effects of renormalization processes on the halo neutrons in ^{10}Li (panel I) and in ^{11}Li (panel II) described in the text (see Ref. [3].)

notice that within the present approach [4] a rather low value ($S = 0.31$) for the spectroscopic factor associated with the ^{11}Be(d,p)^{12}Be(gs) reaction was predicted, in agreement with recent experimental findings [32, 33], and at variance with other calculations (see e.g. [34]).

A detailed comparison with experimental absolute cross sections has been carried out for the reaction p(^{11}Li,^{9}Li)t [35], populating the $3/2^-$ ground state and the first $1/2^-$ excited state of ^{9}Li. The absolute cross sections for transfer to the $3/2^-$ ground state and to the $1/2^-$ first excited state provide a stringent test of the theoretical wavefunction of the ^{11}Li halo, which can be written as

$$|\tilde{0} >_\nu = C|0>_\nu + \alpha|[p_{1/2}, s_{1/2}]_{1^-} \otimes 1^-; 0> + \beta|[s_{1/2}, d_{5/2}]_{2^+} \otimes 2^+; 0>, \quad (3)$$

where the calculated coefficients take the value $C = 0.7$, $\alpha = 0.7$ and $\beta = 0.1$, while $|0>_\nu$ is

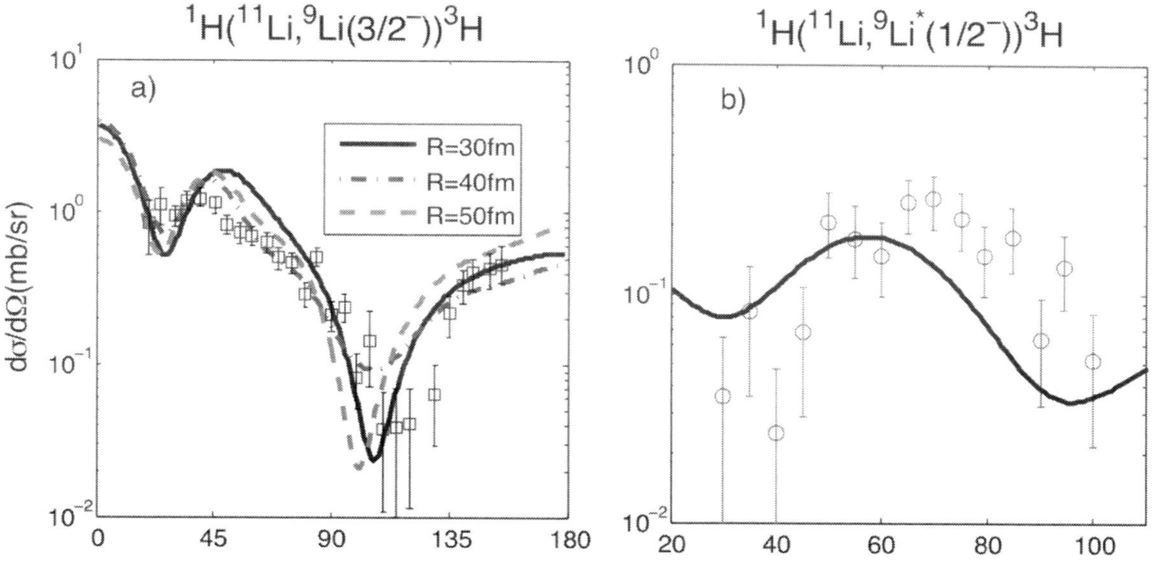

Figure 4. (Color online) Theoretical and experimental absolute differential cross sections associated with the ^{1}H(^{11}Li,^{9}Li)^{3}H two-neutron transfer reaction populating the ground state (a) and the first excited state (b) of ^{11}Li. The three theoretical curves in (a) refer to calculations of the correlated wavefunction $|\tilde{0}>$ (see equation 3) performed in spherical boxes having different radii R.

given by

$$|0>_\nu = \sum_{lj,n_1n_2} a^{lj}_{n_1n_2}[\psi_{n_1lj}\psi_{n_2lj}]_0, \tag{4}$$

where the sum runs over the angular momenta $(lj) = s_{1/2}, p_{1/2}, d_{5/2}$ and over the number of nodes that identify the different wavefunctions in the discretized continuum. The weight of the different angular momenta is given by

$$\sum_{n_1n_2}(a^{lj}_{n_1n_2})^2 = 0.40(\mathrm{s}_{1/2}); 0.59(\mathrm{p}_{1/2}); 0.003(\mathrm{d}_{5/2}). \tag{5}$$

The weight of s and p waves is comparable, in overall agreement with the experimental findings. While the ground state transition provides direct evidence on the s^2 and p^2 admixture, which in the present description arises mainly from the phonon induced pairing interaction, it provides no direct evidence for the phonon exchange correlation (note however the normalization $C^2 = 0.49$ resulting mainly from the coupling to the soft dipole resonance). On the other hand, and in keeping with the fact that the $1/2^-$ first excited state of ^{11}Li can be viewed as the lowest member of the multiplet obtained by coupling the $p_{3/2}$ proton to the quadrupole vibration of the core, the associated absolute differential cross section to the $1/2^-$ state is proportional to the square of the β^2 coefficient in equation (3).

The reaction calculation was performed in the second-order DWBA (see [36] and [37] and references therein). Simultaneous, successive and non-orthogonal contributions to the transfer amplitude were taken into account. Global optical potential were used [35, 38]. The calculation of more reliable optical potentials for the reactions involving halo nuclei, and a better understanding of their connection with elastic scattering represent important open challenges in the field.

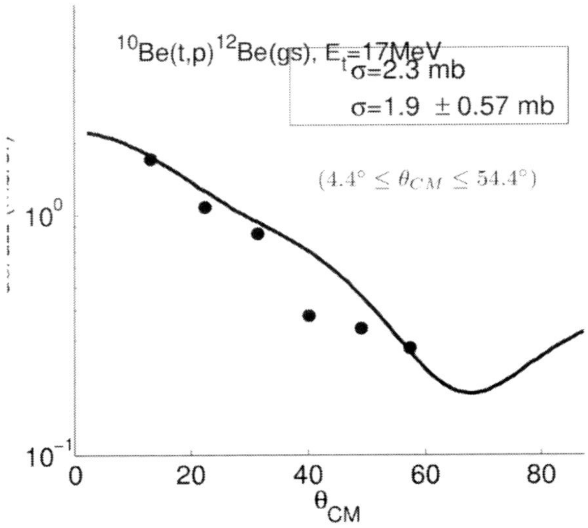

Figure 5. Absolute differential cross section measured [40] in the reaction ^{10}Be(t,p)^{12}Be (gs) at 17 MeV triton bombarding energy (solid dots). The theoretical calculations (continuous solid curve) were obtained making use of the spectroscopic amplitudes associated with the wavefunction of equation (7), and the optical parameters of Refs. [38] and [40]. The integrated cross sections are shown in the inset (see Ref. [6]).

The calculated cross sections [5] for the transfer to the ground state and to the first excited state of ^{11}Li are compared to data in figures 4(a) and (b). In (a) we also show that one obtains convergence respect to the box radius R used in the structure calculation, that is, to calculate the coefficients of the wavefunction (3). It was checked that other channels involving inelastic or breakup processes, which could, in principle, contribute to the transfer to the excited state, turn out to be strongly suppressed as compared to the two-nucleon direct transfer considered here [5, 39].

The wavefunction describing the neutron component of the ^{12}Be ground state can be written as

$$|\tilde{0}\rangle_\nu = C|0\rangle_\nu + \alpha|(p_{1/2}, s_{1/2})_{1^-} \otimes 1^-; 0\rangle + \beta|(s_{1/2}, d_{5/2})_{2^+} \otimes 2^+; 0\rangle + \gamma|(p_{1/2}, d_{5/2})_{3^-} \otimes 3^-; 0\rangle, \quad (6)$$

where $C = 0.87$, $\alpha = 0.10$, $\beta = 0.35$, $\gamma = 0.33$ and (see equations (4) and (5))

$$\sum_{n_1 n_2} (a^{lj}_{n_1 n_2})^2 = 0.19(\mathrm{s}_{1/2}); 0.34(\mathrm{p}_{1/2}); 0.47(\mathrm{d}_{5/2}). \quad (7)$$

The $d_{5/2}$ orbital plays a much more important role in this case than in ^{11}Li. The measurement of the reaction ^{12}Be(p,t)^{10}Be populating the ground and excited states of ^{10}Be would provide a stringent test of the wavefunction but it has not yet been made. Within this context, the absolute theoretical [6] and experimental cross sections associated with the reaction ^{10}Be(t,p)^{12}Be at 17 MeV triton bombarding energy are compared in figure 5.

The wavefunctions (3) and (6) have been obtained without taking into account the effect of the pairing ground state fluctuations of the core. The latter have been included in calculations of the wavefunction of ^{12}Be carried out within the particle-particle RPA using both the Gogny [41] and a monopole pairing force [6], and absolute two-neutron transfer cross sections expected within the pair vibrational scheme around ^{10}Be have also been predicted [6]. In these calculations the coupling with surface degrees of freedom is not explicitly included, and the positions of the single-particle levels are taken from experiment. A calculation taking into account both pairing and surface ground state fluctuations has not yet been performed.

3. Conclusions

Insight in the basic properties of halo nuclei can be obtained taking consistently into account the basic processes associated with vibrations of the system, which renormalize both the single-particle spectrum and the pairing interaction and which can be directly probed with transfer reactions.

References

[1] Tanihata I, Savajols H, and Kanungo R 2013 *Prog. Part. Nucl. Phys.* **68** 215
[2] Broglia R A, Potel G, Barranco F, and Vigezzi E 2010 *J. Phys. G* **37** 064022
[3] Barranco F, Bortignon P F, Broglia R A, Colò G, and Vigezzi E. 2001 *Eur. Jou. Phys. A* **11** 385
[4] Gori G, Barranco F, Broglia R A, and Vigezzi E 2004 *Phys. Rev. C* **69** 041302(R)
[5] Potel G, Barranco F, Vigezzi E, and Broglia R A 2010 *Phys. Rev. Lett.* **105** 172502
[6] Potel G, Idini A, Barranco F, Vigezzi E, and Broglia R A 2012 `ArXiv:1212.2437`; *Yad. Fiz.* in press
[7] Otsuka T, Fukunishi N, and Sagawa H 1993 *Phys. Rev. Lett.* **70** 1385
[8] Sagawa H, Brown B A, and Esbensen H 1993 *Phys. Lett. B* **309** 1
[9] Nunes F M, Thompson I J, and Johnson R C 1996 *Nucl. Phys. A* **596** 171
[10] Nunes F M, Christley J A, Thompson I J, Johnson R C, and Efros V D 1996 *Nucl. Phys. A* **609** 43
[11] Nunes F M, Thompson I J, and Tostevin J A 2002 *Nucl. Phys. A* **703** 593
[12] Hamamoto I 1969 *Nucl. Phys. A* **126** 145
[13] Van de Wiele J, Vdovin A, and Langevin-Joliot H 1996 *Nucl. Phys. A* **605** 173
[14] Winfield J S *et al.* 2001 *Nucl. Phys. A* **683** 48
[15] Deltuva A 2013 *Phys. Rev. C* **88** 011601(R)
[16] Moro A M and Crespo R 2012 *Phys. Rev. C* **85** 054613
[17] Myo T, Kato K, Toki H, and Ikeda K 2007 *Phys. Rev. C* **76** 024305
[18] Ikeda K, Myo T, Kato K, and Toki H 2010 *Clusters in nuclei*, Vol 1, Lectures Notes in Physics Vol 818 (Springer)
[19] Csótó A 1997 `ArXiv:nucl-th/970454v1`
[20] Bertsch G F and Esbensen H 1991 *Ann. Phys.* **209** 327
[21] Zhukov M V, Danilin B V, and Fedorov D V 1993 *Phys. Rep.* **231** 151
[22] Hagino K, Sagawa H, Carbonell J, and Schuck P 2007 *Phys. Rev. Lett.* **99** 022506
[23] Forssen C, Caurier E, and Navratil P 2009 *Phys. Rev. C* **79** 021303(R)
[24] Nörterhäuser W, Neff T, Sánchez R, and Sick I 2011 *Phys. Rev. C* **84** 024307
[25] Liu L, Otsuka T, Shimizu N, Utsuno Y, and Roth R 2012 *Phys. Rev. C* **86** 014302
[26] Bohr A and Mottelson B R 1975 *Nuclear Structure*, Vol. II (Benjamin)
[27] Bohr A and Mottelson B R 1969 *Nuclear Structure*, Vol. I (Benjamin)
[28] Bès D R, Broglia R A, Dussel G G, Liotta R J, and Sofía H M 1976 *Nucl. Phys. A* **260** 1
[29] Bès D R, Broglia R A, Dussel G G, Liotta R J, and Sofía H M 1976 *Nucl. Phys. A* **260** 27
[30] Bès D R, Broglia R A, Dussel G G, Liotta R J, and Perazzo R J 1976 *Nucl. Phys. A* **260** 77
[31] Bortignon P F, Broglia R A, Bes D R, and Liotta R J 1977 *Phys. Rep.* **30** 305
[32] Kanungo R *et al.* 2010 *Phys. Lett. B* **682** 391
[33] Johansen J G *et al.* 2013 *Phys. Rev. C* **88** 044619
[34] Fortune H T and Sherr R 2012 *Phys. Rev. C* **85** 051303(R)
[35] Tanihata I *et al.* 2008 *Phys. Rev. Lett.* **100** 192502
[36] Bayman B F and Chen J 1982 *Phys. Rev. C* **26** 1509
[37] Potel G , Idini A, Barranco F, Vigezzi E, and Broglia R A 2013 *Rep. Prog. Phys.* **76** 106301
[38] An H and Cai C 2006 *Phys. Rev. C* **73** 054605
[39] Vigezzi E, Potel G , Barranco F, and Broglia R A 2011 *J. Phys. G. Conf. Ser.* **312** 092061
[40] Fortune H T, Liu G B, and Alburger D E 1994 *Phys. Rev. C* **50** 1355
[41] Blanchon G, Vinh Mau N, Bonaccorso A, Dupuis M, and Pillet N 2010 *Phys. Rev. C* **82** 034313

Journal of Physics: Conference Series **527** (2014) 012006 doi:10.1088/1742-6596/527/1/012006

Nuclear excitation of Pygmy Dipole Resonance

Edoardo G Lanza[1,2], **Andrea Vitturi**[3,4], **M Victoria Andrés**[5], **Francesco Catara**[2,1]

[1] INFN - Sezione di Catania, 95123 Catania, Italy
[2] Dipartimento di Fisica e Astronomia, 95123 Catania, Italy
[3] Dipartimento di Fisica e Astronomia, 35131 Padova, Italy
[4] INFN - Sezione di Padova, 35131 Padova, Italy
[5] Departamento de Física Atomica, Molecular y Nuclear, 41080 Sevilla, Spain

E-mail: lanza@ct.infn.it

Abstract. Excitations of Pygmy Dipole Resonance (PDR), generated by a nuclear field, are studied within a semiclassical RPA based model. This is possible since the PDR, present in almost all nuclei with neutron excess, has a strong isoscalar character. Our results show that the employ of isoscalar probes finds its best use at incident energies below 50 MeV/u. The structural splitting of the low-lying E1 strength, found in recent $(\alpha, \alpha'\gamma)$ experiments, is corroborated by our cross section calculations.

1. Introduction

Collective states play an important role in the varied panorama of nuclear phenomena. Their role has been strengthened with the advent of the studies of nuclei far from the stability line. In particular, the low lying excitation dipole strength, present in almost all the nuclei with neutron excess, has arisen a lot of attention and many experimental [1] and theoretical [2] studies have been devoted to it. The interest in the Pygmy Dipole Resonance (PDR) states is enhanced by the fact that there is a connection between their strength and the neutron skin as well as with the symmetry energy in the equation of state of nuclear matter [1, 2]. Although there is a controversial debate about whether they can be considered collective, most of the theoretical studies coincide in the description of their transition densities whose behaviour "defines" this new mode: The neutron and proton transition densities are in phase inside the nucleus and at the surface only the neutron part considerably contributes. This property manifests in a strong mixing of isospin character which eventually allows the excitation of these states by means of both isoscalar and isovector probes.

In this paper we will describe the excitation of the PDR via the nuclear interaction at low incident energies within a semiclassical framework and we essentially will concentrate in two aspects. In our previous studies it was shown that these states are not collective, therefore it is not possible to apply the usual standard model to calculated the form factors to be employed in, for instance, distorted-wave Born approximation (DWBA) calculations. Therefore it is of paramount importance the use of a microscopic description of these states both for the transition densities as well as for the form factors. The other aspect regards the comparison with experimental data. Recently, experiments with isoscalar probes have become available [3, 4, 5, 6]. Combining the $(\alpha, \alpha'\gamma)$ coincidence method with high-resolution γ-ray spectroscopy [7] on several semi-magic stable nuclei, a splitting of the low lying dipole strength has come in evidence. The strength separates in two groups: the low lying part is excited by both isoscalar and isovector probes while the high energy group is excited only by electromagnetic probes. Some microscopic nuclear structure calculations show, for the isoscalar and isovector B(E1) response, a qualitative agreement with the experimental findings. Our cross section calculations agree very well with the data reproducing the experimental observed splitting.

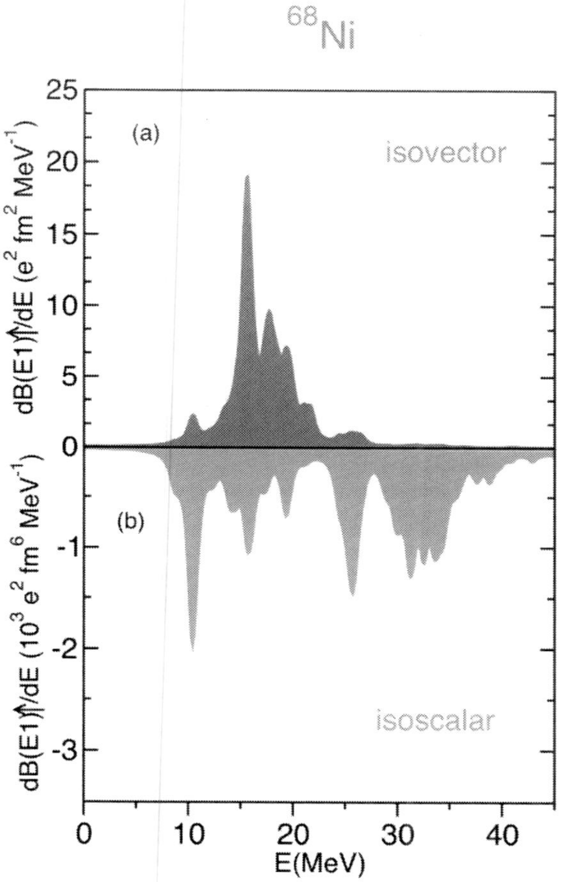

Figure 1. (Color online) RPA strength distributions for isovector (a) and isoscalar (b) response for ^{68}Ni.

2. Semiclassical model

The inelastic cross section calculations are performed within a semiclassical model which has been developed along the years and it has been applied to several nuclear reactions between heavy ions. The idea is that the trajectory of the partners of the reactions can be described classically, while their internal structure and excitation can be described microscopically according to quantum mechanics. These conditions are valid for grazing collisions. Details of the model are given in Ref. [8] and references therein, here we only emphasize the most salient points.

The Hamiltonian describing each of the two colliding nuclei is given by the sum of one term which describes the structure of the nucleus and another one which is responsible, through its mean field, of the excitation of the other partner. The Schrödinger equation for the time dependent wave functions of the states taken into consideration can be written in terms of coupled equations for the probability amplitudes. The solutions of these coupled equations can then be used to calculate cross sections, for each state taken into consideration, as an integral of the excitation probabilities over the impact parameters. The structure of the nucleus under study is obtained by means of an HF + RPA calculation which allows us to select the states with an important EWSR percentage among the RPA states with multipolarities from 0 to 3. For each of these states the transition densities are obtained and they are used to construct the real part of the nuclear optical potential as well as the form factors by the double folding procedure. The classical trajectory is then determined by the Coulomb potential together with the real part of the nuclear potential.

In a previous work we have analyzed in detail the nature of the PDR (see for example Refs. [9, 10]). There we have shown that the dipole responses for an isoscalar and isovector probe are quite different. As an example we show in figure 1 the strength distribution for the isovector (above) and isoscalar (below) response for the isotope ^{68}Ni. The isovector response shows the well known peak of the isovector Giant Dipole Resonance (IVGDR) at about 15 MeV and the small peak at about 10 MeV of the PDR. In the lower frame the response for the isoscalar probe shows a very strong peak in correspondence of the PDR energy and the well known ISGDR peak at around 33 MeV. The transition density for the PDR shows its characteristic pattern: at the surface only the neutron transition density contributes. In order to establish whether these states are collective we have analyzed them in detail. The RPA results tell us that they are formed by many particle-hole configurations and most of them come from the shell occupied by the excess neutrons. But the analysis of the coherence, the other property that has to be satisfied to be called collective, gives different results for the isoscalar and isovector response. While the isoscalar response shows a clear feature of coherence, as in the case of the high lying ISGDR, the isovector one is manifestly not coherent [10].

This implies that the usual standard prescription to calculate the form factor for the collective states cannot be applied in this case. A microscopic construction of the form factors is in order. This can achieved by means of the double folding prescription [11] where a microscopic RPA transition density is double folded with the density of the nucleus responsible of the excitation and with a nucleon nucleon interaction. In our calculation we have chosen an M3Y-Reid type [12]. Examples of the radial form factors are shown in figure 2 where the Coulomb (dashed blue) and the nuclear (full red) contributions are separately presented together with their sum for the system ^{68}Ni + ^{12}C. These states are the results of RPA calculations with SGII interaction [13]. In the upper first frame on the left, we observe the known property of the GMR which can not be excited by the Coulomb interaction. In the other cases, the Coulomb form factor shows clearly its proportionality to the squared root of the B(Eλ). This is manifestly evident if one looks at the dipole form factors, namely the one for PDR and GDR, for which the B(Eλ) values are almost an order of magnitude different. The other thing to be noted is that for the dipole states the nuclear and Coulomb contribution are interfering negatively in the interior region and positively at the surfaces of the nuclei. This depends on the fact that the isoscalar transition densities for the dipole states display a node as it has been shown and discussed in Refs. [8, 14]. For the quadrupole and octupole states one can notice a slight negative interference at the surface region. The interference in the form factors reveals itself also in the inelastic cross section as it will be shown in the following. Another aspect, even more interesting, is the fact that the nuclear form factor for the PDR has a very different slope with respect to the other ones. The prescription of the collective model or the scaling assumption for the radius [15] to calculate the transition densities and hence the form factors seems not to work for the PDR states. For these states a more microscopic description, like the one sketched here, for the construction of the form factor seems to be more appropriate.

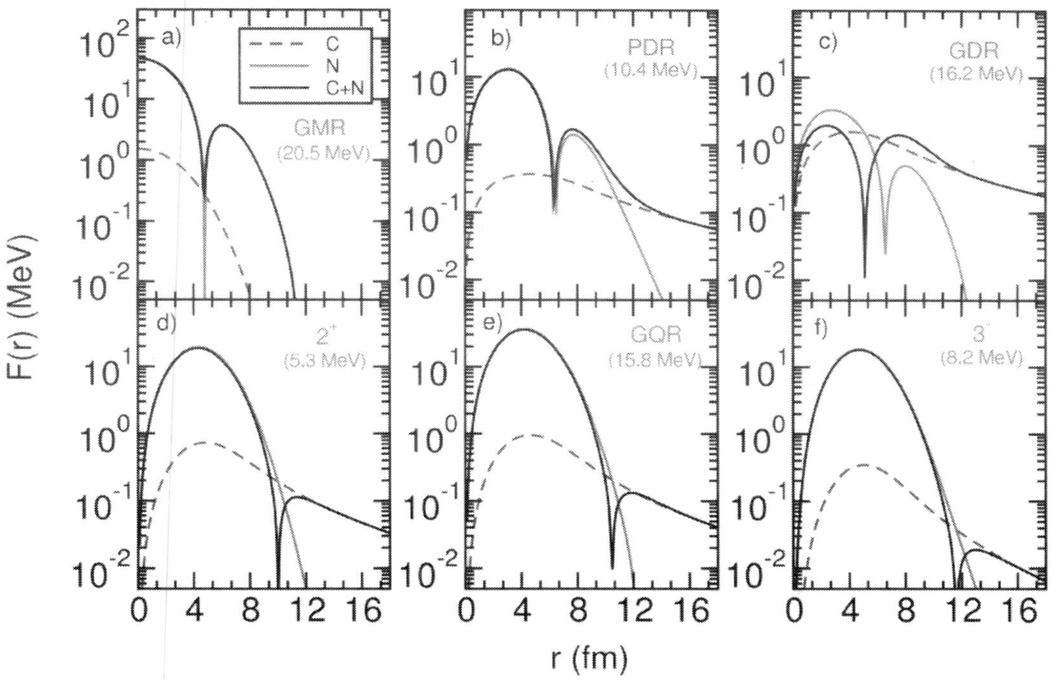

Figure 2. (Color online) Radial form factors, in absolute values, for the states used in the cross section calculations for the system ^{68}Ni + ^{12}C. In each frame is reported the Coulomb (dashed blue) and nuclear (full red) contributions as well as their sum. The states and their energies are also shown.

3. The inelastic cross section

We have performed calculations for the inelastic excitation of various systems with the aim of putting in evidence the appropriate incident energies and/or partners of the reaction that allow the best results for the detection of the PDR via the nuclear interaction. As an example we report here the results referring to calculations performed for the ^{68}Ni + ^{12}C at several incident energies. The dependence of the excitation process on the incident energies for each of the states involved in the reaction can be seen in figure 3 where the inelastic cross sections for each of the states are reported as a function of the incident energy. The contributions due to the two terms of the interactions are separately reported in first two panels of the figure. We note that the PDR is mainly excited by the nuclear interaction as well as the pure isoscalar states. The Coulomb interaction is almost not contributing to the excitation process due to the small electric charge of ^{12}C. The high multipolarities are depleted in Coulomb excitation and that is why the 2^+ low lying state is more excited than the 3^- one. When both interactions are acting (right frame) one can appreciate strong interference effects for the dipole states and this is related to the form factor properties we were discussing above. Therefore, although the Coulomb interaction alone is not producing a strong excitation, because of the small charge of ^{12}C, it is important in the reaction process for the interference effect.

From our results it is evident that the more promising incident energy to study the PDR is around 30 MeV/u. At these energies the value of the cross section for the PDR is smaller than the ones for the other multipole states. But its detection may be possible because they are located in

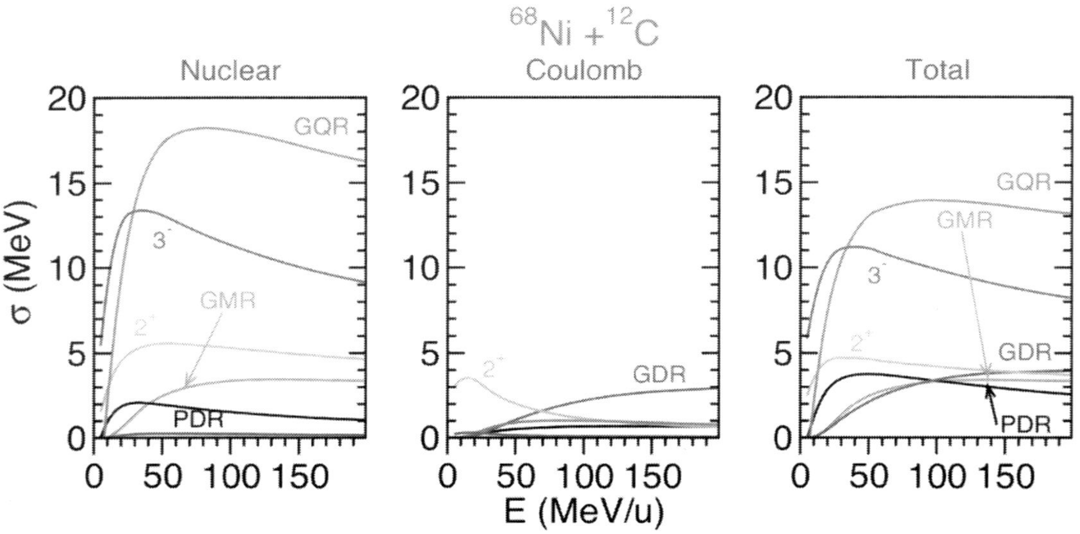

Figure 3. (Color online) Inelastic cross section for the system ^{68}Ni + ^{12}C as a function of the incident energy per nucleon and for all the states involved in the reaction. The contribution of each state is indicated in figure. On the left frame the cross section due to only the nuclear interaction is reported. The middle panel show the contribution of the Coulomb interaction and in the right one are shown the results when both interactions are acting.

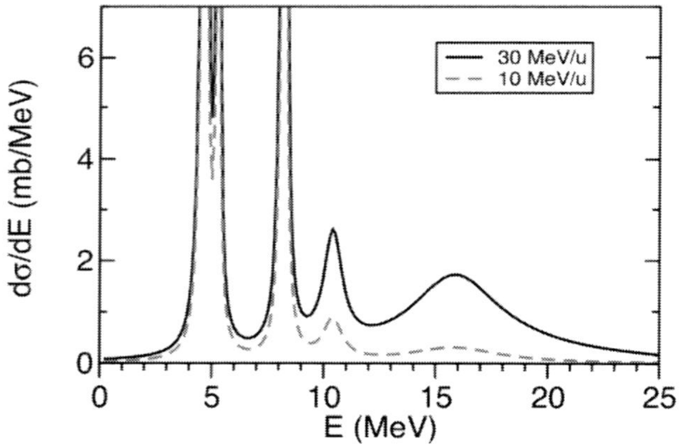

Figure 4. (Color online) Inelastic cross section for the system ^{68}Ni + ^{12}C for two incident energies as a function of the excitation energy. The figure is obtained with a folding procedure as explained in the text.

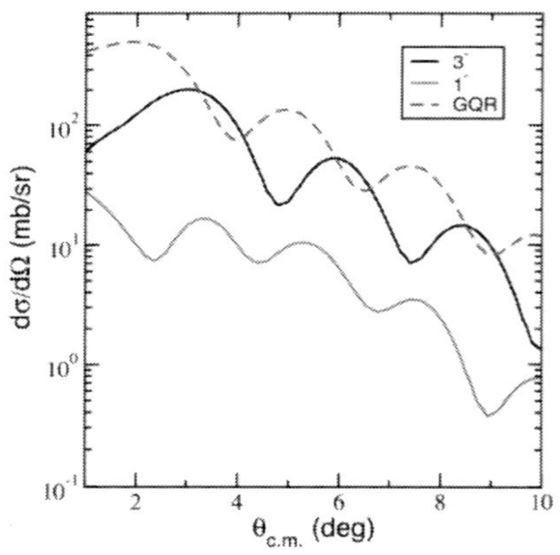

Figure 5. (Color online) Angular distribution for the system ^{68}Ni + ^{12}C at 30 MeV/u for the states indicated in the legend.

the spectrum at different regions of excitation energies. The predicted theoretical shape for the spectrum of the reaction ^{68}Ni + ^{12}C at 30 MeV/u incident energies is shown in figure 4 where the total cross sections for each of the states taken into consideration have been smoothed out

Journal of Physics: Conference Series **527** (2014) 012006 doi:10.1088/1742-6596/527/1/012006

by folding them with a Lorentzian shape with a width of 0.2, 1 and 0.026×$E^{1.9}$ MeV for the low lying states, PDR and giant resonances, respectively. For comparison the result corresponding to 10 MeV/u of incident energy is also reported. As one can see, even in this case the PDR cross section should be measurable, even if in both cases the PDR peak could be masked by the presence of the other states like the giant resonance at higher energies, which are not strongly excited due to the choice of the incident energy, or the low lying octupole state at lower energies. The latter, however, should be lower in energy in the real nuclei, since it is well known that the RPA overestimates the energies of the low lying states of higher multipole. Nevertheless, angular distributions should be able to disentangle the competition between the dipole and octupole cross sections. Indeed, as it is shown in figure 5, the angular distributions for the PDR and the low lying 3^- state have different behaviour at forward angles. For completeness also the angular distribution for the GQR is shown.The angular distributions are calculated for the system ^{68}Ni + ^{12}C at 30 MeV/u incident energies with the DWUCK4 code [16] using our microscopic form factors and potential; for the imaginary part we assume the same geometry of the real part with half strength. We are aware of the fact that for exotic nuclei with neutron excess, this choice may be inaccurate because of the possible incorrect estimation of other channels, as nucleons transfer or breakup, typical surface reactions. Therefore, if one wants to compare, with precise absolute magnitude, calculations of this sort with experimental data, one should use an optical potential which is deduced from elastic scattering experimental data and possibly within the same experimental set-up. On the other hand, our aim here is to show the possibility to discriminate between excitations of states of different multipolarities by means of the shape of the angular distribution at forward angles, and this has been clearly manifested here. Indeed, the shape of the angular distribution depends on the angular momentum of the state while the choice of the correct optical potential, together with the strength deformation parameter, will determine the absolute value of the distribution.

4. Splitting of the PDR

One of the most recent aspects of the problematics related to the low lying dipole states is the so called splitting of the PDR. This comes out when the investigation of the PDR on the same nucleus by means of isovector (IV) as well as isoscalar (IS) probes is carried out in order to put in evidence the strong mixing of isospin character. These studies have been performed on various stable nuclei and here we analyze the α +^{124}Sn reaction at 136 MeV incident energy [5, 6]. The comparison of the high resolution γ ray spectroscopy [7] (nuclear resonance fluorescence (NRF)) with the results obtained applying the $(\alpha, \alpha'\gamma)$ coincidence method on the same nucleus shows a splitting of the low lying dipole strength below the neutron threshold into two groups: The one at lower energy is excited by both isovector and isoscalar probes while the group at higher energy is excited only by the electromagnetic probe. This can be seen in figure 6 where the results from the $(\alpha, \alpha'\gamma)$ coincidence method [6] are plotted (a) together with the NRF results [17] (b). A closer inspection of the figures confirm the existence of the two separate regions. Similar behaviour of the low-lying E1 strength has been reported in the experimental inelastic scattering of ^{17}O off ^{208}Pb at low bombarding energies [18].

The internal structure of the nucleus ^{124}Sn is calculated within the relativistic quasiparticle time blocking approximation (RQTBA) developed in Refs. [19, 20]. The details of the calculations, performed by E. Litvinova, for ^{124}Sn are given in Ref. [6]. The electromagnetic B_{em}(E1) values for the nucleus ^{124}Sn are displayed in frame d) of figure 6. These can be compared directly with the experimental results in the frame b). For the same set of states the isoscalar Bis(E1) has been calculated and its distribution show a pattern which recall the structure of the experimental cross section in frame a) (see figure 2 of Ref. [5]). This qualitative agreement between the experimental $(\alpha, \alpha'\gamma)$ cross section and the B_{is}(E1) distribution is not satisfactory. In fact, while the Coulomb excitation cross section and the B_{em}(E1) distribution

Journal of Physics: Conference Series **527** (2014) 012006 doi:10.1088/1742-6596/527/1/012006

are proportional, the relation between the cross section and the $B_{is}(E1)$ is not so clear. So a calculation of the inelastic cross section for the low lying dipole states with a isoscalar probe is in order.

We have performed a calculation, along the line described above, for the system $\alpha + ^{124}Sn$ at 136 MeV incident energy [21]. In particular, the form factors and the real part of the optical potential have been constructed via the double folding procedure by using the already mentioned transition densities provided by E. Litvinova RQTBA calculations. The reaction calculation has been done in the same angular range of the experimental data and the results are shown in frame b) of figure 6. The gross feature of the experimental cross section is retrieved although there is a shift of about 800 keV (due to the RQTBA strength distributions) and a factor of almost ten bigger in the theoretical calculations, both in the cross section and in the $B_{em}(E1)$. Both features seems to be ascribed to the fact that the coupling to higher configurations have to be included. The RQTBA produces strength functions which are less fragmented with respect to the experimental data. Therefore, it may be more appropriate to compare the integral of the experimental and theoretical cross section. The cumulative sums of these quantities are shown in figure 7. Apart from the global energy shift, the experimental and theoretical total cross section are in good agreement.

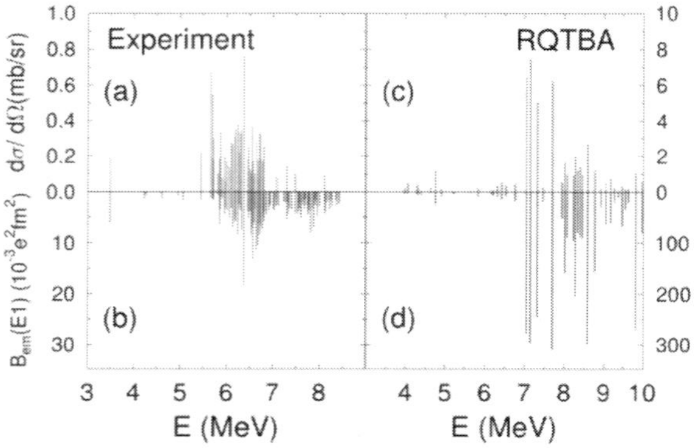

Figure 6. (Color online) Comparison of experimental (a) and RQTBA (c) based cross section for the $(\alpha, \alpha'\gamma)$ reaction and $B_{em}(E1)$ values (b and d) for ^{124}Sn. Experimental values are taken from [6, 17].

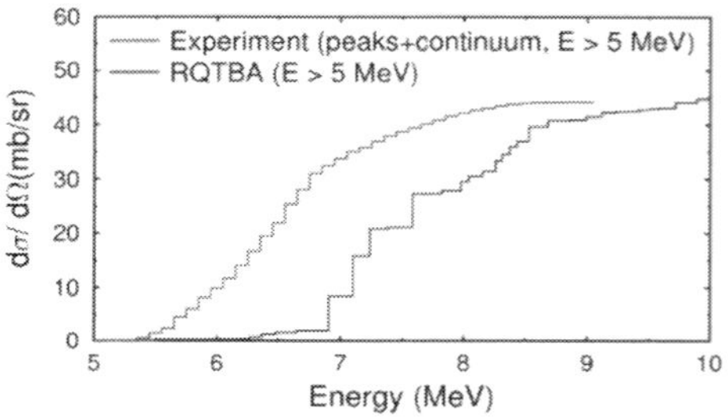

Figure 7. (Color online) Comparison of experimental and RQTBA $(\alpha, \alpha'\gamma)$ cross sections for ^{124}Sn in terms of their cumulative sums.

5. Conclusions

We have presented results where it is shown that the study of the PDR via an isoscalar probe is possible and new features can be envisaged. The study is done within a semiclassical model of the reaction. Important ingredient of this model is the construction of the microscopic form factors that reveals itself of paramount importance for the PDR state which cannot be included among the collective states. Our results show that the proper incident energy to investigate the PDR via isoscalar probes should be below 50 MeV/u. The combination of low incident energies and low charge of the partner of the reaction make the effect of the Coulomb interaction almost negligible, although its contribution to the PDR excitation is important due to a constructive interference effect. To disentangle the low lying dipole state from the other multipole states the use of the DWBA has been proved useful for data taken at very forward angles.

The so called splitting of the PDR has been analyzed with the help of the microscopic structure calculation done within the RQTBA. In the higher part of the low lying dipole peak, the calculated α-scattering cross section is strongly reduced with respect to the isovector channel. Qualitative comparison between the experimental cross section and the isoscalar reduced transition probability was previously done to understand the data. Our proper calculation of the α-scattering excitation cross section confirms the agreement between the experimentally observed splitting of the low-lying E1 strength and the results of the internal structure calculation.

Acknowledgments

We thank Elena Litvinova and Deniz Savran for their essential contribution on the work on the splitting of the PDR.

References

[1] Savran D, Aumann T, and Zilges A 2013 *Prog. Part. Nucl.Phys.* **70**, 210 and references therein
[2] Paar N, Vretenar D, Khan E, and Colò G 2007 *Rep. Prog. Phys.* **70** 691 and references therein
[3] Savran D, van den Berg A M, Harakeh M N, Hasper J, Matic A, Wörtche H J, and Zilges A 2006 *Phys. Rev. Lett.* **97** 172502
[4] Endres J, Savran D, van den Berg A M, Dendooven P, Fritzsche M, Harakeh M N, Hasper J, Wörtche H J, and Zilges A 2009 *Phys. Rev. C* **80** 034302
[5] Endres J *et al.* 2010 *Phys. Rev. Lett.* **105** 212503
[6] Endres J *et al.* 2012 *Phys. Rev. C* **85** 064331
[7] Savran D, van den Berg A M, Harakeh M N, Ramspeck K, Wörtche H, and Zilges A 2006 *Nucl. Instr. and Meth. Phys. Res. A* **564** 267
[8] Lanza E G, Vitturi A, Andrés M V, Catara F, and Gambacurta D 2011 *Phys. Rev. C* **84** 064602
Vitturi A, Lanza E G, Andrés M V, Catara F, and Gambacurta D 2010 *PRAMANA* **75** 73
[9] Lanza E G, Catara F, Gambacurta D, Andrés M V, and Chomaz Ph 2009 *Phys. Rev. C* **79** 054615
[10] Lanza E G, Vitturi A, and Andrés M V 2013 *J. Phys.: Conf. Ser.* **420** 012147
[11] Satchler G R 1983 *Direct Nuclear Reactions* (Oxford University Press)
[12] Bertsch G, Borysowicz J, McManus H, and Love W G 1977 *Nucl. Phys. A* **284** 399
[13] Van Giai N and Sagawa H 1981 *Phys. Lett.* **106B** 379
[14] Catara F, Lanza E G, Nagarajan M A, and Vitturi A 1997 *Nucl. Phys. A* **614** 86
Catara F, Lanza E G, Nagarajan M A, and Vitturi A 1997 *Nucl. Phys. A* **624** 449
[15] Harakeh M N and van der Woude 2001 *Giant Resonances* (New York: Oxford University Press)
[16] Kunz P D *DWUCK4 code for DWBA at http://spot.colorado.edu/ kunz/DWBA.html*
[17] Govaert K, Bauwens F, Bryssinck J, De Frenne D, Jacobs E, Mondelaers W, Govor L, and Ponomarev V Y 1998 *Phys. Rev. C* **57** 2229
[18] Bracco A and Crespi F C L 2013 *EPJ Web of Conferences* **38** 03001
[19] Litvinova E, Ring P, and Tselyaev V 2008 *Phys. Rev. C* **78** 014312
[20] Litvinova E, Ring P, Tselyaev P, and Langanke K 2009 *Phys. Rev. C* **79** 054312
[21] Lanza E G, Vitturi A, Litvinova E, and Savran D 2014 *Phys. Rev. C* **89** 041601(R)

XIV Conference on Theoretical Nuclear Physics in Italy IOP Publishing
Journal of Physics: Conference Series **527** (2014) 012007 doi:10.1088/1742-6596/527/1/012007

Fission at intermediate nucleon energies

S Lo Meo[1,4], **D Mancusi**[2], **C Massimi**[3,4], **G Vannini**[3,4], **A Ventura**[4]

[1] ENEA, Centro Ricerche Clementel, 40129 Bologna, Italy
[2] CEA, Centre de Saclay, Irfu/SPhN, F91191 Gif-sur-Yvette, France
[3] Dipartimento di Fisica e Astronomia dell'Università di Bologna, 40126 Bologna, Italy
[4] INFN, Sezione di Bologna, 40127 Bologna, Italy

E-mail: sergio.lomeo@enea.it

Abstract. In the present work Monte Carlo calculations of fission of actinides and pre-actinides induced by protons and neutrons in the energy range from 100 MeV to 1 GeV are carried out by means of a recent version of the Liège Intranuclear Cascade Model, INCL++, coupled with different evaporation-fission codes, in particular GEMINI++ and ABLA07. Fission model parameters are adjusted on experimental (p, f) cross sections and used to predict (n, f) cross sections, in order to provide a theoretical support to the campaign of neutron cross section measurements at the n_TOF facility at CERN.

1. Introduction

Fission induced by nucleons at intermediate energies is important in both fundamental and applied nuclear physics. From a fundamental viewpoint a theory able to describe nucleon-induced fission over a broad range of target nuclei and a wide interval of projectile energies is still lacking; therefore, Monte Carlo simulations able to reproduce the sparse data on (p, f) and (n, f) reactions at intermediate energies constitute a valuable integration.

From an applied viewpoint, the above mentioned data are very important for energy production, radioactive waste transmutation and radiation shield design for accelerators. In particular, high-accuracy data of neutron-induced fission cross sections are essential to the design of Generation IV reactors and, as far as neutron intermediate energies are concerned, of accelerator-driven subcritical systems (ADS). This is the reason why fission cross sections in the unprecedented range from thermal energies up to about 1 GeV are measured at the n_TOF facility [1] at CERN, exploiting the neutron beam produced by 20 GeV/c protons from the PS accelerator impinging on a lead spallation target.

Fission cross sections for natural lead and bismuth [2], forming the eutectic system that should act as a spallation target and a coolant in an ADS, as well as for actinides of the U-Th cycle and minor actinides [3], have been measured relative to the $^{235}\mathrm{U}(n, f)$ or the $^{238}\mathrm{U}(n, f)$ cross sections, which are considered fission standards at low energy.

In order to obtain absolute fission cross sections of actinides and pre-actinides in the whole energy range measured at n_TOF, experimental ratios have been normalized to an evaluated $^{235}\mathrm{U}(n, f)$ cross section, taken from the ENDF/B-VII library [4] up to 30 MeV and from the JENDL/HE-2007 library [4] from 30 MeV to 1 GeV. It is then natural to investigate whether the choice made for the normalization in the intermediate energy range, from about 100 MeV to 1 GeV, is compatible with recent absolute data of the (p, f) reaction, since it is expected that at energies well above the fission barrier neutrons and protons have a similar behaviour and the corresponding cross sections a similar energy trend. In Ref. [5], (p, f) cross sections for several pre-actinides and actinides, including, of course, ^{235}U, have been measured at nine proton energies in the range from 200 MeV to 1 GeV and are used to calibrate the fission model parameters for our (n, f) calculations.

2. The models

In the energy range of interest, fission induced by nucleons can be seen as a two-stage process: a fast cascade stage, initiated by the high energy projectile, and representable as a succession of two-body collisions, with emission of fast nucleons, light clusters, pions, etc, leaving an excited residual nucleus, and the slow decay stage, where the remnant decays by evaporation, fission, or other mechanisms. In our system of codes, the intranuclear cascade is described by a recent C++ version of the Liège Intranuclear Cascade Model, INCL++ [6], the evaporation-fission model by a C++ version of GEMINI, GEMINI++ [7, 8], or a Fortran version of ABLA07 [9].

INCL++ [6] is a time-like intranuclear cascade model. At the beginning of the cascade stage, the incident nucleon is located with its own impact parameter on the surface of a working sphere, centered on the target nucleus with a radius $R_{max} = R_0 + 8a$, where R_0 and a are the radius and the diffuseness of the target nucleus, respectively. Particles move along straight-line trajectories between collisions in the working sphere and are divided into participants and spectators in the usual sense. Participants that leave the working sphere are considered as ejectiles. Inside the working sphere, nucleons feel a potential that depends on energy and isospin. The depth of the potential well decreases linearly with increasing energy, from ordinary values at the Fermi level to zero at about 200 MeV. The isospin dependence is such that neutron and proton Fermi levels have the same energy.

Collisions are, of course, governed by Pauli blocking, treated in a different way in the first and in the subsequent collisions. The nucleons involved in the first collision are subject to a strict blocking: after the collision, both of them should lie outside the Fermi sphere. In subsequent collisions, the blocking is applied stochastically, with a probability given by the product of final-state blocking factors. A careful definition of the latter allows one to account for surface effects and for the depletion of the Fermi sphere during the evolution of the cascade.

An important novelty of the present version of the code is the introduction of a coalescence model based on phase space, which permits the emission of light clusters, with mass $A \leq 8$, during the cascade stage, in keeping with experimental evidence.

Pions are produced in inelastic nucleon-nucleon collisions through the excitation and subsequent decay of Δ resonances, which sets an upper limit of the order of 3 GeV to the incident nucleon energy for the mechanism of pion production to be valid. The lower energy limit is given by the requirement that the de Broglie wave length of relative motion be much smaller than the range of nuclear forces, which in turn is smaller than the average distance of neighbouring nucleons and is commonly set to 200 MeV, although the model performs reasonably well even at lower energies, as shown in fission calculations of the following section.

An important characteristic of the model is the self-consistent determination of the stopping time of the cascade, which can be simply parametrized as $t_{stop} = 29.8 A_T^{0.16}$ fm/c, with A_T the mass of the target nucleus. At $t = t_{stop}$ many physical quantities, such as the excitation energy of the target nucleus and the average kinetic energy of the ejectiles, do not undergo a rapid variation any more, but only a much slower change. Thanks to this choice of the stopping time, it is not necessary to introduce a pre-equilibrium model describing the intermediate stage between the fast cascade and the evaporation-fission decay.

GEMINI++ [7, 8] is a statistical-model code which follows the decay of a compound nucleus by a series of sequential binary decays until such decays are forbidden by energy conservation or become improbable because of gamma-ray competition. Differently from most statistical-model codes, light-particle evaporation is described by the Hauser-Feshbach formalism, which strictly conserves angular momentum, at the price of higher computational time with respect to the more common Weisskopf-Ewing formalism. An important ingredient of the decay width is the nuclear level density as a function of excitation energy U and angular momentum J, described by a Bethe formula

$$\rho(U,J) \sim (2J+1)\exp\left[2\sqrt{a\,(U,J)\,U}\right] \quad , \tag{1}$$

where the level density parameter $a(U,J)$ includes the damping of shell correction δW according to Ignatyuk [10]

$$a(U) = \tilde{a}(U)\left[1 + \frac{1-e^{-\eta U}}{U}\delta W\right] \quad , \tag{2}$$

with $\frac{1}{\eta} = 18.5$ MeV and the asymptotic level density parameter $\tilde{a}(U)$ is taken of the form

$$\tilde{a}\,(U) = \frac{A}{k_\infty - (k_\infty - k_0)\exp\left(-\frac{\gamma}{k_\infty - k_0}\frac{U}{A}\right)} \quad , \tag{3}$$

with $k_0 = 7.3$ MeV, $k_\infty = 12$ MeV and $\gamma = 0.00493\exp(0.03332A)$, with A the mass number.

The decay width for symmetric fission, dominant at high excitation energy, is given by the well-known Bohr-Wheeler formula, while the decay width for asymmetric fission is derived from Moretto's formalism [11], based on the concept of conditional fission barrier, i.e. a saddle point configuration with fixed asymmetry of mass and charge of the prefragments. Liquid drop barriers are calculated by means of Sierk's finite-range model [12], with shell and pairing corrections taken from Ref. [13].

An important adjustable parameter in the code is the ratio of the asymptotic level density parameter at the saddle point, $\tilde{a}_f$, to the same quantity at ground-state deformation, $\tilde{a}_n$, with a default value of 1.036 [8].

ABLA07 [9] is a statistical code describing the de-excitation of a nucleus in thermal equilibrium by particle evaporation, fission, or, above a prescribed excitation energy per nucleon, multifragmentation. Particle evaporation is treated in an extended Weisskopf-Ewing formalism, where a distribution of orbital angular momenta in the emission of nucleons and clusters is evaluated in semiclassical approximation, based on phase space arguments. An essential ingredient of the decay formalism is the nuclear level density as a function of excitation energy and angular momentum, described by a constant temperature formula at low energy and by a Bethe formula (1) at high energy. The asymptotic level density parameter, $\tilde{a}$ (see formula (2)), is energy-independent and given, in MeV^{-1}, by the original prescription of Ref. [10]

$$\tilde{a} = 0.073A + 0.095B_sA^{2/3} \quad , \tag{4}$$

where B_s is the ratio of the surface of the deformed nucleus to that of a spherical nucleus with the same mass number A. The level density contains a collective enhancement factor, due to nuclear rotations and vibrations, depending on excitation energy. The approach to fission contains elements of dynamics: the time evolution of the fission degree of freedom is treated as a diffusion process, determined by the interaction of collective degrees of freedom with a heat bath formed by the individual nucleons and described by a Fokker-Planck equation whose solution leads to a time-dependent fission width, $\Gamma_f(t)$. An analytical approximation to such a solution and, consequently, to the time dependence of the fission width makes the problem computationally tractable. At low excitation energy, the code uses the standard model of a two-humped fission barrier, whose penetrability is computed in the approximation of full damping of the vibrational resonances in the intermediate well. Like in the GEMINI++ code, liquid drop barriers are computed in the frame of the finite-range model [12] and shell and pairing corrections are taken from Ref. [13].

3. Cross section calculations

In keeping with the philosophy of the authors [6], no parameters have been modified in the INCL++ model, which was already optimized by reproducing a large amount of experimental data connected with the cascade stage, such as total reaction cross sections, double-differential spectra of emitted nucleons and light clusters, etc, while we have taken the freedom to adjust fission parameters in both GEMINI++ and ABLA07. There are essentially two crucial parameters: the height of the fission barrier and the asymptotic level density parameter at the saddle point, $\tilde{a}_f$, or its ratio to the level density parameter at ground state deformation, $\tilde{a}_n$. The barrier height is expected to play a main role at low energy, the level density is important in the whole energy range, in particular at high energy.

While reliable experimental values of absolute (p, f) cross sections in the energy range up to 1 GeV exist for a number of pre-actinides and actinides, (n, f) cross section measurements above 200 MeV are relative to ^{235}U(n, f) or ^{238}U(n, f), which are not experimentally known, but computed by means of some theoretical model. Even if the measured absolute (n, f) cross sections below 200 MeV are systematically lower than the (p, f) cross sections corresponding to the same target nuclei, the effect being more significant for pre-actinides than for actinides, it is expected that, with increasing incident energy, the behaviour of neutrons and protons becomes more and more similar; consequently, (n, f) and (p, f) cross sections are expected to show a similar energy trend and comparable absolute values. Thus, the basic strategy of the present work is to reproduce (p, f) and (n, f) data for the same target with the same model parameters, or, at least, as close as possible in the proton and neutron channels. Where only (p, f) data exist, the model parameters that reproduce them are used to predict (n, f) data.

Figure 1 compares the (p, f) and (n, f) cross sections of ^{235,238}U calculated with the INCL++ + GEMINI++ chain with selected experimental data and with the JENDL/HE-2007 evaluations. The fission model parameters adjusted in GEMINI++ so as to reproduce the (p, f) data of Ref. [5] are the $\tilde{a}_f/\tilde{a}_n$ ratio and the height of the liquid drop fission barrier. For ^{235}U we have chosen $\tilde{a}_f/\tilde{a}_n = 1.050$ and reduced the height of the fission barriers of all residual nuclei by the same amount, ΔB_f = -0.2 MeV, for ^{238}U we have adopted $\tilde{a}_f/\tilde{a}_n = 1.038$ and ΔB_f = -0.3 MeV. The parameters adjusted on the (p, f) cross sections have been used to compute the corresponding (n, f) cross sections: as shown by the figure, the (n, f) data in the energy range from 100 to 200 MeV are satisfactorily reproduced and the predicted cross sections at higher energy are rather close to the corresponding (p, f) cross sections. It is worthwhile to stress that the (n, f) data from Refs. [15, 16] can be considered as absolute cross sections, since the neutron flux is determined by simultaneous measurement of (n, p) scattering events and knowledge of the differential (n, p) cross section in the energy range of interest, while the data of Ref. [17] are relative to the ^{235}U(n, f) cross section, normalized in the intermediate energy range according to the recommendations of Ref. [18].

As for the comparison with JENDL/HE-2007, it is worthwhile to point out that the (p, f) evaluations do not take into account the data of Ref. [5], but are consistent with older data, in conflict with Ref. [5], used by Prokofiev [19] for his (p, f) systematics, which exhibit a steady decrease with increasing energy in the range from 500 MeV to 1 GeV, at variance with Ref. [5], whose data show a large plateau reproduced in our calculations. Two arguments supporting our choice of basic reference data [5] should be mentioned: the responsible for the JENDL evaluations, Tokio Fukahori, co-authored Ref. [5] and reproduced those data in good agreement with our calculations in the range from 200 MeV to 1 GeV; a recent ^{238}U(p, f) measurement [14] in inverse kinematics yields two additional points, shown in figure 1, in excellent agreement with the data of Ref. [5].

As for (n, f) cross sections, JENDL/HE-2007 reproduces the data below 200 MeV very well, but shows in the range from 500 MeV to 1 GeV a fast decrease with increasing energy which is not consistent with the present experimental evidence for (p, f) cross sections. Our calculations

Journal of Physics: Conference Series **527** (2014) 012007 doi:10.1088/1742-6596/527/1/012007

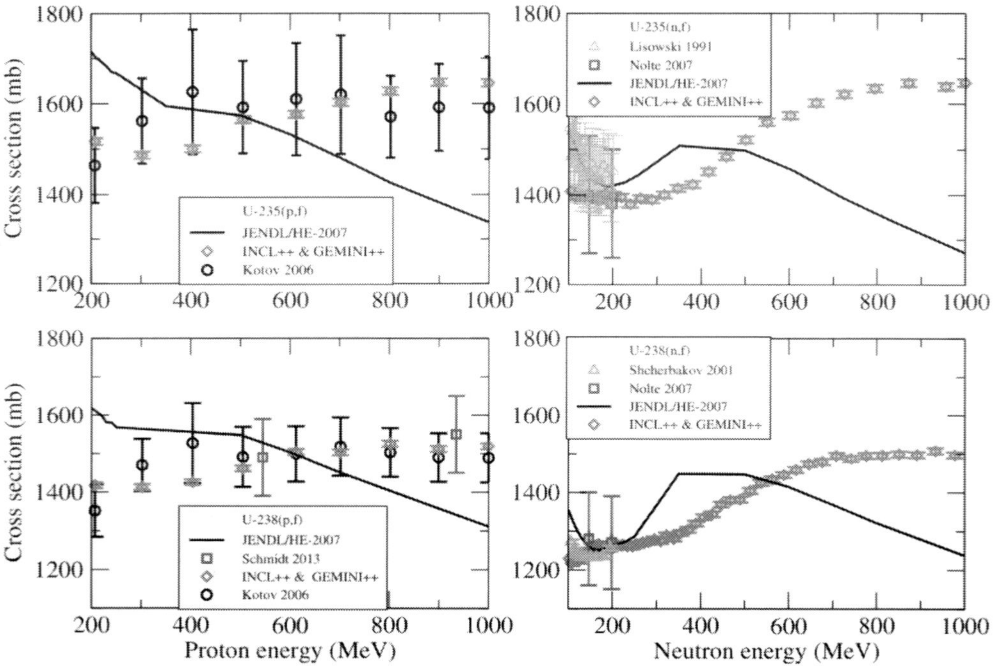

Figure 1. (Color online) Proton- and neutron-induced fission cross sections of ^{235}U and ^{238}U. ^{235}U(p, f): data from Ref. [5]; ^{238}U(p, f): data from Refs. [5, 14]; ^{235}U(n, f): data from Refs. [15, 16]; ^{238}U(n, f): data from Refs. [16, 17].

justify *a posteriori* the use of the intranuclear cascade model in the energy range from 100 to 200 MeV, where the formal conditions of applicability are not so well satisfied. At the present time, the main source of (n, f) data at intermediate energies is the n_TOF experiment [1], where the measurements are relative to the ^{235}U(n, f) and ^{238}U(n, f) cross sections. Final results for ^{234}U and ^{237}Np up to 1 GeV are published in Ref. [3], natural lead and ^{209}Bi are described in Ref. [2], while preliminary data are available for ^{232}Th and ^{233,238}U. In the case of ^{234}U, since no (p, f) data exist in the energy range of interest, we have assumed in our GEMINI++ calculations the same model parameters of ^{235}U, namely $\tilde{a}_f/\tilde{a}_n$ = 1.050 and ΔB_f = -0.2 MeV. Figure 2 shows that theoretical and experimental $\sigma_{nf}(^{234}\mathrm{U})/\sigma_{nf}(^{235}\mathrm{U})$ ratios are in excellent agreement, while the absolute (n, f) cross sections diverge at high energy, owing to the normalization to the JENDL/HE-2007 cross section of ^{235}U, adopted in Ref. [3] . The fission model of ABLA07 is more sophisticated, but less flexible than the one of GEMINI++: the default parameters should be modified very carefully, in order not to spoil the internal consistency. In the present work we have changed only the most crucial quantity, $\tilde{a}_f$, by multiplying it by a constant factor for all fissioning nuclei. The price we pay is that, in general, it is not possible to reproduce (p, f) and (n, f) data of a given target nucleus with the same value of $\tilde{a}_f$: for instance, the best reproduction of ^{235}U(p, f) data from Ref. [5] is obtained by increasing the default $\tilde{a}_f$ value (4) by 6%, while reproducing the ^{235}U(n, f) data below 200 MeV [15, 16] requires an increase of the order of 10%. It is however, expected that acting on other parameters of the fission model this inconvenience can be removed. The $\tilde{a}_f$ values adopted in the proton and neutron channels are closer for pre-actinides than for actinides: for instance, ^{nat}Pb(p, f) data from Ref. [5] are reproduced at best with the default $\tilde{a}_f$ value (4) and ^{nat}Pb(n, f) data from Ref. [2] by increasing it by 1%. The same increase by 1% is adopted in reproducing ^{209}Bi(p, f) [5] and ^{209}Bi(n, f) [2].

Figure 3 compares theoretical and experimental (p, f) cross sections of ^{nat}Pb and ^{209}Bi and figure 4 shows theoretical and experimental ratios to the ^{235}U(n, f) cross section as well as absolute (n, f) cross sections: for the former the agreement is satisfactory, for the latter the results diverge in the energy region from 500 MeV to 1 GeV because of the normalization to the JENDL/HE-2007 evaluation adopted in Ref. [2].

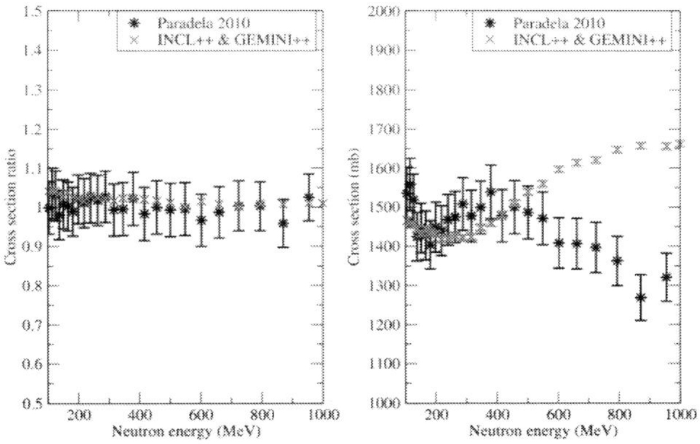

Figure 2. (Color online) Left panel: theoretical and experimental [3] ratio $\sigma_{nf}(^{234}\text{U})/\sigma_{nf}(^{235}\text{U})$. Right panel: theoretical ^{234}U(n, f) cross section compared with the experimental ratio normalized to JENDL/HE-2007.

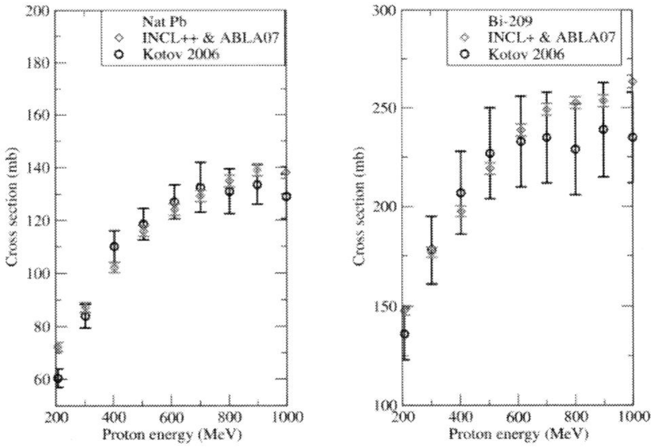

Figure 3. (Color online) (p, f) cross sections of ^{nat}Pb and ^{209}Bi. Experimental data from Ref. [5].

Summing up, experimental (n, f) ratios measured at n_TOF for pre-actinides and actinides are satisfactorily reproduced by using fission model parameters that are similar, or rather close, to the ones adopted in reproducing experimental (p, f) cross sections. Absolute cross sections are larger than those obtained by the n_TOF Collaboration in the energy range from 500 MeV to 1 GeV because of different normalization.

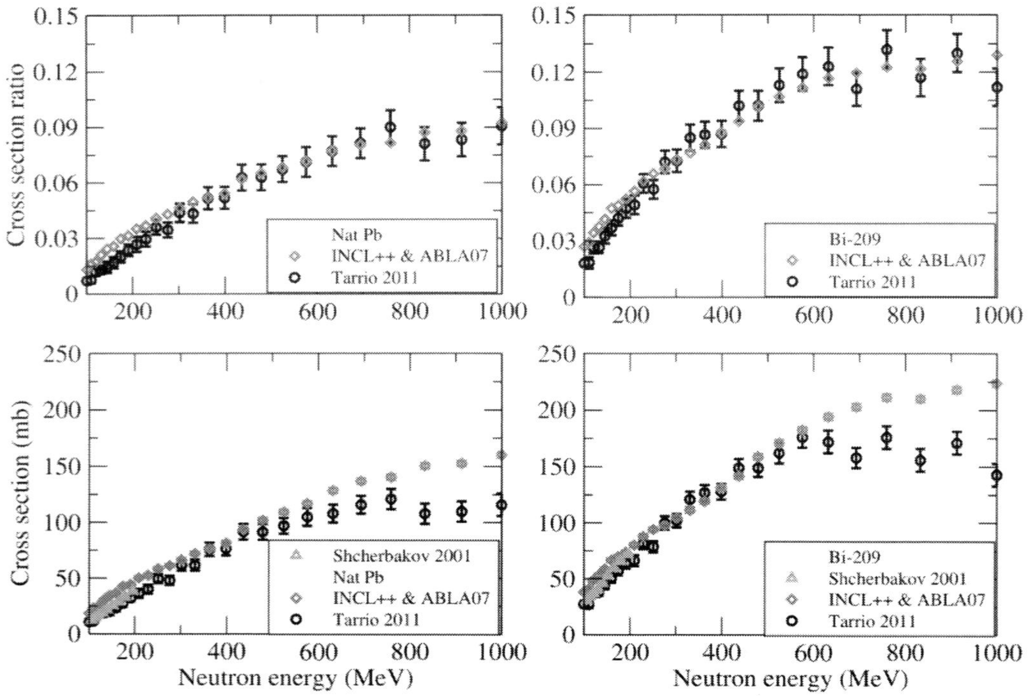

Figure 4. (Color online) (n, f) cross sections of ^{nat}Pb and ^{209}Bi. Upper panels: theoretical vs experimental [2] ratios to ^{235}U(n, f). Lower panels: absolute cross sections.

4. Conclusions and outlook

The preliminary calculations presented in this note indicate that the Liège Intranuclear Cascade Model, INCL, coupled with an appropriate evaporation-fission model like GEMINI or ABLA07, can reproduce the experimental data of fission of actinides and pre-actinides induced by protons and neutrons at intermediate energies with the same, or almost the same model parameters. As a consequence, predicted (n, f) cross sections are rather similar to (p, f) cross sections at high energies, while they are systematically lower than (p, f) cross sections below 200 MeV, in agreement with experiments.

The present study is also intended as a theoretical support to the (n, f) measurements at intermediate energies performed at the n_TOF facility at CERN: in particular, it offers a valid alternative to the normalization of experimental ratios to the JENDL evaluation of the ^{235}U(n, f) cross section adopted in Refs. [2, 3], which does not seem to be consistent with recent data of the (p, f) cross section in the same energy range.

Our preliminary results encourage us to carry out systematic calculations of (p, f) and (n, f) reactions for actinides and pre-actinides and compare our theoretical results with the most recent and reliable experimental data. As a by-product of such a systematic study, it will be of interest to examine the dependence of model parameters that simultaneously reproduce (p, f) and (n, f) data on characteristics of target nuclei, such as the fissility parameter, proportional to Z^2/A. In particular, it has been shown [20] that the $\sigma_{p,f}/\sigma_{n,f}$ ratio at E = 180 MeV decreases exponentially with increasing Z^2/A for a number of pre-actinides and actinides and the experimental result has been interpreted in terms of ratios of fission probabilities, depending, in turn, on neutron and proton fission barriers, B_f^n and B_f^p. In pre-actinides, $B_f^p < B_f^n$, resulting in $\sigma_{p,f}/\sigma_{n,f} > 1$, while, in actinides, $B_f^p \cong B_f^n$, yielding $\sigma_{p,f}/\sigma_{n,f} \sim 1$. Thus, it will be of interest to check this interpretation on the basis of more recent data at other intermediate energies.

Finally, it is our intention to make (n, f) predictions for other isotopes whose measurements are already planned at CERN, not only of fission cross sections, but also of angular distributions of fission fragments, which are also measurable at the n_TOF facility.

Acknowledgments
We are grateful to Dr Sylvie Leray for valuable comments.

References
[1] Guerrero C *et al.* (the n_TOF Collaboration) 2013 *Eur. Phys. J. A* **49** 27
[2] Tarrío D *et al.* (the n_TOF Collaboration) 2011 *Phys. Rev. C* **83** 044620
[3] Paradela C *et al.* (the n_TOF Collaboration) 2010 *Phys. Rev. C* **82** 034601
[4] See, for instance, *www-nds.iaea.org*
[5] Kotov A A *et al.* 2006 *Phys. Rev. C* **74** 034605
[6] Boudard A, Cugnon J, David J-C, Leray S, and Mancusi D 2013 *Phys. Rev. C* **87** 014606
[7] Charity R J *Proceedings of the Joint ICTP-IAEA Advanced Workshop on Model Codes for Spallation Reactions, ICTP Trieste, Italy, 4-8 February 2008, Eds Filges D et al.*, IAEA report INDC(NDS)-530, Vienna, 2008, p. 139
[8] Mancusi D, Charity R J, and Cugnon J 2010 *Phys. Rev. C* **82** 044610
[9] Kelić A, Ricciardi M V, and Schmidt K-H, *Proceedings of the Joint ICTP-IAEA Advanced Workshop on Model Codes for Spallation Reactions, ICTP Trieste, Italy, 4-8 February 2008, Eds Filges D et al.*, IAEA report INDC(NDS)-530, Vienna, 2008, p. 181
[10] Ignatyuk A V, Smirenkin G N, and Tishin A S 1975 *Sov. J. Nucl. Phys.* **21** 255
[11] Moretto L G 1975 *Nucl. Phys. A* **247** 221; Moretto L G and Wozniak G J 1988 *Prog. Part. Nucl. Phys.* **21** 401
[12] Sierk A J 1986 *Phys. Rev. C* **33** 2039
[13] Möller P *et al.* 1995 *Atomic Data Nucl. Data Tables* **59** 185
[14] Schmidt K-H *et al.* 2013 *Phys. Rev. C* **87** 034601
[15] Lisowski P W *et al.* 1991 *Proceedings of the Specialists' Meeting on Neutron Cross-Section Standards for the Energy Region above 20 MeV, Uppsala, Sweden*, NEA report NEANDC-305/U, p. 177
[16] Nolte R *et al.* 2007 *Nucl. Sci. Eng.* **156** 197
[17] Shcherbakov O A *et al.* 2002 *Proceedings of the International Conference on Nuclear Data for Science and Technology, Tsukuba (Japan), 2001, J. Nucl. Sci. Technol. Suppl. 2*, Vol. 1, p. 230
[18] Carlson A D *et al.* 1997 IAEA report INDC(NDC)-368, Vienna, p.23
[19] Prokofiev A V 2001 *Nucl. Instr. Meth. Phys. Res. A* **463** 557
[20] Smirnov A N *et al.* 2002 *Proceedings of the International Conference on Nuclear Data for Science and Technology, Tsukuba (Japan), 2001, J. Nucl. Sci. Technol. Suppl. 2*, Vol. 1, p. 238

Journal of Physics: Conference Series **527** (2014) 012008 doi:10.1088/1742-6596/527/1/012008

Direct reactions with radioactive beams

Angela Bonaccorso
INFN, Sezione di Pisa, 56127 Pisa, Italy
E-mail: bonac@df.unipi.it

Abstract. Since the first measurements in the 80's of total reaction cross sections from radioactive beams up to the most recent and advanced exclusive experiments for dripline unbound nuclei, our understanding of exotic nuclei has greatly evolved. This has been possible through a joint effort in the description of the reaction mechanisms and of the underlying nuclear structure. Most exotic nuclei are weakly bound due to the excess of either neutrons or protons and they breakup easily in peripheral reactions. We will discuss in this paper the mechanism responsible for nuclear and Coulomb breakup. Some open problems will be discussed in detail, such as kinematical effects on core parallel momentum distributions and the possibility and consequences that such cores be themselves weakly bound. A general framework for the reaction theory will be used, based on a time dependent perturbation theory approach. Its eikonal limit will be also mentioned.

1. Introduction

There are many reasons to be interested in the physics of exotic nuclei: they can be found in the crust of neutron's stars where they do not β-decay because of the surrounding electrons; by studying them one could hope to answer the fundamental question of nuclear bindings and relative interactions because they extend our understanding of the nuclear force; they help us checking the limits of validity of structure models.

The present state of the art is that we have very sophisticated structure models and extremely complex experiments with relatively "simple" interpretations for the data. "Ab initio" models have started to bridge the gap between structure and reaction theory (see for example Eurisol Lisbon Meeting report on "Physics of Light Exotic Nuclei" [1]). In this panorama one might wonder about which is the rôle of reaction theory and how simple and/or "complicated" does it need to be. In the following I shall try to show that first of all reaction theory helps understanding the reaction mechanisms. Then it provides guidance in the search for the best *observable* to be measured, in view of the reduced intensity of RIBs. Finally I will discuss the accuracy of methods and numerical implementations.

2. Early experiments and theory

A new chapter in the history of studies of nuclear reactions started around 1985 with the seminal papers by Tanihata I *et al.* [2, 3]. Interaction cross sections, defined as the total reaction cross sections for the change of proton and/or neutron number in the incident nucleus were reported for a series of light nuclei. The measurements had been done at the Bevalac of the Lawrence Laboratory in Berkeley for a series of light nuclei at 790A.MeV on light targets. The systematic showed very large cross sections for some of them. Tanihata *et al.* in view of the high incident energies involved, used a simple geometrical model to interpret the data and deduced unexpected large interaction radii. Some of the "unusual" nuclei were: ^{6}He, ^{11}Be, ^{11}Li, ^{14}Be. Besides the importance of the unexpected results these papers were the first to show that beams of unstable nuclei could be used for nuclear structure studies. A detailed account of subsequent developments can be found in Ref. [4]. Very narrow parallel momentum distributions were also a simple and clear observable measured in the spectrometers [5]-[7]. It was soon clear that the anomalous behavior was due to one or two very weakly bound neutrons whose breakup was dominating the interaction cross section. Interestingly it took twenty five more years to see very clearly such effects in the elastic scattering of ^{11}Be [8].

One and/or two nucleon breakup cross sections from ^{11}Be, ^{11}Li were measured in some pioneering experiments, and discussed in the Anne *et al.* papers [9]. Neutron angular distributions following breakup on three targets of increasing mass were measured. They allowed to distinguish the nuclear and Coulomb breakup mechanisms. At the same time some coincidence experiments [10] with neutrons decaying from excited states of an heavy target gave evidence for the presence of possible projectile-target inelastic excitations associated with breakup along not too peripheral trajectories. These effects were associated with long tails of the core parallel momentum distributions. Up-to-date this interesting phenomenon seen very often on momentum distributions has not been clarified. More recently possible effects of core excitations have been seen and discussed in Refs. [11, 12].

While experiments were planned and performed at various facilities in the world the first theoretical interpretations of the early data started to appear. Hansen and Jonson [13] were the first to propose the presence of a neutron halo as responsible for the large interaction cross sections and narrow momentum distributions. Other works followed, see for example [14, 15] in which the origin of the halo was attributed to the shell inversion between 2s and 1p$_{1/2}$ states caused by core excitations.

3. Interaction cross sections and halo decoupling from the core

As mentioned above Tanihata *et al.* in view of the high incident energies involved, used a simple geometrical model to interpret their first data on interaction cross sections and deduced unexpected large interaction radii. The hypothesis was that projectile and target would not overlap such that $\sigma_R = \pi(R_p^I + R_t^I)^2$. Later on the Glauber model was introduced

$$\sigma_R = \int_0^\infty d\mathbf{b}(1 - |S_{NN}(\mathbf{b})|^2) = \sigma_{ct} + \sigma_{nt} \tag{1}$$

where for the first time the decoupling of core and valence particle, later on identified as "halo" [13], was explicitly considered. It was obtained by first introducing the optical limit of Glauber theory for the phase shift and then writing the total density of the halo nucleus as a sum of the core and valence particle densities, such that

$$|S_{NN}(\mathbf{b})|^2 = e^{-[\sigma_{nn}\int ds\rho_p(|b-s|)\rho_t(s)]} = e^{-[\sigma_{nn}\int ds\rho_c\rho_t]}e^{-[\sigma_{nn}\int ds\rho_n\rho_t]} \; . \tag{2}$$

Equation (2) can be obtained only if $\rho_p = \rho_c + \rho_n$. Such a decoupling of core and valence particle is justified if ρ_c and ρ_n can be considered to span different regions of the projectile-target distance $|\mathbf{b} - \mathbf{s}|$. Thus ρ_n must represent only the peripheral part of the bound state wave function with no significant contribution from the internal region were the particle is in the same potential as the core nucleons. For a true halo this condition is always satisfied. For normally bound nuclei instead some caution has to be taken.

The second factor in equation (2) contains the neutron-target phase shift in the exponent $|S_{nt}|^2 = e^{-[\sigma_{nn}\int ds\rho_n\rho_t]}$. If it can be considered small, then to first order

$$|S_{NN}(\mathbf{b})|^2 \;=\; |S_{ct}|^2|S_{nt}|^2 = |S_{ct}|^2 - |S_{ct}|^2|[\sigma_{nn}\int ds\rho_n\rho_t]. \tag{3}$$

Substitution of this expression in equation (1) leads to the sum of core-target and neutron-target cross sections.

In most halo nuclei there are no bound excited states, therefore the second term in equation (3) was interpreted as the halo breakup probability $P_{b_{up}}(\mathbf{b_c}) \approx \sigma_{nn} \int ds \rho_n \rho_t$ such that one finally obtains

$$\sigma_{nt} = \int_0^\infty d\mathbf{b_c} |S_{ct}(\mathbf{b_c})|^2 P_{b_{up}}(\mathbf{b_c}). \tag{4}$$

Here the no-recoil approximation has been used leading to the substitution of the projectile-target impact parameter $\mathbf{b}$ with the core-target impact parameter $\mathbf{b_c}$. Equations (3)-(4) constituted the building blocks of all subsequent models of halo breakup. In particular equation (4) shows clearly that the condition that the core will be detected intact is of fundamental importance for a correct treatment of breakup and is represented theoretically by $|S_{ct}|^2$. However it was soon realized that the expression for the halo breakup $P_{b_{up}}$ given above was not accurate as it contained the neutron-target interaction only to first order. In the following we shall use more accurate expressions for all breakup cross section calculations, in which an all order treatment of the neutron-target interaction will be used. Recent reports on the experimental and theoretical situation up-to-date can be found in the volume [16].

4. Parallel momentum distributions from breakup

At the early stages of the use of RIBs, besides the interaction cross sections, parallel momentum distributions of the cores following one nucleon breakup, were measured by Tanihata, Kobayashi and collaborators [5] and soon after by several other groups at various facilities [17]-[29]. Based on previous theoretical work [30]-[35] on normal nuclei, it was soon clear that such distributions represented a somewhat distorted "photography " of the Fourier transform of the initial state wave function, and thus could represent a useful way to perform systematic spectroscopic studies of exotic nuclei. The theoretical seminal papers in this field were Refs. [36]-[40]. Following some early publications concentrated on understanding the reaction dynamics correctly to quantify the amount of "distortion" due to the reaction mechanism both experimentally [7] and theoretically [41], an intense activity devoted to spectroscopic studies started with Hansen and collaborators [20] using systematic measurements of parallel momentum distribution and total breakup cross sections to deduce the angular momenta of the initial states and the spectroscopic factors which would make the link between data and the shell model description of exotic nuclei.

The breakup probability in equation (4) which is the sum of the elastic breakup and a second contribution called inelastic breakup or stripping can be obtained if the closure approximation is used [37] in an "all order" eikonal model more refined than in equation (3). This is possible if the nuclei under study do not have bound excited states. Thus the total breakup probability reads

$$P_{-n}(\mathbf{b_c}, k_1) = \int_0^\infty d\mathbf{b_v}(|1 - S(\mathbf{b_v})|^2 + 1 - |S(\mathbf{b_v})|^2)|\tilde{\psi}_i(|\mathbf{b_v} - \mathbf{b_c}|, k_1)|^2 \tag{5}$$

where $S(\mathbf{b_v})$ is the valence nucleon-target S-matrix. $\tilde{\psi}_i$ is the one dimensional Fourier transform of the initial state wave function of the breakup nucleon. $k_1 = (\varepsilon_f - \varepsilon_i - \frac{1}{2}mv^2)/(\hbar v)$, is the neutron (proton) parallel momentum component with respect to the core. Also ε_i is the initial binding energy of the valence particle in the projectile and ε_f its final energy with respect to the target. $\frac{1}{2}mv^2$ is the incident energy per particle at the distance of closest approach. By energy and momentum conservation the core parallel momentum distribution in the laboratory is given by:

$$P_{//} = \sqrt{(T_p + \varepsilon_i - \varepsilon_f)^2 + 2M_r(T_p + \varepsilon_i - \varepsilon_f)}, \tag{6}$$

such that a measure of the latter is considered to give a direct information on the momentum distribution of the valence particle in the initial state of the projectile. Thus inserting equation (5) in equation (4) the cross section differential with respect to the intrinsic parallel momentum in the core is obtained. By using equation (6) and the relative jacobian the cross sections becomes directly comparable to the measured momentum distributions.

Here $P_{//}$ is given in terms of the projectile kinetic energy T_p and of the residual nucleus (A_p-1) mass M_r. Indicating by $M_{p,t,n}$ the projectile, target and neutron mass respectively, then equation (6) is obtained by applying first energy conservation between initial and final states

$$M_p + T_p + M_t = T_r + M_r + M_n + M_t + \varepsilon_f, \tag{7}$$

if the target mass is conserved ($\varepsilon_f << M_t$), M_t cancels out and

$$T_r \;=\; T_p + M_p - (M_r + M_n) - \varepsilon_f = T_p + \varepsilon_i - \varepsilon_f \tag{8}$$

where $-\varepsilon_i \equiv (M_r + M_n) - M_p$ is by definition, the valence neutron separation energy in the projectile. Under the eikonal hypothesis that energy is converted mainly in parallel momentum, inserting (8) in $P_{//} = \sqrt{(T_r + M_r)^2 - M_r^2}$ then equation (6) readily follows.

In this way some kinematical constraints can be satisfied even if an eikonal formalism is used without the need to take the sudden limit [30, 42]. However if the incident energy is very high and the initial separation energy very small, then one can extend the limit of k_1 from $-\infty \to +\infty$. Integrating back on k_1 the momentum distribution the total breakup cross section reads

$$\sigma_{nt} = \Sigma_i (C^2S)_i \int_0^\infty d\mathbf{b_c} d\mathbf{b_v} |S_{ct}(\mathbf{b_c})|^2 (|1 - S(\mathbf{b_v})|^2 + 1 - |S(\mathbf{b_v})|^2) |\psi_i(|\mathbf{b_v} - \mathbf{b_c}|)|^2. \tag{9}$$

This equation is consistent with equation (4.17) of [33], and we have considered that if different initial states contribute to the inclusive breakup cross section (see section IV of Ref. [43]) their contribution should be summed both in the spectra as well as in the integrated cross sections, each weighted by the relative initial wave function spectroscopic factor C^2S. This method has been largely followed in the past years after the seminal applications presented in [20]. Spectroscopic factors have been extracted from the ratio of measured and calculated total breakup cross sections while the angular momentum of the initial state has been deduced from the shape of the momentum distribution. Such approaches derive from the use of shell model initial state wave functions. Recently ab-initio wave functions have also been used [24].

5. Kinematical effects on breakup

The breakup models discussed in section 2 were introduced having in mind very high incident energies and small initial state separation energies. Under such conditions energy and momentum conservation can be neglected. However some dynamical effects are always present [35] and become very important when the separation energies are large and the incident energies comparable to them. This has recently been verified in relation to studies of exotic nuclei having large differences in the proton-neutron separation energies ($\Delta S = |S_n - S_p| \approx 20$MeV) [11]. Such situations are more accurately described in a time dependent formalism. Starting again from equation (4) if the breakup probability is calculated in the target reference frame, then the initial state wave function has to be boosted by a Galilean transformation following which the time dependent exponentials become $e^{-ik_1 z'}$ $e^{ik_2 z}$ where k_1, which has already been defined, and $k_2 = (\varepsilon_f - \varepsilon_i + \frac{1}{2}mv^2)/(\hbar v)$ are the z-components of the neutron momentum in the initial and final state respectively. $\eta^2 = k_1^2 + \gamma_i^2 = k_2^2 - k_f^2$ is the magnitude of the transverse component

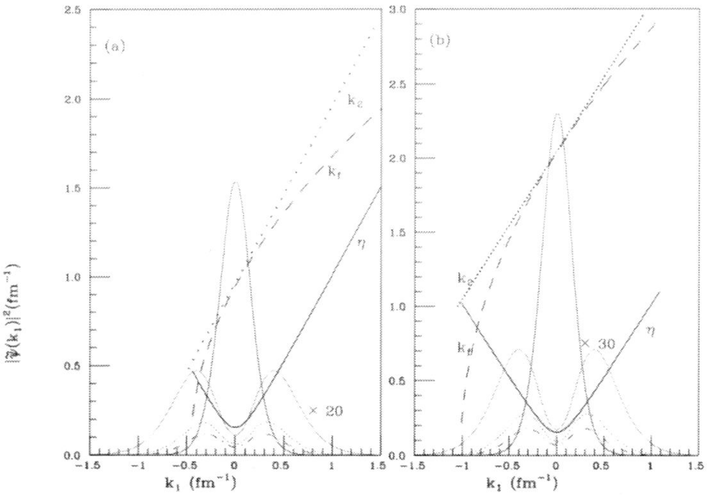

Figure 1. Initial state momentum distributions [35], as a function of k_1. For an s-state full curve peaked at $k_1 = 0$. For a d-state: dotdashed, dotted and thin solid curves. The dotted line is k_2, the dashed line is k_f, and the solid line is η, see text for definitions. (a) $T_p/A_p = 20$MeV, (b) $T_p/A_p = 88$MeV.

$k_\perp = i\hbar\eta$ of the neutron momentum in the initial and final state. Here $\gamma_i = \sqrt{-2m\varepsilon_i}/\hbar$, $k_f = \sqrt{2m\varepsilon_f}/\hbar$ and $k_\perp$ is conserved during the breakup process and it is purely imaginary because the neutron which in the initial state has negative energy [44] is emitted through a potential barrier. Because of this it holds also $k_2 > k_f$. It is straightforward to see from the definitions of these kinematical variables that they satisfy the neutron energy and momentum conservation. In this time dependent formalism that we call transfer-to-the-continuum (TC) [35][41]-[43] [45]-[47], the breakup probability reads

$$\frac{dP}{dk_1} \approx \frac{1}{2}\Sigma_{j_f}(|1 - \bar{S}_{j_f}|^2 + 1 - |\bar{S}_{j_f}|^2)(2j_f + 1)(1 + R)\left[\frac{\hbar}{mv}\right]\frac{1}{k_f}|C_i|^2\frac{e^{-2\eta b_c}}{2\eta b_c}M_{l_f l_i}. \tag{10}$$

$\bar{S}_{j_f}$ are neutron-target S-matrices calculated according to the optical model, including the spin-orbit term of the neutron-target optical potential. The sum over partial waves in equation (10) is indeed a sum over total neutron-target angular momenta. C_i are the initial wave function asymptotic normalization constant. The form factor $\frac{e^{-2\eta b_c}}{2\eta b_c}$ is due to the combined effects of the initial and final wave function Fourier transforms, while $M_{l_f l_i}$ is due the to overlap of the angular parts. Detailed definitions can be found in Ref. [35].

Under the sudden (or adiabatic) hypothesis the parallel momentum distribution of the neutron in the projectile is frozen during the reaction and its shape should be reflected by the final measured distribution. The available neutron final energy is all converted into parallel

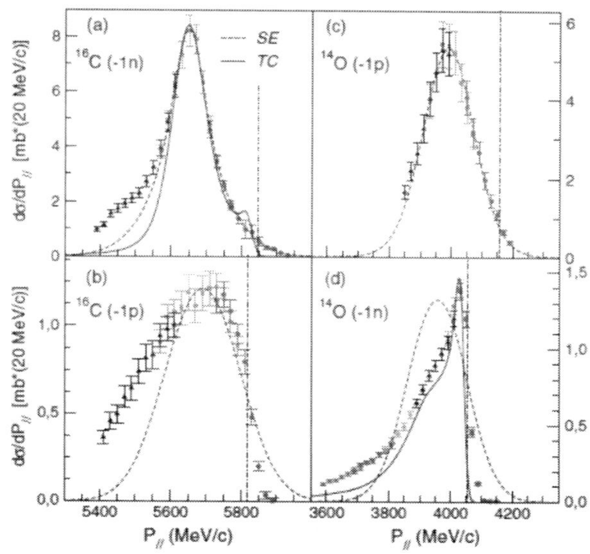

Figure 2. (Color online) Examples of spectra from Refs. [11].

momentum. Interference effects with the transverse distribution are in this way neglected. The sudden hypothesis means also that the whole momentum distribution in the initial state is sampled during the reaction, while the correct kinematical conditions, expressed in the definition of k_1 and k_2, constrain the phase space sampled by the reaction. Also, in the present approach in order to realize the best energy matching conditions for each possible final energy ε_f of the neutron, the reaction mechanism shares the total momentum k_f between the transverse momentum component η and the parallel component k_2, thus allowing the interference (cf. also equation (3) of [41]) between the two corresponding distributions. The factor $M_{l_f l_i}$ in equation (10) contains the interference effects. Thus measured parallel momentum distribution might look deformed as compared to the parallel momentum distribution of the neutron in the initial state.

In order to clarify the importance of the energy sharing between the parallel and transverse components of the neutron final momentum, we show in figures 1a and 1b [35], corresponding to an incident energy of 20A.MeV and 88A.MeV, respectively, the following kinematical variables as functions of k_1, the neutron initial momentum with respect to the projectile: k_2, the neutron final parallel momentum component with respect to the target, by the dotted line; k_f, the magnitude of the total neutron momentum corresponding to each neutron final continuum energy ε_f, by the dashed line; η the neutron transverse momentum component, by the solid line. The minimum values of k_1 correspond to $\varepsilon_f = 0$. Values of all parameters corresponding to values of $\varepsilon_f < 0$ are not accessible by breakup reactions but they would rather correspond to transfer to a final bound state. In both figures there is a region corresponding to very small values of η in which $k_2 \approx k_f$. This is the region of validity of the sudden eikonal approximation. In fact in such conditions since the transverse component of the neutron momentum η is very small, the total momentum k_f is all converted into parallel momentum k_2. In [41] we showed indeed that the condition $k_2 \approx k_f$ was necessary to obtain the eikonal form of the breakup amplitude. In the same figures we show the initial s and d distributions of the parallel neutron momentum as a function of k_1. The shapes of such distributions depend critically from the initial single particle wave function angular momentum through the angular part of the wave function and then of the

Journal of Physics: Conference Series **527** (2014) 012008 doi:10.1088/1742-6596/527/1/012008

momentum distribution. That one neutron breakup could be used to measure such distributions was first suggested in Ref. [45] where normal nuclei were studied.

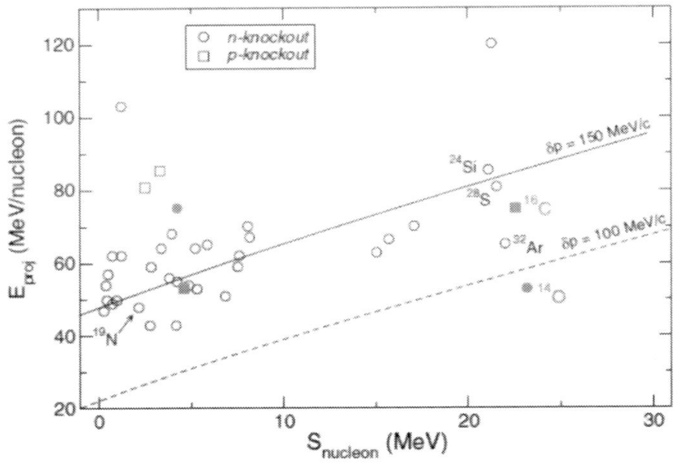

Figure 3. (Color online) Kinematic relation between the energy per nucleon of the projectile and the separation energy of the removed nucleon for a given position of the cutoff. The lines correspond to cutoffs appearing at $\delta p = 100$ and 150 MeV/c with respect to the center of the "sudden" distribution. Data points are taken from [22, 23, 24]. Data points from [11] are indicated with filled (red) symbols.

The initial distributions get wider by increasing the absolute value of the initial binding. Therefore the eikonal approximation is best justified in a range of $k_1 \approx 0$ values and for initial distributions, like the $l_i = 0$ one, concentrated in such a region. Figure 1b shows that such a range increases by increasing the incident energy. On the other hand figure 1a shows that at incident energies around 20A.MeV an important part of the initial neutron momentum distribution corresponding to k_1 values from $-\infty$ to about -0.5 fm^{-1} would not be kinematically allowed. Thus using the frozen limit would give too wide momentum distributions and too large breakup cross sections.

The kinematical effects discussed in this section have recently been seen clearly in some experimental data [11] that we show in figure 2 (d-quadrant). The core parallel momentum spectrum following ^{14}O+^{9}Be one neutron breakup at 53A.MeV is shown. Note that in this case the neutron initial separation energy is S_n=23.2MeV. As expected the spectrum is very deformed and shows a cutoff consistent with equation (6). Furthermore in figure 3 we show the amount of cutoff affecting a series of recent data as a function of initial incident energy and separation energy. δp is easily understood by looking at figure 1 as the absolute value of the lower kinematically allowed limit of k_1 in MeV/c. Such results were already predicted in Ref. [42].

6. Coulomb dissociation and proton breakup

When the target is heavy the effect of the core-recoil, due to the Coulomb interaction between projectile and target, gives rises to a further source of breakup. The dipole approximation for the Coulomb potential is the standard approximation to calculate neutron halo "Coulomb dissociation" as it gives correctly the core-recoil mechanism which causes the reaction [48, 49]. However it is not applicable when the valence particle is a proton. This is due to the fact that in proton breakup there is a very important direct Coulomb proton-target term which needs to be treated correctly, to all orders and all multipolarities as it has been made clear in several publications [50]-[55]. We remind the reader that in calculating breakup, to treat an interaction to all orders means to have in the probability amplitude a final wave function which takes into account all final state interactions of the breakup particle with the target. This is well achieved with the eikonal wave function, as it is easily seen if one expands in series the phase shift. On the other hand to treat the interaction to all multipolarities means that either no multipole expansion is done or all terms are taken into account. The extension of our formalism to take into account the full Coulomb potential for a neutron breakup was made in Ref. [53], while the proton breakup formalism was first introduced in Ref. [54] and then studied in detail in Ref. [55].

The Coulomb potential responsible for proton breakup is

$$V(\mathbf{r},\mathbf{R}) = \frac{V_c}{|\mathbf{R}-\beta_1\mathbf{r}|} + \frac{V_v}{|\mathbf{R}+\beta_2\mathbf{r}|} - \frac{V_0}{R} \tag{11}$$

where $V_c = Z_cZ_te^2$, $V_v = Z_vZ_te^2$ and $V_0 = (Z_v + Z_c)Z_te^2$. β_1 and β_2 are the mass ratios of proton and core, respectively, to that of the projectile. Z_c, Z_t and Z_v are the core, target and proton charge respectively. In this case the differential cross section will get contribution from three terms such that it reads:

$$\frac{d\sigma}{d\mathbf{k}} = \frac{1}{8\pi^3}\int d\mathbf{b}_c|S_{ct}(\mathbf{b}_c)|^2|g^{rec} + g^{dir} + g^{nuc}|^2, \tag{12}$$

where $\mathbf{k}$ is the proton final momentum in the continuum and the density of final states has been introduced. See Ref. [55] for further notation details.

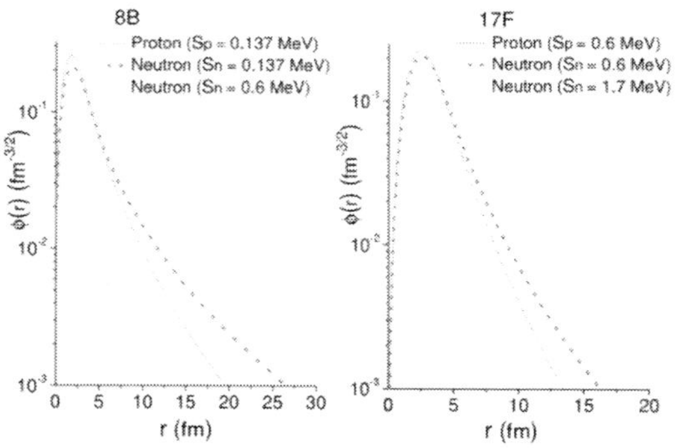

Figure 4. (Color online) Proton vs neutron wave functions for a $p_{3/2}$ single particle state in a ^{8}B and $d_{5/2}$ in ^{17}F as indicated.

In Ref. [55] we followed an idea already presented in Ref. [56] and tried to understand proton breakup in terms of the well known characteristics of neutron breakup. We found that from the point of view of the nuclear breakup a proton behaves exactly as a neutron having an "effective" binding energy

$$\tilde{\varepsilon}_i = \varepsilon_i - \Delta = \varepsilon_i - \frac{Z_pe^2}{R_i} - Z_te^2\left(\frac{1}{2}\left(\frac{1}{|b_c+\beta_2R_i|} + \frac{1}{|b_c-\beta_1R_i|}\right) - \frac{1}{b_c}\right), \tag{13}$$

Journal of Physics: Conference Series **527** (2014) 012008 doi:10.1088/1742-6596/527/1/012008

which through the factors β_1 and β_2 is consistent with equation (11). R_i the position of the projectile top of the Coulomb barrier, b_c is the distance between the center of the two nuclei for which the top of the two Coulomb barriers of projectile and target coincide.

We checked this idea by fitting wave functions for a neutron and a proton in a Woods-Saxon potential. Figure 4 shows two cases for ^{8}B and ^{17}F from Ref. [55]. By applying this method we were able to understand that from the point of view of the nuclear breakup a proton behaves exactly as a neutron of larger binding energy. Our prescription for the effective energy equation (13) proved to be quite accurate but as expected the effective separation energy has to be "incident energy dependent". This is because it contains an average estimate of the Coulomb barrier effect at a fixed typical distance between projectile and target. Such distance is expected to be incident-energy dependent. Indeed the results of Ref. [57] were CDCC calculations were performed for ^{17}F+^{208}Pb at 170MeV, show that at lower incident energies the effective binding energy increases.

On the other hand the Coulomb breakup is dominated by the direct proton-target interaction, as it can be seen from the figures and discussions in Refs. [55] where we showed the utility of using the breakup proton angular distributions to disentangle the reaction mechanism. We finish this section by suggesting that proton halo candidates should be investigated via nuclear breakup and compared to neutron breakup at high incident energy where the effects of the reaction mechanisms will be minimal.

7. Breaking a strongly bound neutron but skipping a weakly bound proton

Table 1. One-nucleon knockout results from ^{14}O on ^{9}Be at 53 A.MeV. Calculated inclusive cross sections σ_{TC} from the TC approach are shown and compared to the measured (σ_{exp}) cross sections. Theoretical spectroscopic factors C^2S from [11]. Reduction factors are indicated and defined as R_f in order to distinguish from the strong absorption radius notation (R_s).

Residue	$S_{p(n)}$ (MeV)	J^π	σ_{exp} (mb)	C^2S	σ_{sp} (mb)	σ_{sp}^{nop} (mb)	σ_{TC} (mb)	σ_{TC}^{nop} (mb)	R_f
^{13}N	4.6	$1/2^-$	58(4)	1.83	34.2		53		0.9
^{13}O	23.1	$3/2^-$	14(1)	3.15	10.9	8.6	34.5	27.1	0.5

There are situations in which a strongly bound neutron is knocked out but a weakly bound proton in the same projectile is not. The final cross section formula for a reaction of this type must contain some information on the strong reaction channel that one thinks is suppressed, thus we suggest the following correction:

$$\sigma_{TC}^{nop} = C^2S \int_0^\infty d\mathbf{b_c} P_{-n}(b_c) e^{-P_{-p}} |S_{ct}(b_c)|^2,$$

where $e^{-P_{-p}} \approx 1 - P_{-p}(b_c)$, is the probability than the weakly bound particle in the projectile does not breakup. This is equivalent to calculating a Dynamic Polarization Potential (DPP) from phase shift following [58]

$$|S_{NN}(b_c)|^2 = e^{-4\delta_I(b_c)} = e^{-4\delta_I^V(b_c)} e^{-4\delta_I^S(b_c)} = |S_{ct}(b_c)|^2 e^{-P_{-p}(b_c)}.$$

This improved formalism has been applied to the reaction previously studied in Refs. [11, 59]. Using the Dispersive Optical Potential neutron-target S-matrices of Ref. [60] we obtained the results shown in table 1. The single particle neutron/proton wave functions were calculated in the Woods-Saxon potential given in Ref. [59]. The proton wave function was calculated according to Ref. [55], as a neutron wave function having an effective binding energy such as to fit the exact proton wave function of Ref. [59]. The results of table 1 for the reduction seen in the spectroscopic factors are consistent with those of Ref. [59] which makes one thinking that transfer and knockout lead to the same results about reduction factors if almost all uncertainties due to the difference in wave functions (structure information) are removed and if the weakly and strongly bound particle breakup are treated consistently, including kinematical cutoff and the effect of the DPP on the core.

8. Conclusions

In this contribution I have discussed some direct reactions used to study exotic nuclei and I have tried to enlighten open problems for the future. My final remarks are as follows. From simple halo nucleus beams we have gone to more complicated RIBs, therefore reaction theory needs to accommodate such developments and simple eikonal models of knockout need to be improved by including kinematics and a better treatment of the core-target scattering in particular when strong reaction channels such as breakup would be expected for the core. From the experimental point of view, kinematical complete experiments are necessary with reconstruction of the target final state and core angular distributions if and when possible. It would be nice to have more proton breakup experiments since we now understand better the dynamics of their reactions. The past has been characterized by studies at high incident energy and for weakly bound projectiles. In the future more and more strongly bound nuclei will be studied at lower energies at ISOL-type facilities. Then the issues raised here will be particularly relevant.

Acknowledgements

Some unpublished results presented in this paper have been obtained in collaboration with Charity R J, Flavigny F, Obertelli A and Salvioni G. I thank Barbieri C for providing his overlap functions.

References

[1] http://www.df.unipi.it/ angela/lisbon_report.pdf
[2] Tanihata I *et al.* 1985 *Phys. Lett. B* **160** 380
[3] Tanihata I *et al.* 1985 *Phys. Rev. Lett.* **55** 2676
[4] Tanihata I 1995 *Prog. Part. Nucl. Phys.* **35** 505
[5] Kobayashi T *et al.* 1988 *Phys. Rev. Lett.* **60** 2599
[6] Tanihata I 2011 *Nucl. Phys. A* **685** 80c
[7] Kelley J H et al 1995 *Phys. Rev. Lett.* **74** 30
[8] Di Pietro A *et al.* 2010 *Phys. Rev. Lett.* **105** 22701
[9] Anne R *et al.* 1990 *Phys. Lett. B* **250** 19; 1993 *Phys. Lett. B* **304** 55; 1994 *Nucl. Phys. A* **575** 125
[10] Périer Y *et al.* 1999 *Phys. Lett. B* **459** 55
[11] Flavigny F *et al.* 2012 *Phys. Rev. Lett.* **108** 252501
[12] Louchart C *et al.* 2011 *Phys. Rev. C* **83** 011601
[13] Hansen P G and Jonson B 1997 *Euro Phys. Lett.* **4** 409
[14] Zhukov M V *et al.* 1993 *Phys. Rep.* **231** 151
[15] Vinh Mau N 1995 *Nucl. Phys. A* **592** 33
[16] Proceedings Nobel Symposium NS152, Fahlander C, Jonson B eds 2013 *Phys. Scr. T* **152** 1
[17] Orr N *et al.* 1992 *Phys. Rev. Lett.* **69** 2050
[18] Blank B *et al.* 1991 *Z. Phys. A* **340** 41
[19] Kobayashi T *et al.* 1989 *Phys. Lett. B* **232** 51
[20] Hansen P G and Tostevin J A 2003 *Ann. Rev. Nucl. Part. Sci.* **53** 219
[21] Riisager H *et al.* 1992 *Nucl. Phys. A* **540** 365
[22] Sauvan E *et al.* 2000 *Phys. Lett. B* **491** 1
[23] Gade A *et al.* 2008 *Phys. Rev. C* **77** 044306

[24] Grinyer G *et al.* 2011 *Phys. Rev. Lett.* **106** 162502; 2012 *Phys. Rev. C* **86** 024315
[25] Hansen P G, Jensen A S, and Jonson B 1995 *Ann. Rev. Nucl. Part. Sci.* **45** 591
[26] Schwab W *et al.* 1995 *Z. Phys. A* **350** 238
[27] Pecina I *et al.* 1995 *Phys. Rev. C* **52** 191
[28] Negoita F *et al.* 1996 *Phys. Rev. C* **54** 1787
[29] Bazin D *et al.* 1995 *Phys. Rev. Lett.* **74** 3569; 1998 *Phys. Rev. C* **57** 2156
[30] Bonaccorso A and Brink D M 1998 *Phys. Rev. C* **58** 2864
[31] Serber R 1947 *Phys. Rev.* **72** 1008
[32] Baur G, Rösel F , Trautmann D, and Shyam R 1984 *Phys. Rep.* **111** 33
[33] Hussein M S and McVoy K W 1985 *Nucl. Phys. A* **445** 124
[34] Bertulani C A and Hussein M S 1990 *Phys. Rev. Lett.* **64** 1099
[35] Bonaccorso A 1999 *Phys. Rev. C* C **60** 054604; 1999 *Nucl. Phys. A* **649** 315c
[36] Bertsch G F, B. A. Brown, and Sagawa H 1989 *Phys. Rev. C* **39** 1154
[37] Yabana K, Ogawa Y, and Suzuki Y 1992 *Nucl. Phys. A* **539** 295; 1992 *Phys. Rev. C* C 45 2909
[38] Bertulani C A and Mc Voy K W 1992 *Phys. Rev. C* **46** 2638
[39] Sagawa H and Takigawa N 1994 *Phys. Rev. C* **50** 985
[40] Hencken K, Bertsch G F, and Esbensen H 1996 *Phys. Rev. C* **54** 3043
[41] Bonaccorso A and Brink D M 1998 *Phys. Rev. C* **57** R22
[42] Bonaccorso A and Bertsch G F 2001 *Phys. Rev. C* **63** 044604
[43] Bonaccorso A and Brink D M 1994 *Phys. Rev. C* **49** 329
[44] Bonaccorso A, Brink D M, and Lo Monaco L 1987 *J. Phys. G* **13** 1407
[45] Bonaccorso A and Brink D M 1991*Phys. Rev. C* **44** 1559
[46] Bonaccorso A and Brink D M 1988 *Phys. Rev. C* **38** 1776
[47] Bonaccorso A and Brink D M 1991 *Phys. Rev. C* **43** 299
[48] Margueron J, Bonaccorso A, and Brink D M 2002 *Nucl. Phys. A* **703** 105
[49] Margueron J, Bonaccorso A, and Brink D M 2003 *Nucl. Phys. A* **720** 337
[50] Esbensen H, Bertsch G F, and Bertulani C A 1995 *Nucl. Phys. A* **581** 107
[51] Esbensen H and Bertsch G F 2002 *Phys. Rev. C* **66** 044609
[52] Esbensen H, Bertsch G F, and Snover K A 2005 *Phys. Rev. Lett.* **94** 042502
[53] García-Camacho A, Bonaccorso A, and Brink D M 2006 *Nucl. Phys. A* **776** 118
[54] García-Camacho A, Blanchon G, Bonaccorso A, and Brink D M 2007 *Phys. Rev. C* **76** 014607
[55] Ravinder Kumar and Bonaccorso A 2011 *Phys. Rev. C* **84** 014613; 2012 *Phys. Rev.* C **86** 061601(R)
[56] Bonaccorso A, Brink D M, and Bertulani C A 2004 *Phys. Rev. C* **69** 024615
[57] Kucuk Y and Moro A M 2012 *Phys. Rev. C* **86** 034601
[58] Bonaccorso A and Carstoiu F 2002 *Nucl. Phys. A* **706** 322
[59] Flavigny F *et al.* 2013 *Phys. Rev. Lett.* **110** 122503
[60] Bonaccorso A and Charity R J 2014 *Phys. Rev. C* **89** 024619

Journal of Physics: Conference Series **527** (2014) 012009 doi:10.1088/1742-6596/527/1/012009

News on the equation of state from heavy ion reactions

M Colonna[1], **V Baran**[2], **M Di Toro**[1,3]

[1]Laboratori Nazionali del Sud, INFN, via Santa Sofia 62, I-95123, Catania, Italy
[2]Physics Faculty, University of Bucharest, Romania
[3]Physics-Astronomy Department, University of Catania, Italy

E-mail: colonna@lns.infn.it

Abstract. We discuss recent results on heavy ion reactions between charge asymmetric systems, from low up to intermediate energies. The theoretical framework is provided by mean-field based transport approaches. We focus on isospin sensitive observables, aiming at extracting information on the density dependence of the isovector part of the nuclear effective interaction and of the nuclear symmetry energy. For reactions close to the Coulomb barrier, we explore the structure of collective dipole oscillations, rather sensitive to the low-density behavior of the symmetry energy. At intermediate energies, where regions with higher density and momentum are reached, we discuss collective flows and their sensitivity to the momentum dependence of the isovector interaction channel, which determines the splitting of neutron and proton effective masses.

1. Introduction

The behavior of nuclear matter in several conditions of density, temperature and N/Z asymmetry is of fundamental importance for the understanding of many phenomena involving nuclear systems and astrophysical compact objects. This information can be accessed by mean of heavy ion collision experiments, where transient states of nuclear matter spanning a large variety of regimes can be created. Actually this study allows one to learn about the corresponding behavior of the nuclear effective interaction, which provides the nuclear Equation of State (EOS) in the equilibrium limit. Over the past years, measurements of experimental observables, like isoscalar collective vibrations, collective flows and meson production, have contributed to constrain the EOS of symmetric matter for densities up to five time the saturation value [1]. More recently, the availability of exotic beams has made it possible to explore, in laboratory conditions, new aspects of nuclear structure and dynamics up to extreme ratios of neutron (N) to proton (Z) numbers, also opening the way to the investigation of the EOS of asymmetric matter, which has comparatively few experimental constraints. Indeed, the isovector part of the nuclear effective interaction and the corresponding symmetry energy of the EOS (Asy-EOS) are largely unknown as soon as we move away from normal density, especially in the high density regime.

Nevertheless, this information is essential in the astrophysical context, for the understanding of the properties of compact objects such as neutron stars, whose crust behaves as low-density asymmetric nuclear matter [2] and whose core may touch extreme values of density and asymmetry. Moreover, the low-density behavior of the symmetry energy also affects the

Journal of Physics: Conference Series **527** (2014) 012009 doi:10.1088/1742-6596/527/1/012009

structure of exotic nuclei and the appearance of new features involving the neutron skin, which are currently under intense investigation [3].

Over the past years, several observables which are sensitive to the Asy-EOS and testable experimentally, have been suggested [4, 5, 6, 7]. In this article we will review recent results on dissipative collisions in a wide range of beam energies, from just above the Coulomb barrier up to the AGeV range, on the basis of transport theories of the Stochastic Mean Field (SMF) type. Low energies will bring information on the symmetry term around (below) normal density, while intermediate energies will probe high density regions.

2. Transport theories and symmetry energy

Nuclear reactions are modeled by solving transport equations based on mean field theories, with short range (2p-2h) correlations included via hard nucleon-nucleon elastic collisions and via stochastic forces, selfconsistently evaluated from the mean phase-space trajectory, see [8, 5]. Stochasticity is essential in order to get distributions as well as to allow for the growth of dynamical instabilities.

In the energy range up to a few hundred A MeV, the appropriate tool is the so-called Boltzmann-Langevin (BL) equation [8]:

$$\frac{df}{dt} = \frac{\partial f}{\partial t} + \{f, H\} = I_{coll}[f] + \delta I[f], \quad (1)$$

where $f(\mathbf{r}, \mathbf{p}, t)$ is the one-body distribution function, the semi-classical analog of the Wigner transform of the one-body density matrix, $H(\mathbf{r}, \mathbf{p}, t)$ the mean field Hamiltonian, I_{coll} the two-body collision term incorporating the Fermi statistics of the particles, and $\delta I[f]$ its fluctuating part. Here we follow the approximate treatment of the BLE introduced in Ref. [9], the Stochastic Mean Field (SMF) model. The numerical procedure to integrate the transport equations is based on the test-particle method.

Effective interactions (associated with a given EOS) can be considered as an input of the transport code and from the comparison with experimental data one can finally get some hints on the properties of the nuclear interaction.

We recall that the symmetry energy E_{sym} appears in the energy density functional $\epsilon(\rho, \rho_i) \equiv \epsilon(\rho) + \rho\frac{E_{sym}}{A}(\rho_i/\rho)^2 + O(\rho_i/\rho)^4 + ..$, expressed in terms of total ($\rho = \rho_p + \rho_n$) and isospin ($\rho_i = \rho_p - \rho_n$) densities. E_{sym} gets a kinetic contribution directly from basic Pauli correlations and a potential part, $C(\rho)$, from the highly controversial isospin dependence of the effective interactions:

$$\frac{E_{sym}}{A} = \frac{E_{sym}}{A}(kin) + \frac{E_{sym}}{A}(pot) \equiv \frac{\epsilon_F}{3} + \frac{C(\rho)}{2\rho_0}\rho \quad (2)$$

(ρ_0 denotes the saturation density). The nuclear mean-field, consistently derived from the energy functional, can be written as:

$$U_q = A\frac{\rho}{\rho_0} + B(\frac{\rho}{\rho_0})^{\alpha+1} + C(\rho)\frac{\rho_n - \rho_p}{\rho_0}\tau_q + \frac{1}{2}\frac{\partial C}{\partial \rho}\frac{(\rho_n - \rho_p)^2}{\rho_0}, \quad (3)$$

where $\tau_q = +1(-1)$ for $q = n(p)$. The isoscalar section is fixed requiring that the saturation properties of symmetric nuclear matter, with a compressibility modulus around $K = 200$ MeV [1], are reproduced (which corresponds to the Skyrme SKM* effective interaction). The corresponding values of the coefficients are $A = -356.8$ MeV, $B = 303.9$ MeV, $\alpha = 1/6$.

The sensitivity of the simulation results is tested against different choices of the density dependence of the isovector part of the effective interaction. We employ three different parameterizations of $C(\rho)$: the asysoft, the asystiff and asysuperstiff respectively, see [5] for a detailed description. The value of the symmetry energy, E_{sym}/A, at saturation, as well as the

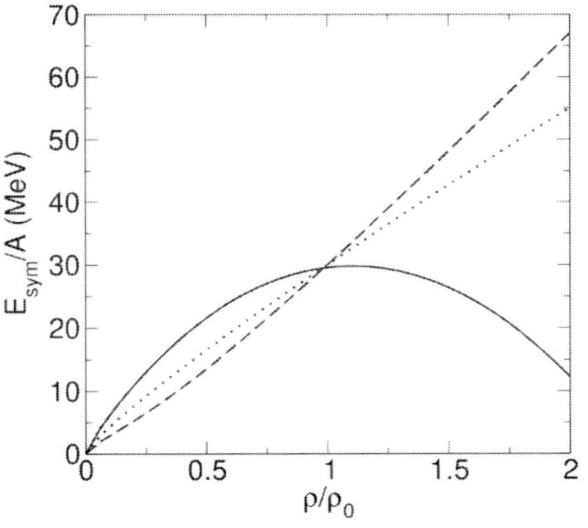

Figure 1. Three effective parameterizations of the symmetry energy : asystiff (dotted line), asysoft (full line) and asysuperstiff (dashed line).

slope parameter, $L = 3\rho_0 \frac{dE_{sym}/A}{d\rho}|_{\rho=\rho_0}$, are reported in table 1 (first two columns) for each of these Asy-EOS. Just below the saturation density the asysoft mean field has a weak variation with density while the asysuperstiff shows a rapid decrease, see figure 1.

For protons, the Coulomb interaction is also included in the simulations.

Surface terms are not explicitly included in the mean-field potential, however surface effects are accounted for by considering finite width wave packets for the test particles employed in the numerical resolution of equation (1). The width is tuned to reproduce the surface energy of nuclei in the ground state [10]. This method also induces the presence of a surface term in the symmetry energy. We have checked that properties connected to surface effects, such as the neutron skin of neutron-rich nuclei, are in reasonable agreement with calculations of other models employing similar interactions [11, 12].

Momentum-dependent effective interactions may also be implemented into equation (1) and will be considered in the following for the study of observables which are particularly sensitive to this ingredient. In particular, we will discuss results related to the momentum dependence of the isovector channel of the interaction, leading to the splitting of neutron and proton effective masses (see section 4).

3. Collective excitations in neutron-rich systems

One of the important tasks in many-body physics is to understand the emergence of collective features as well as their structure in terms of the individual motion of the constituents. The experimental characterization and theoretical description of collective excitations appearing in charge asymmetric and exotic systems is a challenge for modern nuclear physics.

3.1. New exotic collective excitations

Recent experiments of Coulomb dissociation have provided several evidences about the existence of new collective excitations in neutron-rich systems, but the available information is still incomplete and their nature is a matter of debate. In particular, many efforts have been devoted to the study of the Pygmy Dipole Resonance (PDR), identified as an unusually large concentration of the isovector dipole response at energies below the values corresponding to the Giant Dipole Resonance (GDR). The latter is one of the most prominent and robust collective motions, present in all nuclei, whose centroid position varies, for medium-heavy nuclei, as $80A^{-1/3}$ MeV. From a comparison of the available data for stable and unstable Sn isotopes

Table 1. The symmetry energy at saturation (in MeV), the slope parameters, neutron rms radius, protons rms radius, neutron skin thickness of ^{132}Sn for the three Asy-EOS.

asy-EoS	E_{sym}/A	L(MeV)	R_n(fm)	R_p(fm)	ΔR_{np}(fm)
asysoft	29.9	14.4	4.90	4.65	0.25
asystiff	28.3	72.6	4.95	4.65	0.30
asysupstiff	28.3	96.6	4.96	4.65	0.31

a correlation between the fraction of pygmy strength and isospin asymmetry was noted [13]. In general the exhausted sum-rule increases with the proton-to-neutron asymmetry. This behavior was related to the symmetry energy properties below saturation and therefore connected to the size of the neutron skin [14, 3].

In spite of the theoretical progress in the interpretation of this mode and new experimental information [15, 16, 17, 18], a number of critical questions concerning the nature of the PDR still remains. Here we want to address the important issue related to the collective nature of the PDR in connection with the role of the symmetry energy.

A microscopic, self-consistent study of the collective features and of the role of the nuclear effective interaction upon the PDR can be performed within the Landau theory of Fermi liquids. This is based on two coupled Vlasov kinetic equations (see equation (1), neglecting two-body correlations) for neutron and proton one-body distribution functions $f_q(\mathbf{r}, \mathbf{p}, t)$ with $q = n, p$, and was applied quite successfully in describing various features of the GDR, including pre-equilibrium dipole excitation in fusion reactions [19]. However, it should be noticed that within such a semi-classical description shell effects are absent, certainly important in shaping the fine structure of the dipole response [20]. By solving numerically the Vlasov equation, in the absence of Coulomb interaction, Urban [21] evidenced from the study of the total dipole moment D a collective response around 8.6 MeV which was identified as a pygmy mode. Here, considering in the transport simulations also the Coulomb interaction, we explore the isoscalar character of the mode by a comparative analysis, employing the three different density parametrizations of the symmetry energy introduced above (see figure 1).

One of the most important goals is to address the connection between the development of the neutron skin and the emergence of a low-energy E1 response, a subject under intense debate during the last years. Since, as in the case of the GDR, the evolution with mass of the low-energy E1 response provides an additional insight into the nature of the mode, we shall consider the systems ^{48}Ca, ^{68}Ni, ^{86}Kr and ^{208}Pb, as well as a chain of Sn isotopes, $^{108,116,124,132,140}Sn$.

From the one-body distribution functions one obtains the local densities in the ground state: $\rho_q(\vec{r}, t) = \int \frac{2d\mathbf{p}}{(2\pi\hbar)^3} f_q(\mathbf{r}, \mathbf{p}, t)$ as well as the quadratic radii $\langle r_q^2 \rangle = \frac{1}{N_q} \int d\mathbf{r}\; r^2 \rho_q(\mathbf{r}, t)$ and the width of the neutrons skin $\Delta R_{np} = \sqrt{\langle r_n^2 \rangle} - \sqrt{\langle r_p^2 \rangle} = R_n - R_p$.

The integration of the transport equations is based on the test-particle (t.p.) method, with a number of 1300 t.p. per nucleon, ensuring in this way a good spanning of the phase-space.

An efficient method to extract the values of R_n and R_p is by observing their time evolution after a gentle perturbation. Both quantities perform small oscillations around equilibrium values and we remark that the numerical simulations keep a very good stability of the dynamics for at least 2000 fm/c. Using this procedure we obtain for the charge mean square radius of ^{208}Pb a value around $R_p = 5.45$ fm, to be compared with the experimental value $R_{p,exp} = 5.50$ fm. For Sn isotopes we display the mass dependence of R_n, R_p in figure 2 (a) and of ΔR_{np} respectively in figure 2 (b). Values for ^{132}Sn are also reported in table 1. For the charge radii the predictions from the three Asy-EOS virtually coincide and we notice a good agreement with the experimental

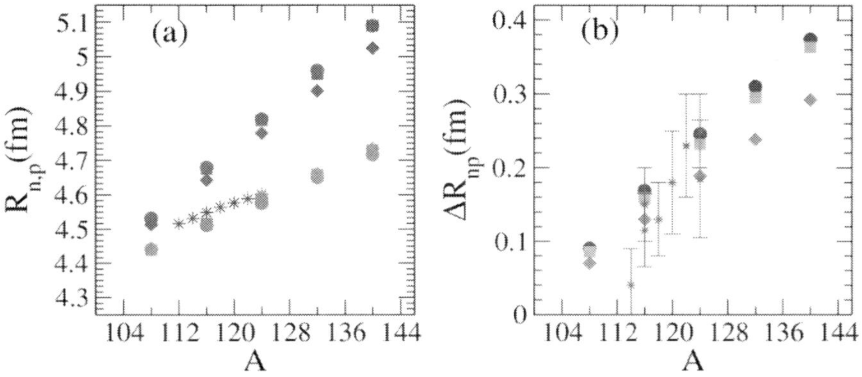

Figure 2. (Color online) (a) The neutron (blue) and proton (red) mean-square radius for Sn isotopes: asysoft (diamonds), asystiff (squares) and asysuperstiff (circles) EOS. The stars are experimental data from Refs. [22]. (b) The neutron skin size as a function of mass for Sn isotopes: asysoft (magenta diamonds), asystiff (green squares), asysuperstiff (maroon circles). The stars and the error bars (blue) are experimental data from Ref. [23].

data reported in [22]. For all adopted parametrizations the predicted values of the neutron skin thickness are within the experimental errors bars, see the data presented in [23] for the stable Sn nuclei. In the case of ^{208}Pb we find $\Delta R_{np} = 0.19$ fm for asysoft, $\Delta R_{np} = 0.25$ fm for asystiff and $\Delta R_{np} = 0.27$ fm for asysuperstiff, whereas for ^{68}Ni the corresponding values are $\Delta R_{np} = 0.17, 0.19, 0.20$ fm. We see that the neutron skin thickness increases with the slope parameter L, an effect related to the tendency of the system to stay more isospin symmetric even at lower densities when the symmetry energy changes slowly below saturation, as in the case of the asysoft EOS.

3.2. Results of transport approaches

We study the E1 response considering a GDR-like initial condition [24], determined by the instantaneous excitation $V_{ext} = \eta\delta(t - t_0)\hat{D}$ at $t = t_0$ [25]. This situation corresponds to a boost of all neutrons against all protons while keeping the Center of Mass (CM) at rest. Here $\hat{D}$ is the dipole operator. If $|\Phi_0\rangle$ is the state before perturbation then the excited state becomes $|\Phi(t_0)\rangle = e^{i\eta\hat{D}}|\Phi_0\rangle$ and the value of η can be related to the initial expectation value of the collective dipole momentum $\hat{\Pi}$, $\langle\Phi(t_0)|\hat{\Pi}|\Phi(t_0)\rangle = \hbar\eta\frac{NZ}{A}$. Here $\hat{\Pi}$ is canonically conjugated to the collective coordinate $\hat{X}$ which defines the distance between the CM of protons and neutrons, i.e. $[\hat{X}, \hat{\Pi}] = i\hbar$ [26]. Then the strength function $S(E) = \sum_{n>0} |\langle n|\hat{D}|0\rangle|^2\delta(E_n - E_0)$, directly related to the excitation probability in unit time (where E_n are the excitation energies of the states $|n\rangle$ and E_0 is the energy of the ground state $|0\rangle = |\Phi_0\rangle$), is obtained from the imaginary part of the Fourier transform of the time-dependent expectation value of the dipole momentum $D(t) = \frac{NZ}{A}X(t) = \langle\Phi(t)|\hat{D}|\Phi(t)\rangle$, as $S(E) = \frac{Im(D(\omega))}{\pi\eta\hbar}$. We consider the initial perturbation along the z-axis and follow the dynamics of the system until $t_{max} = 1830$ fm/c in each case. At $t = t_0 = 30$ fm/c we extract the collective momentum and determine η. For the three Asy-EOS the E1 strength functions of ^{208}Pb and ^{140}Sn are represented in figure 3. As a test of the quality of our method we compared the numerically estimated value of the first moment $m_1 = \int_0^\infty ES(E)dE$ with the value predicted by the Thomas-Runke-Kuhn (TRK) sum rule

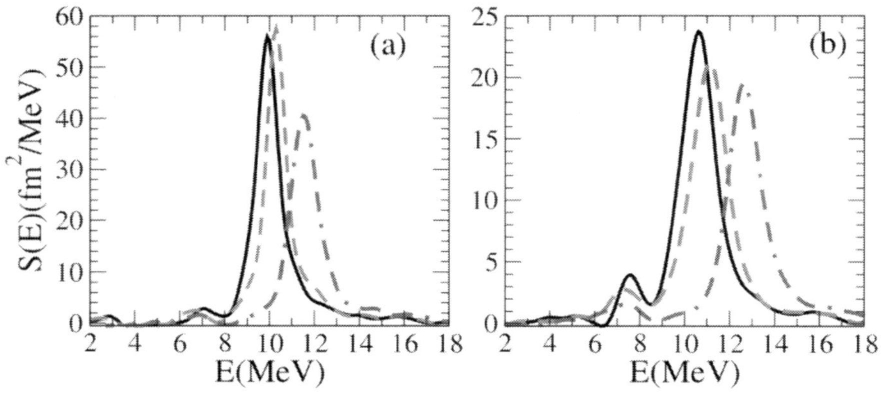

Figure 3. (Color online) The strength function for ^{208}Pb (a) and ^{140}Sn (b) for asysoft (blue, dot-dashed line), aystiff (red, dashed line) and asysuperstiff (black, solid line) EOS.

$m_1 = \frac{\hbar^2}{2m}\frac{NZ}{A}$. In all cases the difference was of only few percentages. The energy peak of the PDR for ^{208}Pb, see figure 3(a), is located around 7 − 7.5 MeV in good agreement with recent experimental data which indicate $E_{PDR,Pb} = 7.36$ MeV [27]. For ^{68}Ni we obtain 9.8 MeV, quite close to the recent reported data $E_{PDR,Ni} = 9.9$ MeV [28]. We observe that the GDR energy centroid is underestimated in comparison with experimental data, a fact probably related to the choice of the interaction which has not an effective mass [29]. In any case, a clear dependence with the slope parameter L manifests as a consequence of the isovector nature of the mode. This feature shows that the symmetry energy behavior below saturation is also affecting the dipole oscillations of finite systems. Figure 4 displays the position of the PDR energy centroid, which is independent of the Asy-EOS, as a function of mass for all studied systems (blue circles). In addition, we represent the position of the PDR energy peak as it results from the power spectrum analysis of the pygmy dipole $D_y(t)$ after a pygmy-like initial condition, see [24] (red diamonds), and the available experimental data (maroon square) [30]. The differences between the two methods are within a half of MeV. An appropriate parametrization, obtained from the fit of the numerical results is $E_{PDR} = 41A^{-1/3}$, quite close to what is expected in the harmonic oscillator shell model (HOSM) approach [26] and in agreement with some recent experimental data. Whereas the isovector residual interaction pushes up the value of the GDR energy it seems that the PDR energy centroid is not much affected by this part of the interaction, indicating the isoscalar-like nature of this mode. This feature may explain the better agreement with experimental observations in comparison with the GDR case. Let us mention that for Ni, Sn and Pb isotopic chains, the isotopic dependence of the PDR energy was also investigated by Paar *et al.* [38], within a HFB and RQRPA treatment. The presence of a collective mode with energy centroid around 10 MeV for ^{68}Ni, 8 MeV for ^{132}Sn and 7.5 MeV for ^{208}Pb was predicted. A comparison with our results shows a good agreement between the two theoretical approaches.

We can now determine the Energy Weighted Sum Rule (EWSR) exhausted by the PDR by integrating over the low-energy resonance region:

$$m_{1,y} = \int_{PDR} ES(E)dE \tag{4}$$

The dependence of the moment $m_{1,y}$ on the neutron skin thickness is shown in figure 5, where the information concerning all mentioned systems for the three Asy-EOS is included. While below 0.15 fm the EWSR acquired by the PDR manifests a saturation tendency, above this value a

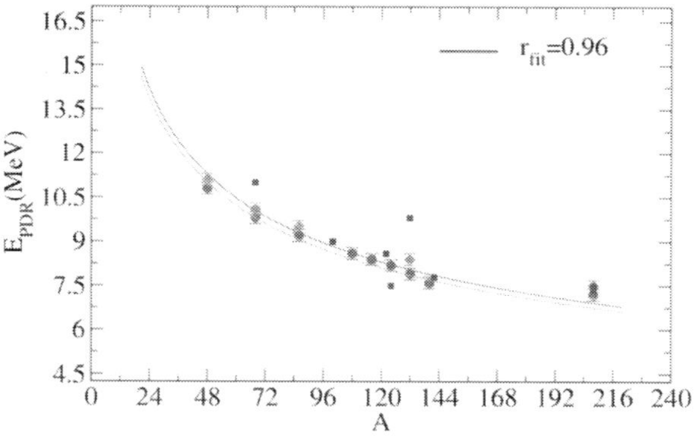

Figure 4. (Color online) The energy centroid of PDR as a function of mass (blue circles and red diamonds). The solid lines correspond to parametrizations $40A^{-1/3}$ (green), $41A^{-1/3}$ (blue) and $42A^{-1/3}$ (yellow). The maroon squares are experimental data from Ref. [30]. r_{fit} refers to the correlation coefficient.

linear correlation clearly manifests. For the same system, when we pass from asysuperstiff to asysoft parametrisation, the neutron-skin shrinks and correspondingly the value of $m_{1,y}$ decreases. This behavior is in agreement with the results reported in [39] with a self-consistent RPA approximation based on relativistic energy density functionals. Moreover we notice that the variation rate appears to be system independent, obtaining an increase of 15 MeV fm^2 of the exhausted EWSR to a change of 0.1 fm of the neutron skin width. Such features suggest that the acquired EWSR should not differ too much even for different nuclei if they have close values of the neutrons skin thickness. We would like to mention that these findings look qualitatively in agreement with those of Inakura [40], based on systematic RPA calculations.

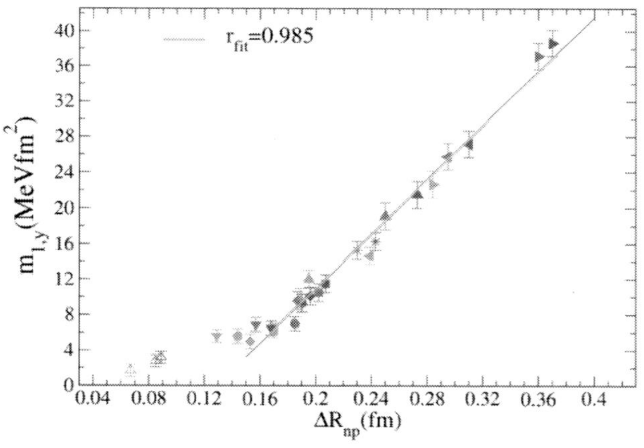

Figure 5. (Color online) The EWSR exhausted by PDR as a function of neutron skin in ^{108}Sn (empty triangles up), ^{116}Sn (triangles down), ^{124}Sn (stars), ^{132}Sn (triangles left), ^{140}Sn (triangles right), ^{48}Ca (circles), ^{68}Ni (squares), ^{86}Kr (diamonds), ^{208}Pb (full triangles up), for asysoft (green), asystiff (red) and asysuperstiff (blue) EOS. r_{fit} refers to the correlation coefficient. The error bars are related to the uncertainties in defining the integration domain for the PDR response.

To conclude this section, we have discussed recent open questions [41] regarding the nature of the PDR. By performing a systematic analysis which includes several nuclear systems we obtained new results aiming at contributing to a more complete picture of the PDR dynamics. In a microscopic transport approach, a low-energy dipole collective mode is evidenced as an ubiquitous feature of all investigated nuclei. The analysis leads to a dependence of the PRD energy centroid with mass described by $E_{PDR} = 41A^{-1/3}$, in agreement with recent experimental information. This indicates a close connection with the characteristic frequency of the HOSM, $\hbar\omega_0 = 41A^{-1/3}$, and a weak influence of the residual interaction in the isovector sector [42]. Such behavior points out the isoscalar-like nature of this mode. From our calculations a universal, linear correlation of the EWSR exhausted by PDR with the neutron skin thickness occurs [43].

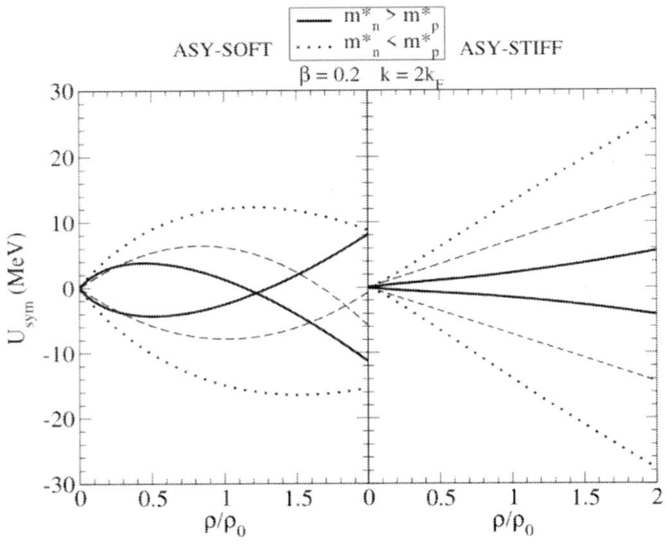

Figure 6. Density dependence of neutron(upper)-proton(lower) potentials for an asymmetry $\beta = (N-Z)/(N+Z) = 0.2$ for the asysoft (left) and asystiff (right) choices. Dashed: No momentum dependence. Momentum dependent potentials at momentum $k = 2k_F$: solid lines for the $m^*_n > m^*_p$ case, dotted lines for the opposite $m^*_n < m^*_p$ choice.

It appears as a very specific signature, showing that the neutrons which belong to the skin play an essential role in shaping the E1 response in the PDR region. We consider that the new findings presented here can be useful for further systematic experiments searching for this quite elusive mode. A precise estimate of the EWSR acquired by the PDR can provide indications about the neutron skin size which in turn will add more constraints on the slope parameter L of the symmetry energy.

4. Asy-EOS at supra-saturation density: collective flows

Reactions with neutron-rich systems at intermediate energies (100-500 AMeV) are of interest in order to have high momentum particles and to test regions of high baryon and isovector density during the reaction dynamics. In such a context, it is important to consider momentum dependent effective interactions, which essentially lead to the concept of effective masses. If also the isovector component of the interaction is momentum dependent, one observes different effective masses (i.e. effective mass splitting) for neutrons and protons. The problem of the precise determination of the Momentum Dependence in the Isovector channel ($Iso - MD$) of the nuclear interaction is still very controversial and it would be extremely important to get more definite experimental information, looking at observables which may also be sensitive to the mass splitting.

Transport codes are usually implemented with different (n, p) momentum dependent interactions, see for instance [44, 45]. This allows one to follow the dynamical effect of opposite n/p effective mass (m^*) splitting while keeping the same density dependence of the symmetry energy [46]. Neutron and proton symmetry potentials, corresponding to different symmetry energy parametrizations and effective mass splitting, are represented as a function of the density in figure 6. One can observe an interesting competition between the two ingredients: above normal density, for instance, the asysoft parametrization, coupled to the $m^*_n < m^*_p$ splitting, leads to similar potentials as obtained in the asystiff case, with $m^*_p < m^*_n$.

Let us consider semicentral ($b/b_{max} = 0.5$) collisions of $^{197}Au + {}^{197}Au$ at 400 AMeV [47]. In the interacting zone baryon densities about $1.7 - 1.8\ \rho_0$ can be reached in a transient time of the order of 15-20 fm/c. The system is quickly expanding and the freeze-out time is around 50 fm/c. A rather abundant particle emission is observed over this time scale. Here it is very interesting to study again the collective response of the system. Collective flows are very good candidates since they are expected to be rather sensitive to the momentum dependence of the mean field, see [48, 5]. The transverse flow, $V_1(y, p_t) = \langle \frac{p_x}{p_t} \rangle$, where $p_t = \sqrt{p_x^2 + p_y^2}$ is the transverse momentum and y the rapidity along the beam direction, provides information on the anisotropy of nucleon emission on the reaction plane. Very important for the reaction dynamics is also the elliptic flow, $V_2(y, p_t) = \langle \frac{p_x^2 - p_y^2}{p_t^2} \rangle$. The sign of V_2 indicates the azimuthal anisotropy of the emission: on the reaction plane ($V_2 > 0$) or out-of-plane ($squeeze - out$, $V_2 < 0$) [48].

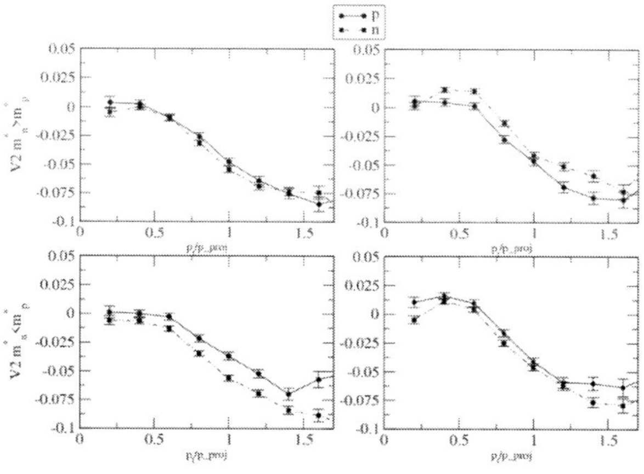

Figure 7. Proton (thick) and neutron (thin) V_2 flows in a semi-central reaction $Au + Au$ at 400 AMeV. Transverse momentum dependence at midrapidity, $| y_0 |< 0.3$. Upper curves for $m_n^* > m_p^*$, lower curves for the opposite splitting $m_n^* < m_p^*$. Left: asystiff. Right: asysoft.

In figure 7, we plot the elliptic flow of emitted neutrons and protons, for different asy-stiffness and effective mass splitting choices. We are now exploring density regions mainly above normal density, therefore we expect a larger neutron repulsion in the asystiff case, corresponding to the larger symmetry energy value (see figure 1). Indeed in figure 7 we observe a larger (negative) squeeze-out for neutrons in the asystiff case (compare left and right panels). Moreover, the $m_n^* < m_p^*$ case will favor the neutron repulsion, leading to a larger squeeze-out for neutrons, compare top and bottom panels. In particular, in the asysoft case (on the right) we see an inversion of the neutron/proton squeeze-out at mid-rapidity ($|y_0| = |y/y_{beam}| < 0.3$) for the two effective mass-splittings.

Actually, we observe a rather interesting interplay between the effects linked to the symmetry energy and to the mass splitting, connected to the behavior of neutron and proton potentials described above: a larger (smaller) neutron effective mass may compensate the larger (smaller) neutron repulsion corresponding to the asystiff (asysoft) case. Indeed the $m_n^* < m_p^*$ case, with the soft Asy-EOS, yields very similar results of the $m_p^* < m_n^*$ case with the stiff Asy-EOS.

It seems to be difficult to conclude on the properties of the effective interaction (asystiffness and MD) just from the analysis of one single observable. However, coupling the flow information to the study of other observables, it would be possible to reach more definitive constraints of the effective interaction. For instance, in the considered beam energy range, the N/Z content of the particle emission looks particularly sensitive just to the sign of mass splitting, rather than to the asy-stiffness [47]. Indeed pre-equilibrium particles emerge from different density regimes and the effect of the density dependence of the isovector interaction is averaged out. This is illustrated in figures 8-9, where we plot the ratio of neutrons and protons emitted at mid-rapidity, as a function of transverse momentum or energy.

Hence it would be very interesting the combine the information coming from particle flows and yields. Recent experimental analyses look very promising in this direction [49, 50]. Due to the difficulties in measuring neutrons, one could also investigate the difference between light isobar (like ^{3}H vs. ^{3}He) flows and yields, also shown in figures 8-9. We still expect to see effective

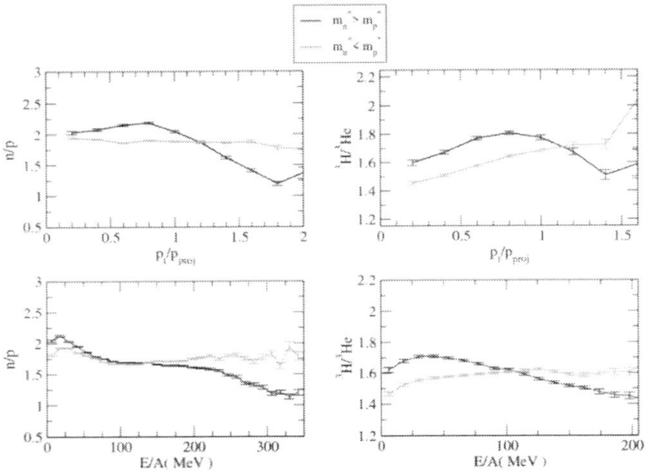

Figure 8. $^{197}Au + ^{197}Au$ at 400 AMeV, central collisions. Isospin content of nucleon (left) and light ion (right) emissions vs. p_t at midrapidity, $\mid y_0 \mid < 0.3$, (upper) and kinetic energy (lower), for all rapidities, for the two nucleon mass splitting choices. Asysoft interaction.

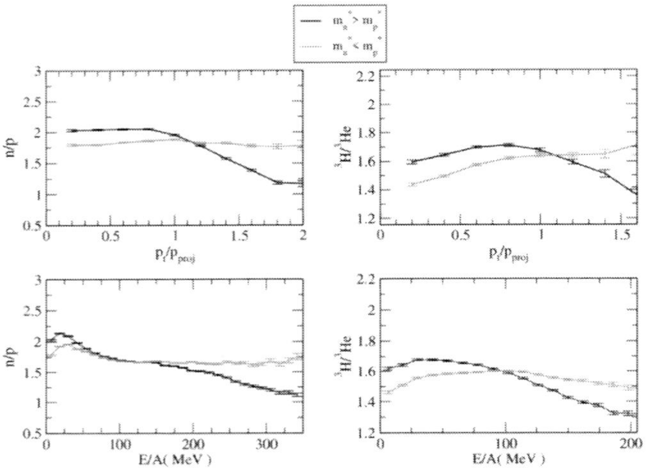

Figure 9. Same as in figure 8, but for the asystiff interaction.

mass splitting effects [47].

5. Conclusions

Heavy ion reactions, from low up to intermediate energies, allow one to explore several regimes of nuclear matter, opening the possibility to constrain the nuclear EOS and, in particular, the largely debated density behavior of the symmetry energy.

Information on the low density region can be accessed in reactions from low to Fermi energies, where collective excitations and fragmentation mechanisms are dominant. Results on isospin sensitive observables have been presented. In particular, we have concentrated our analysis on the features of collective dipole oscillations in exotic systems, showing an interesting correlation between the strength exhausted by the "pygmy" dipole resonance and the extension of the neutron skin, which in turn reflects the behavior of the symmetry energy below normal density. The conclusions drawn from this analysis seem to be compatible with the information extracted from fragmentation reactions and related observables, for which new experimental evidences have recently appeared, pointing towards a stiff behavior of the symmetry energy around normal density [51, 52].

The greatest theoretical uncertainties concerns the high density domain, that has the largest impact on the understanding of the properties of neutron stars. This regime can be explored in terrestrial laboratories by using intermediate energy and relativistic heavy ion collisions of charge asymmetric nuclei. Collective flows, light particle and meson production are promising observables.

More global analyses, also based on new sensitive observables, together with the comparison and the improvement of the available theoretical models, are expected to improve our present knowledge, providing further stringent constraints.

Acknowledgements
This work for Baran V was supported by a grant of the Romanian National Authority for Scientific Research, CNCS - UEFISCDI, project number PN-II-ID-PCE-2011-3-0972.

References

[1] Danielewicz P, Lacey R, and Lynch W G 2002 *Science* **298** 1592
[2] Lattimer J M and Prakash M 2007 *Phys. Rep.* **442** 109
[3] Piekarewicz J 2006 *Phys. Rev. C* **73** 044325
[4] *Isospin Physics in Heavy-ion Collisions at Intermediate Energies*, Eds. Li B A and Schröder W U, Nova Science Publishers (2001, New York)
[5] Baran V, Colonna M, Greco V, and Di Toro M 2005 *Phys. Rep.* **410** 335
[6] Colonna M and Tsang M B 2006 *Eur. Phys. J. A* **30** 165, and references therein
[7] Li B A, Chen L W, and Ko C M 2008 *Phys. Rep.* **465** 113
[8] Chomaz P, Colonna M, and Randrup J 2004 *Phys. Rep.* **389** 263
[9] Colonna M *et al.* 1998 *Nucl. Phys. A* **642** 449
[10] Guarnera A, Colonna M, and Chomaz Ph 1996 *Phys. Lett. B* **373** 267
[11] Baran V, Frecus B, Colonna M, and Di Toro M 2012 *Phys. Rev. C* **85** 051601(R)
[12] Paar N, Nicsik T, Vretenar D, and Ring P 2005 *Phys. Lett. B* **606** 288
[13] Klimkiewicz A *et al.* 2007 *Phys. Rev. C* **76** 051603(R)
[14] Yoshida S and Sagawa H 2004 *Phys. Rev. C* **69** 024318; 2006 *Phys. Rev. C* **73** 044320
[15] Savran D *et al.* 2008 *Phys. Rev. Lett.* **100** 232501
[16] Wieland O *et al.* 2010 *Phys. Rev. Lett.* **102** 092502; Wieland O and Bracco A 2011 *Prog. Part. Nucl. Phys.* **66** 304; Toft H K *et al.* 2010 *Phys. Rev. C* **81** 064311
[17] Tonchev A P *et al.* 2010 *Phys. Rev. Lett.* **104** 072501
[18] Makinaga A *et al.* 2010 *Phys. Rev. C* **82** 024314
[19] Baran V, Rizzo C, Colonna M, Di Toro M, and Pierroutsakou D 2009 *Phys. Rev. C* **79** 021603(R)
[20] Roca-Maza X, Pozzi G, Brenna M, Mizuyama K, and Colò G 2012 *Phys. Rev. C* **85** 024601
[21] Urban M 2012 *Phys. Rev. C* **85** 034322
[22] De Jager C W *et al.* 1987 *At. Data Nucl. Data Table* **36** 495; Audi G and Wapstra A H 1995 *Nucl. Phys. A* **595** 409
[23] Krasznahorkay A *et al.* 1999 *Phys. Rev. Lett.* **82** 3216
[24] Baran V, Frecus B, Colonna M, and Di Toro M 2012 *Phys. Rev. C* **85** 051601
[25] Calvayrac F, Reinhard P G, and Suraud E 1997 *Ann. Phys.* **225** 125
[26] Abrosimov V I and Davydovs'ka O I 2009 *Ukr. J. Phys.* **54** 1068; Baran V *et al.* 2012 *Rom. J. Phys.* **57** 36
[27] Tamii A *et al.* 2011 *Phys. Rev. Lett.* **107** 062502
[28] Rossi D 2014 *INPC2013 Proceedings* and private communication
[29] Suraud E, Pi M, and Schuck P 1988 *Nucl. Phys. A* **482** 187c
[30] For ^{68}Ni from [31]; for ^{100}Mo from [32]; for ^{122}Sn from [33]; for ^{124}Sn from [34]; for ^{132}Sn from [35]; for ^{142}Nd from [36]; for ^{208}Pb from [27, 37]
[31] Wieland O *et al.* 2009 *Phys. Rev. Lett.* **102** 092502
[32] Rusev G *et al.* 2006 *Eur. Phys. J. A* **27** 171

[33] Toft H K *et al.* 2011 *Phys. Rev. C* **83** 044320
[34] Endres J *et al.* 2010 *Phys. Rev. Lett.* **105** 212503
[35] Adrich P *et al.* 2005 *Phys. Rev. Lett.* **95** 132501
[36] Angell C T *et al.* 2012 *Phys. Rev. C* **86** 051302(R)
[37] Kondo T *et al.* 2012 *Phys. Rev. C* **86** 014316
[38] Paar N *et al.* 2005 *Phys. Lett. B* **606** 288
[39] Vretenar D *et al.* 2012 *Phys. Rev. C* **85** 044317
[40] Inakura T *et al.* 2011 *Phys. Rev. C* **84** 021302(R)
[41] Savran D, Aumann T, and Zilges A 2013 *Prog. Part. Nucl. Phys.* **70** 210
[42] Lanza E *et al.* 2011 *Phys. Rev. C* **84** 064602
[43] Baran V *et al.* 2013 *Phys. Rev. C* **88** 044610
[44] Li B A, Das Champak B, Das Gupta S, and Gale C 2004 *Nucl. Phys. A* **735** 563
[45] Rizzo J, Colonna M, and Di Toro M 2005 *Phys. Rev. C* **72** 064609
[46] Rizzo J *et al.* 2008 *Nucl. Phys. A* **806** 79
[47] Giordano V *et al.* 2010 *Phys. Rev. C* **81** 044611
[48] Danielewicz P 2000 *Nucl. Phys. A* **673** 375
[49] Russotto P *et al.* 2011 *Phys. Lett. B* **697** 471
[50] Cozma M D 2011 *Phys. Lett. B* **700** 139
[51] De Filippo E *et al.* (Chimera Collaboration) 2012 *Phys. Rev. C* **86** 014610
[52] Tsang M B *et al.* 2012 *Phys. Rev. C* **86** 015803

Journal of Physics: Conference Series **527** (2014) 012010 doi:10.1088/1742-6596/527/1/012010

Study of nucleonic matter with a consistent two- and three-body perturbative chiral interaction

L Coraggio[1], **J W Holt**[2], **N Itaco**[1,3], **R Machleidt**[4], **L E Marcucci**[5,6], **F Sammarruca**[4]

[1] Istituto Nazionale di Fisica Nucleare, Sezione di Napoli
Complesso Universitario di Monte S. Angelo, Via Cintia - I-80126 Napoli, Italy
[2] Department of Physics, University of Washington
Seattle, Washington 98195, USA
[3] Dipartimento di Fisica, Università di Napoli Federico II,
Complesso Universitario di Monte S. Angelo, Via Cintia - I-80126 Napoli, Italy
[4] Department of Physics, University of Idaho
Moscow, Idaho 83844, USA
[5] Istituto Nazionale di Fisica Nucleare, Sezione di Pisa
Largo Bruno Pontecorvo 3 - I-56127 Pisa, Italy
[6] Dipartimento di Fisica "Enrico Fermi", Università di Pisa,
Largo Bruno Pontecorvo 3 - I-56127 Pisa, Italy

E-mail: luigi.coraggio@na.infn.it

Abstract. We calculate perturbatively the energy per nucleon in infinite nuclear matter with a chiral N^3LO (next-to-next-to-next-to-leading order) two-body potential plus a N^2LO three-body force (3BF). The 3BF low-energy constants which cannot be constrained by two-body observables are chosen such as to reproduce the $A = 3$ binding energies and the triton Gamow-Teller matrix element. This enables to study the nuclear matter equation of state in a parameter-free approach.

1. Introduction

The study of infinite nuclear matter is a very interesting topic for its connections with the properties of other physical systems. As a matter of fact, for instance, the equation of state (EOS) of pure neutron matter (PNM) may be related to the features of supernova explosions and neutron stars, while the compressibility and the symmetry energy of symmetric nuclear matter (SNM) are linked to giant dipole excitations and to the neutron-proton radii in atomic nuclei, respectively. This is why nowadays a great effort is currently being made to theoretically study infinite nuclear matter using high-precision nucleon-nucleon (NN) potentials based on chiral perturbation theory (ChPT) [1–3], that may provide a clear link between nuclear physics and quantum chromodynamics (QCD) (see for instance [4–11]).

In the framework of ChPT, nuclear two- and many-body forces are generated on an equal footing [3, 12, 13]. As a matter of fact, most interaction vertices that appear in the three-nucleon forces (3NF) and in the four-nucleon forces (4NF) are present already at the two-body level. In this context, a major issue is the choice of the low-energy constants (LECs) related to the interaction vertices. The LECs corresponding to the two-nucleon-force (2NF) vertices are fitted

Journal of Physics: Conference Series **527** (2014) 012010 doi:10.1088/1742-6596/527/1/012010

to two-nucleon data, and consistency requires that these values are not changed for the same vertices appearing in the 3NF, 4NF, ... As regards the other LECs, it is worth to note that the calculation of the EOS for pure neutron matter, with chiral 3NF up to N^{2}LO, depends only on constants that are fixed in the two-nucleon system. This comes out since, in this case, the contact interaction, V_E, and the 1π-exchange term, V_D, that appear in the N^{2}LO three-body force, vanish [14]. This is not the case in SNM, whose EOS calculation is influenced by the intermediate-range 1π-exchange component V_D and the short-range contact interaction V_E of the 3NF. Therefore, the LECs of V_E and V_D (known as c_E and c_D), which are not constrained by two-body observables, affect the calculation of the ground-state energy of SNM. Their values should be fixed taking into account only $A \leq 3$ observables, to avoid biases induced by additional many-body contributions. Since the reproduction of the observed $A = 3$ binding energies is not sufficient to fully constrain both c_D and c_E, recently, the attention has been focused on the fitting of the triton half-life, specifically the Gamow-Teller matrix element [15, 16], as first suggested by Gårdestig and Phillips [17].

In the present paper we study the EOS of both SNM and PNM employing two- and three-nucleon chiral forces with consistent LECs, aiming to ascertain the possibility to obtain realistic nuclear matter predictions with chiral interactions constrained by the properties of the two- and the three-nucleon systems without any additional adjustment. We employ a chiral potential with a cutoff parameter $\Lambda = 414$ MeV [18] and calculate, including 3NF effects, the energy per nucleon for infinite nuclear matter at nuclear densities in the framework of many-body perturbation theory. It is worth pointing out that this NN potential has been already employed in perturbative many-body calculations for finite nuclei providing a successfull reproduction of their spectroscopic properties [18, 19]. The effects of the N^{2}LO 3NF are taken into account via a density-dependent two-body potential $\overline{V}_{NNN}$ obtained by summing one nucleon over the filled Fermi sea [20, 21], and that is added to the chiral N^{3}LO potential V_{NN}. The LECs c_D and c_E of the N^{2}LO chiral three-nucleon force are fixed so to reproduce to the binding energies of $A = 3$ nuclei and the ^{3}H-^{3}He Gamow-Teller matrix element.

The paper is organized as follows. In section 2, we briefly describe the procedure we have followed to choose the LECs of the 1π-exchange term V_D and contact interaction V_E, and the perturbative calculation of the properties of infinite nuclear matter is outlined. Our results and some concluding remarks are presented in sections 3 and 4, respectively.

2. Outline of calculations

2.1. Chiral potential

In the last decade NN potentials that are able to reproduce accurately the NN data have been derived in the framework of chiral perturbation theory (ChPT) [1–3]. In this work, we employ a chiral N^{3}LO NN potential with a cutoff $\Lambda = 414$ MeV (published in Ref. [18]), and using $n = 10$ in the regulator function $f(p', p) = \exp[-(p'/\Lambda)^{2n} - (p/\Lambda)^{2n}]$, i.e., a smooth, but rather steep cutoff function is applied.

In addition to this 2NF, we consider the contributions of a N^{2}LO 3NF. At this order the chiral 3NF is built up by three terms: a two-pion exchange term, a one-pion exchange plus a 2N-contact interaction, and a pure 3N-contact interaction. The two-pion exchange 3NF contains only LECs that are already present in the NN potential, while the last two terms

$$V_D = -\frac{c_D}{f_\pi^2 \Lambda_\chi} \frac{g_A}{8 f_\pi^2} \sum_{i \neq j \neq k} \frac{\vec{\sigma}_j \cdot \vec{q}_j}{q_j^2 + m_\pi^2} (\boldsymbol{\tau}_i \cdot \boldsymbol{\tau}_j)(\vec{\sigma}_i \cdot \vec{q}_j) \tag{1}$$

and

$$V_E = \frac{c_E}{f_\pi^4 \Lambda_\chi} \frac{1}{2} \sum_{j \neq k} \boldsymbol{\tau}_j \cdot \boldsymbol{\tau}_k \, . \tag{2}$$

involve two new parameters c_D and c_E, which do not appear in the 2N problem. There are many ways to fix these two parameters. In the present work we have adopted a procedure that has been recently introduced to constrain c_D and c_E [15–17]. This procedure is based on the observation that the LEC c_D appears also in a two-nucleon contact term in the NN axial current operator derived in chiral EFT up to N^{2}LO. Therefore, c_D can be fixed to reproduce the accurate experimental value of the triton β-decay half-life, and in particular of its Gamow-Teller component (GT). More precisely, we have first calculated the ^{3}H and ^{3}He wave functions within the hyperspherical harmonics method (see Ref. [22] for a review), using the chiral 2NF plus 3NF presented above. The LECs c_D and c_E are then determined by fitting the $A = 3$ experimental binding energies and the observed triton GT value. The values obtained are: c_D = -0.40 GeV^{-1} and c_E = -0.07 GeV^{-1}.

2.2. Nuclear matter calculations

We calculate the ground-state energy (g.s.e.) per nucleon of infinite nuclear matter within the framework of many-body perturbation theory, expressing the energy as a sum of Goldstone diagrams up to third order in the interaction.

In order to take into account the effects of the N^{2}LO 3NF, a density-dependent two-body potential $\overline{V}_{NNN}(k_F)$ is obtained by summing one nucleon over the filled Fermi sea and added to the chiral N^{3}LO potential V_{NN}. It is worth pointing out that, in the evaluation of the diagrams, the matrix elements of $\overline{V}_{NNN}(k_F)$ have been multiplied by a factor 1/3 in the first-order Hartree-Fock (HF) diagram, and by a factor 1/2 in the calculation of the self-consistent single-particle energies. This is done to take care of the correct combinatorial factors of the normal-ordering at the two-body level of the 3NF [14].

In figure 1 we report the diagrams we have included in our calculation with V_{NN} and $\overline{V}_{NNN}(k_F)$ vertices. The contribution of the third-order particle-hole (ph) diagram (diagram (e) in figure 1), whose calculation is cumbersome, has been included, at present, only for the SNM EOS.

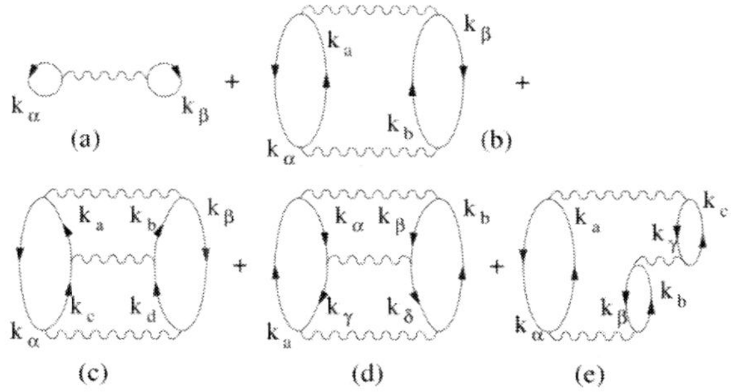

Figure 1. First-, second-, and third-order diagrams of the Goldstone expansion included in our calculations.

The analytic expressions of first-, second-, and third-order particle-particle (pp) and hole-hole (hh) contributions may be found in Ref. [23], while the implicit expression of the third-order ph diagram is reported in Ref. [24].

Padé approximants give an estimate of the value to which a perturbative series may converge [25], therefore we have calculated the [2|1] Padé approximant [25]

$$E_{[2|1]} = \mathcal{E}_0 + \mathcal{E}_1 + \frac{\mathcal{E}_2}{1 - \mathcal{E}_3/\mathcal{E}_2} , \qquad (3)$$

$\mathcal{E}_i$ being the ith order energy contribution in the perturbative expansion of the g.s.e., in order to study its convergence properties.

3. Results

As mentioned in the previous section, we calculate the energy per particle of infinite nuclear matter in the framework of many-body perturbation theory, including contributions up to third-order in the interaction. Therefore, it is worth studying the convergence of the g.s.e. perturbative expansion.

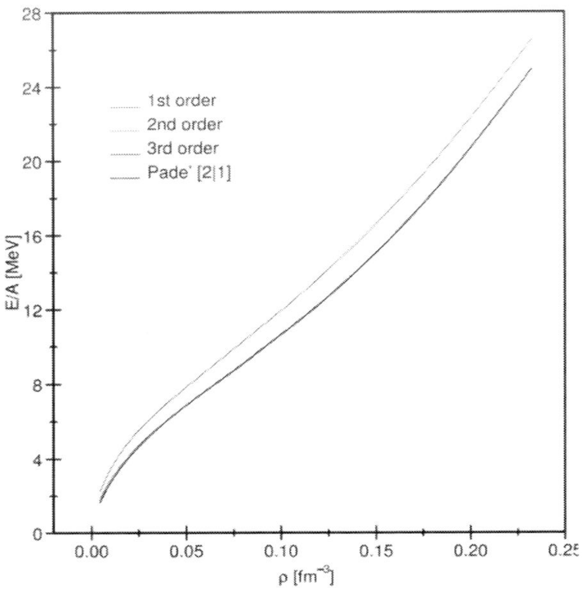

Figure 2. (Color online) PNM energy per particle. The first, second, and third order in the perturbative expansion and the Padé approximant [2|1] are shown as a function of density.

In figure 2 and 3 we show, respectively, the PNM and SNM EOS as a function of density, calculated at various orders in the perturbative expansion. From figure 2 it can be seen that the PNM energy per nucleon calculated at second order is almost indistinguishable from the one computed at third order and consequently from the [2|1] Padé approximant for the whole range of density considered. As regards the SNM, we see from figure 3 that the EOS calculated at second order does not differ much from the one computed at third order, the latter being placed almost on the top of the [2|1] Padé approximant. This is a clear indication that the adopted chiral N^{3}LO NN+N^{2}LO NNN potential has a satisfactory perturbative behavior for both PNM and SNM calculations.

The calculated PNM and SNM EOS and the corresponding symmetry energy are shown in figure 4 and 5, respectively. The inspection of these figures shows how our results reproduce well the empirical SNM saturation point and the value of the symmetry energy at saturation density, the latter quantity being related to the isospin dependence of nuclear forces.

Journal of Physics: Conference Series **527** (2014) 012010 doi:10.1088/1742-6596/527/1/012010

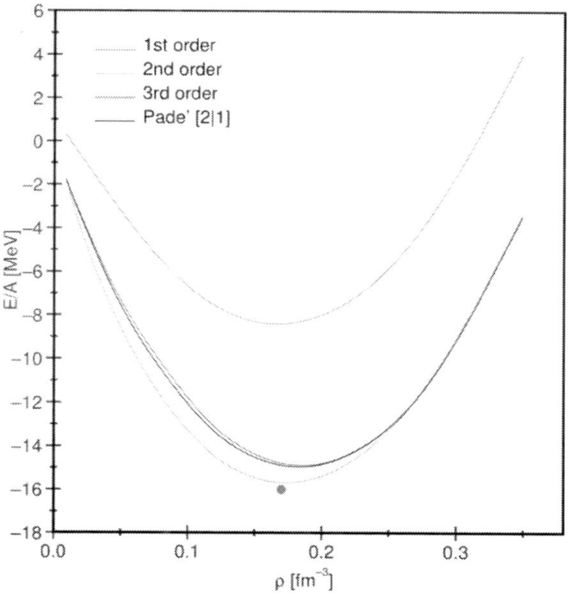

Figure 3. (Color online) Same as in figure 2, but for the SNM energy per particle.

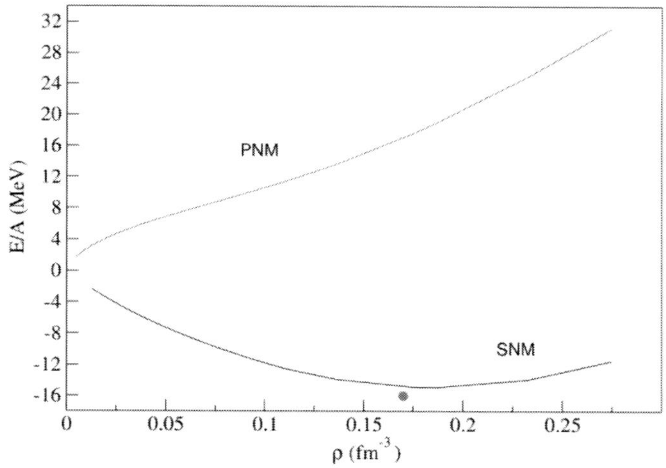

Figure 4. (Color online) Results obtained for the PNM (red line) and the SNM (black line) energy per nucleon at third-order in perturbation theory.

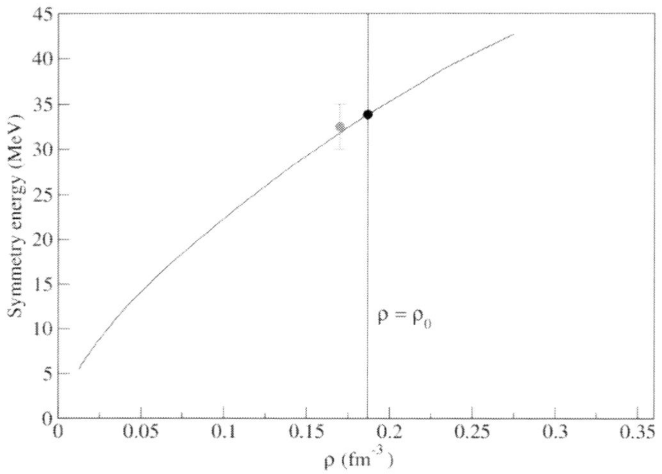

Figure 5. (Color online) Calculated symmetry energy as a function of density. The red point represent the empirical value at saturation density [26, 27].

Two other interesting physical quantities related to PNM and SNM at saturation density are the incompressibility K_0 and the slope of the symmetry energy L.

The empirical value of the SNM incompressibility is determined from experimental data on Giant Monopole Resonance in finite nuclei, at present its accepted value is $K_0 = 230 \pm 30$ MeV [26]. From our calculations we obtain a compressibility $K_0 = 279$ MeV, a value that is very close to the empirical one.

The quantity L determines most of the behavior of the symmetry energy in the proximity of saturation. Its empirical determination - centered at $L = 70$ MeV - is indirect and mainly based on the analysis of heavy-ion collisions at intermediate energies and nuclear structure measurements [27]. Our calculated value for L is 68 MeV, in good agreement with the empirical value.

4. Concluding remarks

In this paper we have studied the properties of infinite nuclear matter employing a chiral potential. This has been done within the framework of the perturbative Goldstone expansion, using a chiral N^{3}LO NN and N^{2}LO NNN potential with a sharp cutoff $\Lambda = 414$ MeV. The LECs involved in the potential have been chosen consistently for the two- and three-body components, and in particular the 3NF LECs c_D and c_E have been fixed as to reproduce the experimental $A = 3$ binding energies and Gamow-Teller matrix element in triton β-decay. Our results for PNM and SNM EOS turn out to be in good agreement with the empirical properties of infinite nuclear matter. The ability to provide realistic nuclear matter predictions employing (consistent) two- and three-body interactions whose LECs are constrained by the properties of the two- and the three-nucleon systems is the main outcome of our study, and this is certainly a very important point that should be further investigated.

Acknowledgements

This work was supported in part by the U.S. Department of Energy under Grant No. DE-FG02-03ER41270 and No. DE-FG02-97ER-41014. We thank Norbert Kaiser for helpful discussions concerning the calculation of the third-order particle-hole diagram.

References

[1] Entem D R and Machleidt R 2003 *Phys. Rev. C* **68** 041001(R)
[2] Epelbaum E, Glöckle W, and Meissner U G 2005 *Nucl. Phys. A* **747** 362
[3] Machleidt R and Entem D R 2011 *Phys. Rep.* **503** 1
[4] Sammarruca F, Chen B, Coraggio L, Itaco N, and Machleidt R 2012 *Phys. Rev. C* **86** 054317
[5] Holt J W, Kaiser N, and Weise W 2013 *Phys. Rev. C* **87** 014338
[6] Tews I, Krüger T, Hebeler K, and Schwenk A 2013 *Phys. Rev. Lett.* **110** 032504
[7] Krüger T, Tews I, Hebeler K, and Schwenk A 2013 *Phys. Rev. C* **88** 025802
[8] Gezerlis A, Tews I, Epelbaum E, Gandolfi S, Hebeler K, Nogga A, and Schwenk A 2013 *Phys. Rev. Lett.* **111** 032501
[9] Carbone A, Polls A, and Rios A 2013 *Phys. Rev. C* **88** 044302
[10] Baardsen G, Ekström A, Hagen G, and Hjorth-Jensen M 2013 *Phys. Rev. C* **88** 054312
[11] Kohno M 2013 *Phys. Rev. C* **88** 064005
[12] Weinberg S 1992 *Phys. Lett. B* **295** 114
[13] van Kolck U 1994 *Phys. Rev. C* **49** 2932
[14] Hebeler K and Schwenk A 2010 *Phys. Rev. C* **82** 014314
[15] Gazit D 2008 *Phys. Lett. B* **666** 472
[16] Marcucci L E, Kievsky A, Rosati S, Schiavilla R, and Viviani M 2012 *Phys. Rev. Lett.* **108** 052502
[17] Gårdestig A and Phillips D R 2006 *Phys. Rev. Lett.* **96** 232301
[18] Coraggio L, Covello A, Gargano A, Itaco N, Kuo T T S, Entem D R, and Machleidt R 2007 *Phys. Rev. C* **75** 024311
[19] Coraggio L, Covello A, Gargano A, and Itaco N 2010 *Phys. Rev. C* **81** 064303
[20] Holt J W, Kaiser N, and Weise W 2009 *Phys. Rev. C* **79** 054331
[21] Holt J W, Kaiser N, and Weise W 2010 *Phys. Rev. C* **81** 024002
[22] Kievsky A, Rosati S, Viviani M, Marcucci L E, and Girlanda L 2008 *Journal of Physics G: Nuclear and Particle Physics* **35** 063101
[23] Coraggio L, Holt J W, Itaco N, Machleidt R, and Sammarruca F 2013 *Phys. Rev. C* **87** 014322
[24] MacKenzie J J 1969 *Phys. Rev.* **179** 1002–1010
[25] Baker G A and Gammel J L 1970 *The Padé Approximant in Theoretical Physics* (*Mathematics in Science and Engineering* vol 71) (Academic Press, New York)
[26] Dutra M, Lourenço O, Sá Martins J S, Delfino A, Stone J R, and Stevenson P D 2012 *Phys. Rev. C* **85** 035201
[27] Tsang M B, Stone J R, Camera F, Danielewicz P, Gandolfi S, Hebeler K, Horowitz C J, Lee J, Lynch W G, Kohley Z, Lemmon R, Möller P, Murakami T, Riordan S, Roca-Maza X, Sammarruca F, Steiner A W, Vidaña I, and Yennello S J 2012 *Phys. Rev. C* **86** 015803

Journal of Physics: Conference Series **527** (2014) 012011 doi:10.1088/1742-6596/527/1/012011

Use of the Sumudu transform to extract response functions from Quantum Monte Carlo calculations

Francesco Pederiva, Alessandro Roggero, Giuseppina Orlandini
Dipartimento di Fisica, Università di Trento, and INFN-TIFPA, via Sommarive, 14, 38123 Trento, Italy

E-mail: pederiva@science.unitn.it, roggero@science.unitn.it, orlandin@science.unitn.it

Abstract. We review an *ab-initio* method for calculating the dynamical structure function of an interacting many–body quantum system. The method consists in coupling a generalized integral transform approach with imaginary time Quantum Monte Carlo calculations. The strength of the method has been tested on the excitation spectrum of bulk atomic ^{4}He. The peculiar form of the kernel as a representation of the delta-function has allowed to minimize the ill-posedness of the integral transform inversion. In fact it has been possible to obtain, at a considerable degree of reliability, both position and width of the collective excitations in the maxon–roton region, as well as the second collective peak. What we stress here is the ability of such a δ-function-like kernel, for which one can control position and width, to maintain in the transformed space the characteristics of the collective structures. The application to the coherent and incoherent density excitation spectrum of liquid ^{4}He is discussed.

1. Introduction

The computation of microscopic dynamic quantities is routinely performed for classical systems by numerically solving Newton equations and collecting the information of interest from the generated trajectories. In quantum mechanics this process is not possible, since there is no reliable method for solving the time-dependent Schroedinger equation for a general many-body system. Information about the dynamical evolution must be obtained by the computation of spectral function relative to some excitation operator. It is possible, by a simple mathematical manipulation, to show that some *integral transform* of the spectral function can be easily obtained as a ground-state expectation value. While this fact is extremely valuable, since there are many reliable *ab-initio* methods available for ground state calculations, the price to pay is the ill-posedness of the inversion problem that needs to be faced to extract the excitation spectrum.

In this paper we will discuss the problem of extending in a reliable way a class of Quantum Monte Carlo (QMC) algorithms to make it possible to determine the Dynamical Structure Function (DSF) for a generic quantum many-body system and a generic excitation operator. As already mentioned, we will follow the general approach of computing an Integral Transform (IT) of the DSF with generalized kernels. As an example, we will present an application to the study of the coherent and incoherent density excitation spectrum in ^{4}He.

At present, QMC calculations provide benchmark-quality results for the study of a variety of many-body systems as, for example, in quantum chemistry, physics of ultra-cold gases, and nuclear physics. One of the most severe limitation of QMC methods is their inability to treat dynamical properties in a similarly reliable way. This failure is essentially due to the fact that the propagators required to implement the time evolution are usually not normalizable, and cannot be sampled. However, by Monte Carlo methods it is possible to implement an imaginary-time evolution, which yields the projection of some generic state belonging to an arbitrary Hilbert space on the component along the eigenstate of the Hamiltonian with lowest eigenvalue. Quantities that do not directly translate into imaginary-time language, are affected by statistical noise when analytically continued to real time. This noise can be hardly reduced and make calculations unfeasible.

The standard approach to the problem is to compute the Laplace transform of the DSF as an imaginary-time autocorrelation function and then to attempt a numerical inversion. However, this inversion is an exponentially ill-posed problem [1]. Sophisticated regularization techniques [2, 3] are needed to correctly extract the physical information. In this context, the reliability of the calculation becomes strongly dependent on the employed inversion scheme. As an example, one of the most powerful inversion schemes, the Maximum Entropy Method [4, 5], cannot resolve the (measured) double peaked structure of $S(\mathbf{q},\omega)$ in superfluid ^{4}He [6], corresponding to a higher energy collective roton mode. In a recent paper this structure was eventually resolved inverting the imaginary time correlation function in a Path Integral Ground State calculation by using a falsification method based on a genetic algorithm [7].

For strongly interacting few-body systems the problem of computing various DSF of density and current excitations is solved by using a generalized integral transform approach, i.e. the Lorentz Integral Transform (LIT) method [8, 9]. The success of this approach, so far applied in nuclear physics, is based on the specific choice of a Lorentzian function as kernel of the IT. On the one hand this choice allows to calculate the transform with bound state techniques, such as expansions on localized many-body basis functions, even when the response is defined in the continuum. On the other hand, and most important, the fact that the kernel is a representation of the δ-function allows for a reliable and stable inversion. So far, the application of this technique has been limited to a small number of particles (up to N=6 [10] and 7 [11]), due to the huge computational costs of the diagonalizations needed to calculate the LIT. In the following we will discuss how it is possible to extend the same ideas to many-body system. In general, QMC methods do not allow for computing generic integral transforms, due to the fact that not all kernels easily translate into functions that can be sampled in coordinates space. However, we will see that it is possible to write a kernel that is suitable for QMC evaluation, and at the same time retains the main features of the Lorentz kernel, i.e. being a peaked function (and a δ-function in some limit), and having a controllable width.

2. Integral transforms

At zero temperature the contribution to the response of a system of interacting particles due to a perturbative probe transferring momentum $\mathbf{q}$ and energy ω to it, can be expressed using a spectral representation

$$\begin{aligned} S_{\hat{O}}(\mathbf{q},\omega) &= \textstyle\sum_\nu |\langle\Psi_\nu|\hat{O}(\mathbf{q})|\Psi_0\rangle|^2\delta(E_\nu-\omega) \\ &= \langle 0|\hat{O}^\dagger(\mathbf{q})\delta(\hat{H}-\omega)\hat{O}(\mathbf{q})|0\rangle, \end{aligned} \tag{1}$$

where $|\Psi_0\rangle$ is the ground state of the system, $|\Psi_\nu\rangle$ are the final states of the reaction, $\hat{O}$ is an excitation operator, $\delta(\hat{H}-\omega)$ is the spectral-density of the hamiltonian and the summation is extended to all discrete and continuum spectrum states in the set.

Journal of Physics: Conference Series **527** (2014) 012011 doi:10.1088/1742-6596/527/1/012011

The cost of a direct calculation of $S_{\hat{O}}(\mathbf{q},\omega)$ becomes rapidly not affordable when the number of particles or the energy transfer ω increases. The latter problem is due to the fact that to account for continuum states one would need to solve the many-body scattering problem. It is instead possible to consider an integral transform of $S_{\hat{O}}(\mathbf{q},\omega)$ with a generic kernel $K(\sigma,\omega)$

$$\Phi(\mathbf{q},\sigma) = \int K(\sigma,\omega)\, S_{\hat{O}}(\mathbf{q},\omega)\, d\omega. \tag{2}$$

Substituting the expression (1) for $S_{\hat{O}}(\mathbf{q},\omega)$ one obtains:

$$\begin{aligned}\Phi(\mathbf{q},\sigma) &= \sum_{\nu} \langle \Psi_0 | \hat{O}^{\dagger}(\mathbf{q}) | \Psi_\nu \rangle K(\sigma,\omega) \langle \Psi_\nu | \hat{O}(\mathbf{q}) | \Psi_0 \rangle \\ &= \langle \Psi_0 | \hat{O}^{\dagger}(\mathbf{q})\, K(\sigma,\hat{H})\, \hat{O}(\mathbf{q}) | \Psi_0 \rangle \end{aligned} \tag{3}$$

Equation (3) can be viewed as a generalization of the energy-weighted sum rules, which now depends on a continuous parameter σ. Provided that the kernel and the excitation operator have suitable analytic properties, the right hand side of equation (3) can be calculated using bound-state type methods. This is the case both for the Stieltjes kernel [12, 13], and for the Lorentz kernel. However, while in the former case the inversion of the transform is as problematic as in the case of the Laplace kernel, in the latter case even a rather simple regularization procedure allows to obtain accurate and stable results. The reason can be easily understood. In the case of the Laplace or the Stieltjes kernels, or, in general, kernels that are significantly non-zero over a wide σ range, the information about $S_{\hat{O}}(\mathbf{q},\omega)$ in the ω domain is spread in a large σ domain. On the contrary the Lorentz kernel, as well as any function that is a δ-function representation, keeps that information in an arbitrarily narrow σ domain, governed by the width of the kernel. In the δ function limit of the kernel no inversion is needed. Once the transform is available we can obtain $\mathcal{R}(\omega)$ via a proper inversion of the transform, using a suitable regularization procedure.

3. Sumudu transform

As we already mentioned, the way to the success for a method based on IT is using kernels that are δ-function representations. While there is a large number of them, very few have a practical implementation. In the past the use of Gaussian kernels has been investigated in different fields from condensed matter [14, 15], to non perturbative QCD [16, 17], but with limited results.

Quite fortunately, it is possible to recast one specific δ-function representation in the imaginary-time propagation language, typical of QMC methods.

Consider the following family of integral kernels built out of the so-called Sumudu transform:

$$K_P(\sigma,\omega;a,b) = \frac{1}{\sigma}\left[\left(\frac{b}{a}\right)^{-a\frac{\omega}{\sigma}} - \left(\frac{b}{a}\right)^{-b\frac{\omega}{\sigma}}\right]^P, \tag{4}$$

where the parameters $b > a$ are integer numbers. This kernel function converges to a scaled delta function $\delta(\omega-\sigma)$ in the $P \to \infty$ limit, independent of the choice of a and b. For a finite P the kernel is still centered around $\omega = \sigma$ but has a finite width that depends on σ and the integers P,a and b. This property makes the choice of the resolution in a given energy range extremely flexible.

The simplest possible choice of the parameters is $a = 1$ and $b = 2$. In this case, using a binomial expansion, and rewriting powers as exponential functions, leads to a more transparent form of the kernel:

$$K_P(\sigma,\omega) = \sum_{k=0}^{P} \binom{P}{k} (-1)^k e^{-\ln(2)(P+k)\frac{\omega}{\sigma}}, \tag{5}$$

Journal of Physics: Conference Series **527** (2014) 012011 doi:10.1088/1742-6596/527/1/012011

By operating the substitution $\omega \to \hat{H}$ according to equation (3), we are lead to a simple linear combination of imaginary-time propagators ($\hbar$=1), taken at imaginary-time points $\tau_{Pk} = \ln(2)(P+k)/\sigma$.

4. Integral transforms in projection Monte Carlo methods

Projection QMC methods are all based on the implementation of such an imaginary-time propagation. The underlying idea is to solve the corresponding integral equation:

$$\Psi(R,\tau) = \int dR' G(R,R',\tau)\Psi(R',0). \tag{6}$$

This can be achieved, for instance, sampling a representation in coordinate space of the Green's function $G(R,R',\tau)$ to propagate a set of configurations representing in turn an expansion of the function $\Psi(R,\tau)$ in eigenstates of the position operator (Diffusion Monte Carlo methods). Alternatively, it is possible to break up the Green's Function in a product of short time propagators in coordinate space:

$$\begin{aligned}\Psi(R,\tau) &= \int dR' \dots dR^n G(R,R^n,\Delta\tau)\times \\ &\quad \times G(R^n,R^{n-1},\Delta\tau)\cdots G(R'',R',\Delta\tau)\psi(R',0),\end{aligned} \tag{7}$$

with $\tau = n\Delta\tau$. This formulation is implemented in the so-called Path-Integral Ground State methods [18], and in the Reptation Monte Carlo (RMC) algorithm [19], where the whole path $\{R,R',R'',\dots,R^n\}$ is sampled from the product of the short-time propagators G, possibly modified with the use of a suitable importance function Ψ_T to be determined in a variational calculation. The fact that estimating $\Phi(\sigma)$ reduces to the computation of an imaginary time correlation function makes this second formulation more convenient and straightforward. In particular, path-based methods yield estimates that never depend on the (necessary) importance function used to improve the convergence of the calculation.

In order to evaluate the transform (2) within a QMC approach, we need to compute the imaginary-time correlation functions, and then construct the corresponding linear combinations. When needed, a smaller width of the kernel could be achieved using a large value of P or reducing the value of the ratio b/a. In both cases the imaginary time correlation function needs to be evaluated for long imaginary time, severely increasing the computational time. In any case, the guiding criterion is that the kernel should have a width at least comparable to the distance between the structures of the DFT.

5. Application to liquid helium

As a first application to a realistic physical problem, we have computed the density excitation response in bulk atomic ^{4}He at $T = 0$. The system is modeled as a periodic box containing $N = 64$ or $N = 125$ ^{4}He atoms. The chosen interaction is the HFDHE2 potential [20, 21], which quantitatively reproduces the binding energy of bulk ^{4}He up to the freezing point. Calculations were performed at the experimental saturation density ($n_0 = 0.02186$ Å^{-3}). The density excitation operator is defined as:

$$\hat{O}(\mathbf{q}) \equiv \hat{\rho} = \sum_{i=1}^{N} e^{i\mathbf{q}\cdot\mathbf{r}_i}, \tag{8}$$

and the transformed DSF in equation (2) becomes

$$\Phi(\sigma) = \langle\Psi_0| \sum_{i,j=1}^{N} e^{i\mathbf{q}\cdot\mathbf{r}_i(0)} K_P(\sigma,\hat{H}) e^{-i\mathbf{q}\cdot\mathbf{r}_j(\sigma)}|\Psi_0\rangle.$$

Computations have been performed by means of a RMC algorithm, as described in Ref. [19]. The variational importance function includes two- and three-body correlations expanded in a basis set [22] and optimized using a variational Monte Carlo procedure. Ground state properties are well reproduced: the ground-state energy per particle is $\epsilon_0^{RMC} = -7.23 \pm 0.01$ K, in good agreement with previous calculations using the same potential [19], and with the experimental value $\epsilon_0^{exp} = -7.17$ K. In the computation of the Sumudu transform, we have found by numerical testing that the values of $P = 2$, $a = 1$, and $b = 2$ satisfy this criterion in the range of energy and momenta explored.

As it is customary in neutron spectroscopy one can distinguish the contribution coming from the so-called *coherent* part, given by the terms with $i \neq j$, related to collective excitations, and an *incoherent* part with $i = j$ that essentially picks up contributions from single particle excitations. We have obtained results for both the full and for its incoherent part, in the most studied region of the spectrum: the low momentum phonon-maxon-roton part $q \approx 0.3 \div 3.5$ Å^{-1}. The computed static structure factor $S(\mathbf{q})$ is consistent with experimental data and previous calculations [19, 22].

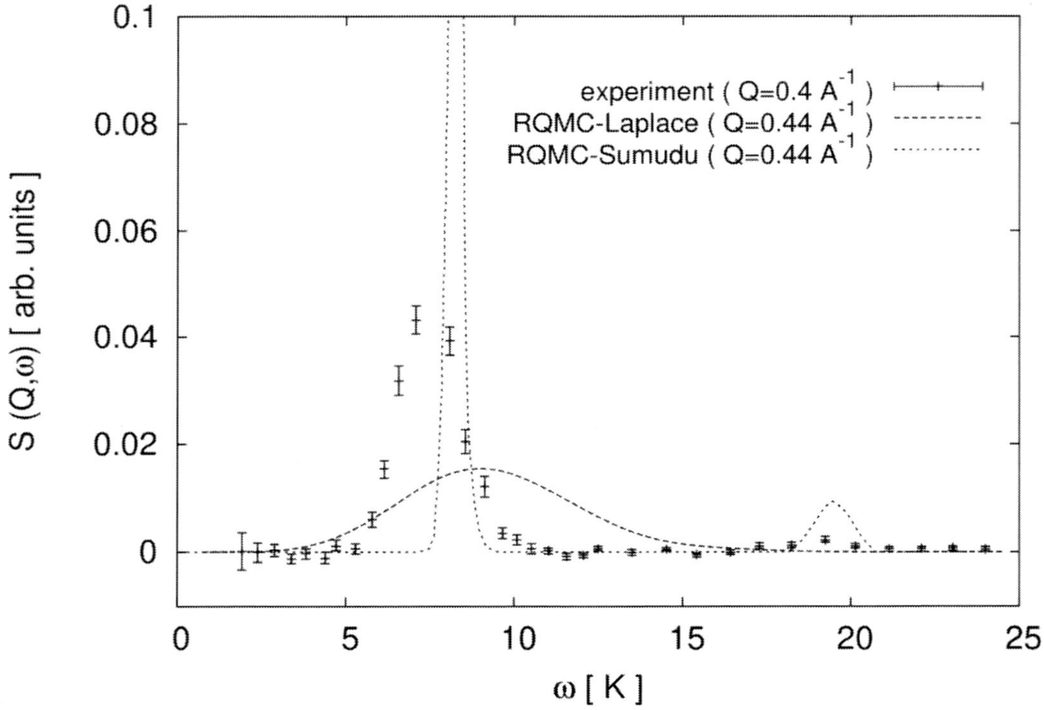

Figure 1. A typical result for the response function (Q, ω) in liquid ^{4}He. Points with errorbars are experimental results at $Q = 0.4$Å^{-1}. The thick dotted line is the result obtained using a Laplace kernel. The dotted line is the result computed by using the Sumudu kernel. Theoretical calculations refer to $Q = 0.44$Å^{-1}, value determined by the size of the simulation box.

When looking at the results on $S(\mathbf{q}, \omega)$, the difference between the inversion of the Laplace transform and the transform defined in equation (4) is clearly seen in figure 1, where we compare the results of the inversion obtained from RMC data with both kernels. Momentum discretization due to the finite simulation cell does not allow for a strict comparison to the experimental results. The small shift of the peak might indeed be due to the 0.04 Å^{-1} difference

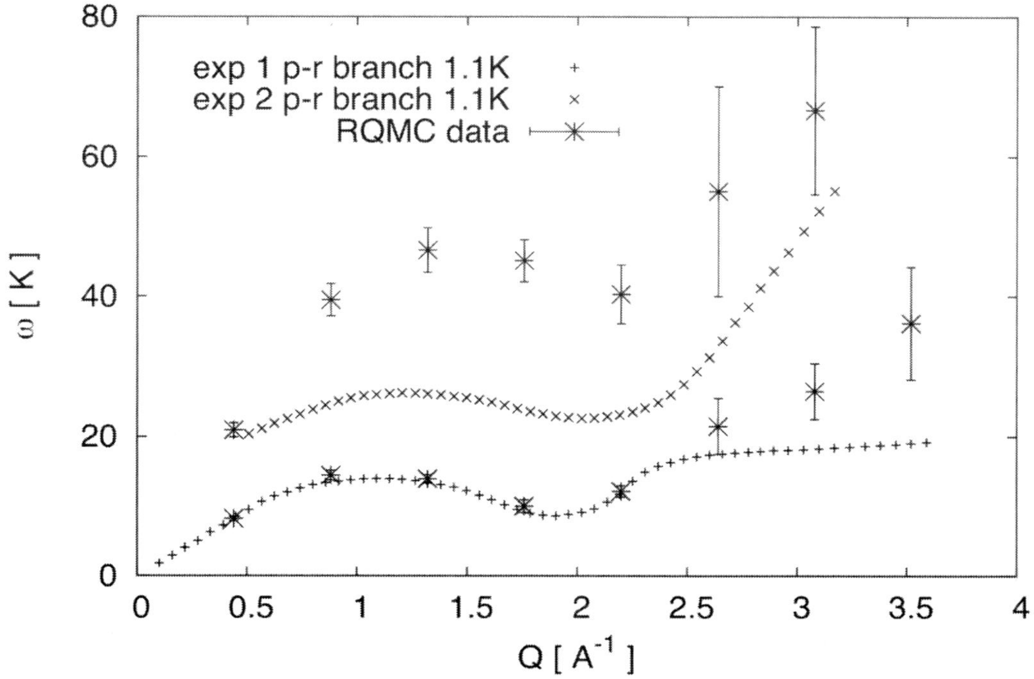

Figure 2. Dispersion of the collective modes in liquid ^{4}He at equilibrium density and $T = 0$. Points with errorbars are the computed values. Errorbars are estimates of the width of the peaks. $+$ and $\times$ are the corresponding experimental data from Ref. [24] at $T = 1.1$K.

in the momentum transfer. However, the new kernel permits to retrieve the information on the second peak and gives a much more realistic height and width of the one-phonon peak. It should be pointed out that the width of the peak in the experiments is essentially due to instrumental resolution.

As regards the inversion procedure, we have used both the Entropy Maximization Maximum Likelihood (EMML) [23] and the Simultaneous Algebraic Reconstruction Technique (SMART) [23]. Error bars in the figures are obtained taking the maximum spread between the half-width position obtained from the two methods. We found that for $q \leq 2.4\text{Å}^{-1}$ both methods converge to the same solution, confirming the robustness of the result.

In figure 2 we plot the excitation spectrum obtained using the new transform. The experimental low-lying part [24] is extremely well reproduced up to $q \approx 2.6\text{Å}^{-1}$, where the dispersion does not bend over around 2Δ but continues to raise. In this region (the so-called Pitaevskii plateau), the peak corresponding to the collective excitation is mixing in the single particle excitations spectrum. The resolution that can be reached by the kernel with $P = 2$ is no longer sufficient, and EMML and SMART give different results indicating that the statistical uncertainty in the QMC data is too large to allow a consistent reconstruction. The dispersion corresponds to the centroid of the overall spectrum.

The two-phonon branch is clearly visible and well resolved. As it happens in Ref. [7], it only qualitatively compares to the experiment. The difference might be due either to the fact that measurements are taken at finite temperature, or to a larger sensitivity of the second peak to the details, for instance, of the interatomic interaction (e.g. the presence of effective three-body

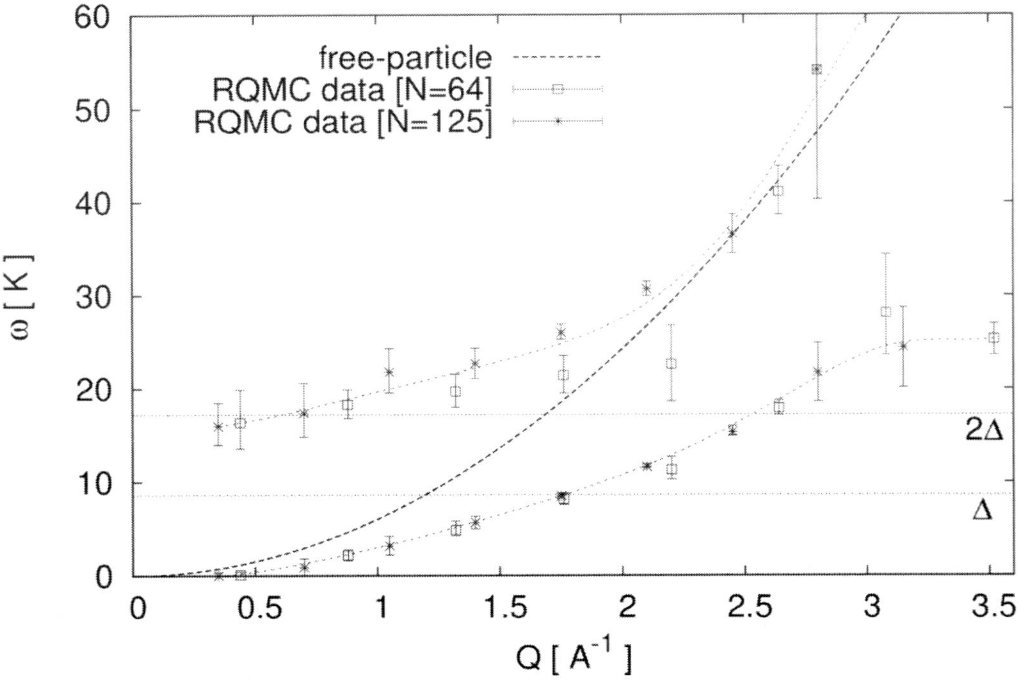

Figure 3. Incoherent Dynamic Structure factor computed by means of the Sumudu integral transform. Empty squares are simulations performed in a simulation box with N=64 atoms, stars refer to a box with N=125 atoms. Dotted lines are plotted as a guid to the eye. The dashed line is the free-particle excitation spectrum. Δ is the roton gap, and the lines at energy Δ and 2Δ are drawn as reference.

forces).

The most useful feature of these calculations is that the resolution is enough to allow for separately compute the incoherent part of the full response function, in order to study single-particle excitations.

In figure 3 we have plotted the calculated excitation spectrum of single-particle excited states. The spectrum shows at least two distinct branches. A lower energy excitation starts from $Q \approx 0.5A^{-1}$ and propagates with a velocity resulting slightly higher than the superfluid critical velocity $Ve/Vc \approx 1.57$. A second branch can be observed starting at an energy slightly below two times the roton energy, tending asymptotically to the free particle spectrum. Interestingly enough, the lower energy branch crosses the collective excitation spectrum exactly at the roton minimum, thereby reinforcing the picture of the roton as a single particle excitation of an atom exiting the superfluid. All these single-particle excitations might be related to the quantum correlations induced by the particle-particle interaction. An extensive experimental analysis of the single particle excitations properties in superfluid ^{4}He is unfortunately not available.

6. Conclusions

We have proposed a method to reliably extract well resolved spectra from numerical calculations implementing imaginary time propagation of an initial state, such as DMC or RMC. The computations might be easily extended to the $T \neq 0$ case by using standard PIMC methods.

The application to the study of the collective and single-particle excitations in ^{4}He shows the robustness and the higher resolution power of this technique. The limit to the accuracy of the spectra is in principle limited only by the available computer power available.

Acknowledgements

We thank W. Leidemann, V. Efros, G.V. Chester, M.H. Kalos, J.Carlson and S. Gandolfi for useful and stimulating discussions about this subject. Authors are members of LISC, Interdisciplinary Laboratory for Computational Science, a joint venture of FBK and the University of Trento.

References

[1] Hadamard J 1952 *Lectures on Cauchy's problem in linear partial differential equations* (Dover, New York)
[2] Tikhonov A and Arsenin V 1977 *Solutions of Ill-posed Problems* (J Wiley, New York)
[3] Talbot J 1979 *J. Inst. Maths. Applics.* **23** 97
[4] Jaynes E T 1963 *Information Theory and Statistical Mechanics* in *Statistical Physics* ed Ford K (Benjamin, New York) p 181
[5] Silver R N, Sivia D S, and Gubernatis J E 1990 *Phys. Rev. B* **41** 2380
[6] Boninsegni M and Ceperley D M 1996 *J. Low Temp. Phys.* **104** 339
[7] Vitali E, Rossi M, Reatto L, and Galli D E 2010 *Phys. Rev. B* **82** 174510
[8] Efros V D, Leidemann W, and Orlandini G 1994 *Phys. Lett. B* **338** 130
[9] Efros V D, Leidemann W, Orlandini G, and Barnea N 2007 *J. Phys. G* **34** R459
[10] Bacca S, Marchisio M A, Barnea N, Leidemann W, and Orlandini G 2002 *Phys. Rev. Lett.* **89** 052502
[11] Bacca S, Arenhoevel H, Barnea N, Leidemann W, and Orlandini G 2004 *Phys. Lett. B* **603** 159
[12] Efros V D 1985 *Sov. J. Nucl. Phys.* **41** 949
[13] Efros V D, Leidemann W, and Orlandini G 1993 *Few-Body Syst.* **14** 151
[14] Munehisa T and Munehisa Y 2000 *Calculations for a delta function of Hamiltonian by the Suzuki-Trotter decomposition* RIKEN Review N 29
[15] Kornilovitch P E 2001 `arXiv:cond-mat/0010100v2`
[16] Bertlmann R A, Launer G, and de Rafael E 1985 *Nucl. Phys. B* **250** 61
[17] Orlandini G, Steele T, and Harnett D 2001 *Nucl. Phys. A* **686** 261
[18] Sarsa A, Schmidt K E, and Magro W R 2000 *J. Chem. Phys.* **113** 1366
[19] Baroni S and Moroni S 1998 *Proceedings of the NATO ASI Quantum Monte Carlo Methods in Physics and Chemistry* (Cornell University) p 12
[20] Aziz R A, Carley J S, Nain V P S, Taylor W L, and McConville G T 1979 *J. Chem. Phys.* **70** 4330
[21] Withlock P A, Kalos M H, Lee M A, and Chester G V 1981 *Phys. Rev. B* **2** 115
[22] Fantoni S, Moroni S, and Senatore G 1995 *Phys. Rev. B* **52** 547
[23] Byrne C L 1998 *IEEE Transactions on Image Processing* Vol 7, N 1
[24] Woods A and Cowley R 1971 *Can. J. Phys.* **49** 177

Journal of Physics: Conference Series **527** (2014) 012012 doi:10.1088/1742-6596/527/1/012012

The Quark Gluon Plasma and relativistic heavy ion collisions in the LHC era

Francesco Becattini

Dipartimento di Fisica e Astronomia, Università di Firenze, Via G. Sansone 1, I-50019, Sesto Fiorentino (Firenze)

E-mail: becattini@fi.infn.it

Abstract. In this talk we briefly review some of the achievements in the study of the QCD matter at high temperature produced in the nuclear collisions at very high energies. We focus on its bulk properties: hydrodynamical and thermodynamical.

1. Introduction

Quark Gluon Plasma (QGP) is a phase of strongly interacting matter (see figure 1) whose existence is predicted by Quantum-Chromo-Dynamics (QCD) to occur at sufficiently high temperatures and/or baryon densities. In the QGP phase, quarks and gluons are deconfined, i.e. they can move over distances much larger than a typical hadron size. This peculiar QCD prediction follows from a property of the theory known as asymptotic freedom: the interaction of the fundamental fields become weaker as the energy density increases. Therefore, for a strongly interacting system, there should be a critical temperature (at vanishing chemical potentials) separating the phase where quarks and gluons are deconfined from the hadron gas. It is by now established that, at least for $\mu_B = 0$, the transition from hadron gas to QGP is a continuous one, that is a crossover [1].

The experimental quest for Quark Gluon Plasma has been going on for more than 30 years by now. It was soon proposed that QGP could have been produced in high energy collisions of heavy nuclei. The main reason of colliding large nuclei is that, in order to obtain a proper deconfinement, the region of space where high energy density is achieved should be much larger than a typical hadron size. Therefore, for a fixed nucleon beam energy, it seems more likely to produce a deconfined QGP in collisions of heavy nuclei (several fm's of diameter) rather than, e.g., in proton-proton collisions, although, as we will see, there are striking similarities between the two systems.

A vaste experimental programme of heavy ion collisions at several centre-of-mass energies was then initiated; from few GeV (Au-Au collisions at BNL-AGS) to an intermediate range of 6-17 GeV (Pb-Pb collisions at CERN-SPS) up $O(100)$ GeV at BNL-RHIC, started in 2000 and finally, from 2009, collisions of Pb nuclei at the LHC at $\sqrt{s_{NN}}$ of few TeV. It is now widely believed that the threshold for QGP production has been overcome at RHIC and probably at SPS.

The main issue of the whole experimental programme is how to prove QGP formation, i.e. to find a signature thereof in the final observable particles produced in the collision events. Indeed, most of the effort of theorists in this field has been spent into the search of the most powerful signatures or probes, besides the description of the collision itself. In this talk, I will just focus on soft hadronic probes and will not dwell on the so-called hard and electromagnetic probes, which would deserve a separate chapter.

Because of the complexity of the process of collision of two nuclei at high energy and the lack (or unfeasibility) of QCD first-principle calculations at the energy scale of interest, most phenomenology in this field has to resort to different approaches and models. One of the most successful models in relativistic heavy ion physics is the hydrodynamical model, which should be rather called "hydrodynamical description" because it is based on a general physical scheme rather than on peculiar assumptions for the specific process (like for instance the string model in pp collisions). The hydrodynamical description has indeed provided much insight and understanding of the collision process and bears out the QGP formation.

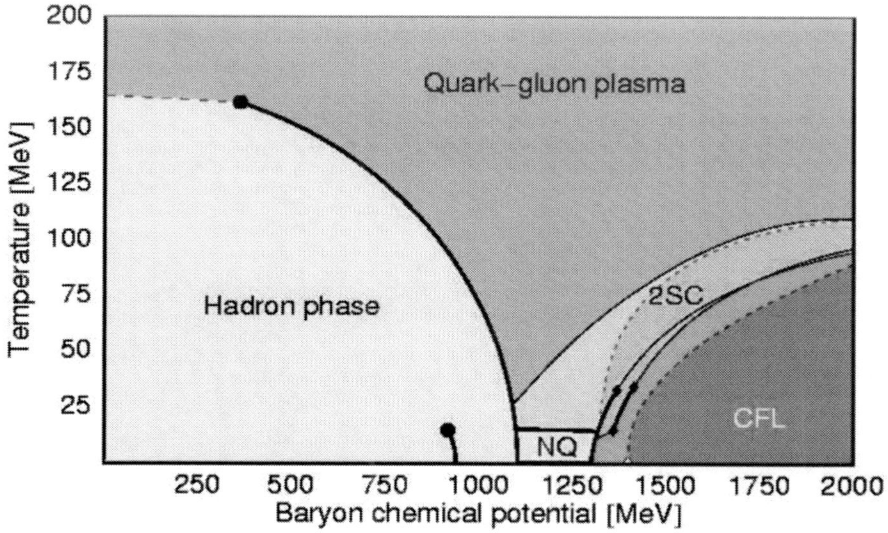

Figure 1. (Color online) QCD phase diagram in the $T - \mu_B$ plane.

2. The QGP as a nearly ideal fluid

For hydrodynamics to be a sensible description of the QGP evolution, two requirements must be met:

(i) the system should be close to local thermodynamical equilibrium for a sufficiently long time, that is local equilibrium must be achieved at some early stage, in the QGP phase;

(ii) the typical microscopic interaction scale (e.g. the mean free path in a kinetic framework) should be much smaller than the length over which temperature and energy density vary, hence much smaller of the size of the system.

The processes leading to the local thermalization in the plasma phase are non-perturbative and still subject of investigation. But if this assumption holds, much of the life of the plasma can be reproduced through the hydrodynamical continuity equations, until the phase transition into hadron gas takes place. The second requirement is generally fulfilled in relativistic heavy ion collisions as the fireball is estimated to have a radius of several fm's while the microscopic scale should be smaller than 1 fm (see later).

The hydrodynamical-based calculations try to reproduce measured particle spectra at low momenta and to determine the initial conditions (e.g. the initial energy density profile) adjusting the pertaining parameters, as well as the parameters (temperature and baryon chemical potential) at which the hydrodynamical regime ceases. The equation of state is an interpolation of lattice QCD results at high temperature and hadron-resonance gas at temperatures lower than critical one.

Mostly adopted initial conditions include a density profile calculated from the geometrical overlap of the two incoming nuclei and longitudinal flow profile according to Bjorken's scaling hypothesis with vanishing transverse flow. Most calculations indicate an early equilibration time of the order of 1 fm/c both at SPS and RHIC energies, with an initial temperature well above the critical value of ≈ 160 MeV.

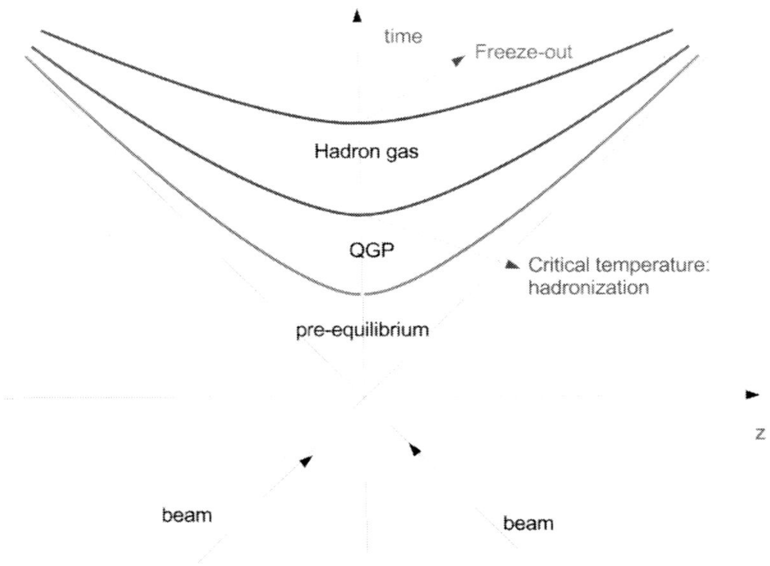

Figure 2. (Color online) Sketch of a relativistic nuclear collision process at high energy. Thermalization is achieved after some time, then plasma expands and hadronizes at the critical QCD temperature. Later, the hadron gas expands and freezes out; for the details of this process, see text.

Of special interest is the so-called *elliptic flow*, which shows as an anisotropy of differential azimuthal particle momentum spectrum in the reaction plane in peripheral collisions. Because of the non-zero impact parameter in peripheral collisions, the overlapping region of the two incoming nuclei has an initial almond shape, unlike in central collisions where it is spherical. If local thermal equilibrium sets in early, pressure gradient is significantly higher in the reaction plane than in the orthogonal direction and this drives an enhancement of the collective flow along the reaction plane, resulting in an enhancement of particle momentum. The magnitude of the elliptic flow is usually gauged with the second coefficient of the Fourier expansion of the $dN/dp_T d\varphi$ spectrum as a function of φ for fixed p_T, the so-called $v_2(p_T)$.

Necessary conditions for reaching high values of v_2 are an early equilibration time of and low values of viscosity, that is a nearly ideal fluid. The fact that QCD matter apparently behaves like an ideal fluid gave rise to the folklore that QGP is a strongly interacting fluid, in stark contrast to the naive expectation of a gas of weakly interacting quarks and gluons. Indeed, lattice QCD calculations (for instance the stress-energy tensor trace anomaly) already indicate that the interaction energy is very significant around T_c, about 1/3 of the total energy density. The estimated low ratio viscosity/entropy η/s from hydrodynamical fits reinforces this conclusion. This can be understood by simply reminding the familiar textbook formula of viscosity from kinetic theory:

$$\eta = \frac{1}{3}\rho v_T \lambda$$

where ρ is the mass density, v_T the particle mean thermal speed and λ its mean free path. We can rewrite the above formula with the mean thermal momentum p_T and particle density n:

$$\eta = \frac{1}{3} n m v_T \lambda = \frac{1}{3} p_T n \lambda$$

and make an educated guess of its extension to the relativistic domain by replacing the particle density n with the entropy density s (which measures the density of microscopic degrees of freedom). Therefore, one has:

$$\frac{\eta}{s} = \mathcal{O}(p_T \lambda)$$

that is the viscosity over entropy ratio is of the order of the ratio between the mean free path and the mean particle wavelength. For usual fluids, this is much larger than 1 [1]. If it is 1 or smaller than 1, as it seems to be the case for the QGP [2] where $\eta/s \approx 1/2\pi$ the mean free path is less than the thermal particle wavelenght, implying a breakdown of the kinetic and particulate description. Particularly, one has, at $T = 160$ MeV:

$$\lambda = \mathcal{O}\left(\frac{\eta}{s}\frac{1}{p_T}\right) \approx \frac{1}{2\pi T} \simeq 0.2 \text{ fm}$$

Indeed, the QGP near the critical temperature is a remarkable instance of a fluid which has no underlying kinetic substratum (in other words, no colliding particle or quasi-particle excitations), a features shared with the so-called unitary Fermi gas [3].

For this system, a kinetic description in terms of independent colliding particles is therefore questionable. Still, hydrodynamics and hydrodynamical description hold, including transport coefficients. Indeed, it is possible to formulate hydrodynamics using the mean value of the fundamental quantum stress-energy tensor operator:

$$T^{\mu\nu} = \mathrm{tr}(\widehat{\rho}\widehat{T}^{\mu\nu})$$

and to show that all transport coefficients can be defined in terms of the two-point functions of this tensor at equilibrium through the Kubo formulae [4].

3. Freeze-out in an expanding hadron gas

The QGP expands and cools until it reaches the critical crossover temperature, where it hadronizes. The hadron gas keeps expanding for a while until particles do not interact anymore: the system is *frozen out*. In an expanding sytem of interacting particles freeze-out occurs when the mean scattering time $\tau_{\rm scatt}$ exceeds the mean collision time $\tau_{\rm exp}$:

$$\tau_{\rm scatt} = \frac{1}{n\sigma\langle v\rangle} > \tau_{\rm exp} = \frac{1}{\partial\cdot u} \tag{1}$$

[1] This situation should be contrasted with a familiar fluid, liquid water at $T = 0^o$C. The mean free path of water molecules can be estimated to be 0.3 nm, which is approximately equal to the square root of the cross section $\sqrt{\sigma}$, yet the mean thermal wavelength is of the order of 5 pm, mostly due to the large molecular mass.

Journal of Physics: Conference Series **527** (2014) 012012 doi:10.1088/1742-6596/527/1/012012

u being the hydrodynamical velocity field and $\langle v\rangle$ is the mean velocity of particles. If the cross-section σ is the inelastic one, the freeze-out is called *chemical*, whereas if it includes elastic processes, the freeze-out is called *kinetic*. Chemical freeze-out of course precedes the kinetic as the inelastic cross section is smaller than the total.

We can obtain a gross approximation of the expansion time with the ratio $V/\dot{V}$ where $V(t)$ is the volume of the fireball at the time t. For a fireball which is spherical in shape with a radius R, this is $R/3\dot{R}$ and if the radius increases at approximately the mean particle velocity $\langle v\rangle$, we have the condition:

$$\frac{1}{n\sigma\langle v\rangle}=\frac{R}{3\langle v\rangle}\implies\frac{1}{n\sigma}=\frac{R}{3} \tag{2}$$

For a given number of particles N within the volume, this inequality yields the radius at which freeze-out occurs as a function of N and of the average cross-section:

$$R_{\rm fo}=\sqrt{\frac{N\sigma}{4\pi}} \tag{3}$$

and the density at which freeze-out occurs, which decreases with N according to:

$$n_{\rm fo}=\frac{N}{\frac{4\pi}{3}R_{\rm fo}^3}=3\sqrt{\frac{4\pi}{N}}\frac{1}{\sigma^{3/2}} \tag{4}$$

For instance, for $N=1000$, typical of heavy ion collisions, and $\sigma=30\,\rm mb=3\,fm^2$ one has $R_{\rm fo}\simeq 15$ fm and $n_{\rm fo}\simeq 0.06\ \rm fm^{-3}$, which are in the right ballpark taking into account the drastic approximations made. The estimates (3) and (4) are obviosuly crude, but it tells us that the freeze-out radius, for each particle, approximately scales with the square root of the number of scattering centers a particle can interact with and the related cross section.

An interesting question, which is relevant for a hadronic gas, is whether multi-body collisions are significant at some stage in the expansion process. One can write a simple generalization of the two-body collision rate formula:

$$\frac{dR}{d^4x}=n_1n_2\sigma v_{rel}=n_1n_2\sigma\frac{\sqrt{(p_1\cdot p_2)^2-m_1^2m_2^2}}{\varepsilon_1\varepsilon_2} \tag{5}$$

where p_i are the particle four-momenta and m_i their masses, n_i their densities and v_{rel} the relativistic relative velocity, to a three-body collision assuming that the particle 3 collides with a cluster (12) made of particles 1 and 2:

$$\left.\frac{dR}{d^4x}\right|_3=n_3n_{(12)}\sigma_{3(12)}v_{rel}P_{(12)} \tag{6}$$

where $n_{(12)}$ is the density of clusters, v_{rel} a suitable extension of the relativistic relative velocity function for the 2-body problem and $P_{(12)}$ the cluster formation probability. The latter can be assumed to be proportional to $n_1n_2\sigma_{12}^{3/2}$, i.e. proportional to the probability that 1 and 2 find themselves within a range of $\approx\sqrt{\sigma}$. Replacing $\sigma_{3(12)}$ and σ_{12} with a typical hadronic cross section σ one has:

$$\left.\frac{dR}{d^4x}\right|_3\approx\frac{n_1n_2n_3}{\varepsilon_1\varepsilon_2\varepsilon_3}\sigma^{5/2}f_{rel}(m_1,m_2,m_3,\sqrt{s},...) \tag{7}$$

where f is a function of relativistic invariants pertaining to a three-particle system. This approximated formula can be extended to N-body collisions:

$$\left.\frac{dR}{d^4x}\right|_N\approx\left(\prod_{i=1}^N\frac{n_i}{\varepsilon_i}\right)\sigma^{(3N-4)/2}f_{rel}(m_1,\ldots,m_N,\sqrt{s},\ldots) \tag{8}$$

Binary collisions prevail when the left hand side of (5) largely exceeds the one in (7). Assuming that the ratio of the relativistic functions f_{rel} averages to $\mathcal{O}(1)$, this happens when:

$$n\sigma^{3/2} \ll 1 \implies \lambda = \frac{1}{\sigma n} \gg \sqrt{\sigma} \tag{9}$$

This is a physically very clear condition: the particle mean free path λ should be much larger than its effective interaction distance as determined by the square root of the cross section. This is by no means a trivial requirement; for an expanding hadron gas to have a stage where binary collisions prevail before freeze-out, this implies, according to equation (2):

$$R_{\rm fo} = \sqrt{\frac{N\sigma}{4\pi}} = 3\lambda \gg 3\sqrt{\sigma} \tag{10}$$

that is the system should contain a *large* number of particles, $N \gg \mathcal{O}(100)$. If this condition is not fulfilled, the system decouples when ternary and, in general, N-nary collisions are still relevant. This is an important conclusion, to be emphasized: *in relativistic ion collisions, for the decoupling to occur in the binary collision regime, the number of particles should largely exceed* $\mathcal{O}(100)$. Thus, this may happen only for *heavy* ion collisions.

There is one more important condition to be addressed: for the freeze-out to be in the kinetic regime, meaning a system of quasi-free colliding particles, it is necessary that the mean free path of particles largely exceeds their mean (thermal) wavelength, a condition that we have seen is not fulfilled in the plasma phase:

$$\lambda = \frac{1}{\sigma n} \gg \frac{1}{\langle p \rangle} = \frac{1}{m\langle v \rangle} = \frac{1}{\sqrt{3Tm}} \tag{11}$$

where the last formula assumes the non-relativistic regime of most hadrons, which is a good approximation for a hadron gas in the actual observed range of temperatures. Hence, writing $n = N/(4\pi R^3/3)$ and using (3):

$$\frac{4\pi}{3}R^3 \gg N\sigma\frac{1}{\sqrt{3Tm}} \implies R \gg R_{\rm fo}^{2/3}\left(\frac{3}{Tm}\right)^{1/6} \tag{12}$$

with the assumption that the number of particles N does not change till freeze-out. The last inequality dictates that we have a kinetic stage in hadron gas expansion, that is $R < R_{\rm fo}$ provided that the radius at freeze-out is very much larger than the thermal wavelength of the hadrons. As a numerical example, for pions at $T = 160$ MeV the last factor equals 1.3 fm$^{1/3}$ and for $R_{\rm fo} = 15$ fm, the above inequality demands $R \gg 8$ fm. We stress again that these numbers are crude approximations and only have an illustrative purpose, they should not be compared as such with actual measurements of the system size with e.g. pion interpherometry.

Another relevant question is whether multi-body collisions can occur in what we can properly call a kinetic regime. According to (9), multi-body collisions are relevant if the mean free path is of the order of the square root of the cross-section. Therefore, we are led to demand that:

$$\lambda \approx \sqrt{\sigma} \gg \sqrt{\frac{1}{3mT}} \tag{13}$$

Is this condition fulfilled? For pions at the typical temperature $T = 160$ MeV, the right hand side is about 1 fm, which is not much smaller than the square root of the typical hadronic cross-section. Under this circumstance, the hadron gas is in the very same condition of a strongly interacting fluid as described at the end of the previous section.

Journal of Physics: Conference Series **527** (2014) 012012 doi:10.1088/1742-6596/527/1/012012

4. Hadronization and beyond: the statistical model

Among the models which have been proposed to account for the quantitative features of hadron production in relativistic heavy ion collisions, the statistical hadronization model can be considered as of today a reference. Besides its remarkable success in reproducing particle multiplicities with few parameters, this model is used in nearly all hydrodynamical calculations to turn the fluid cells into hadrons through the Cooper-Frye prescription, enforcing local thermodynamical equilibrium. Furthermore, it has been shown that this model is able to reproduce hadronic multiplicities in elementary collisions, from e^+e^- to pp and from low to high energy [5]. Therefore, it turns out that the statistical model grasps a peculiar universal feature of the hadronization process, hence of QCD in the non-perturbative regime, regardless of the kind of collision.

Indeed, the statistical model has proved to be successful in relativistic heavy ion collisions throughout the explored centre-of-mass energy range [5]. The interpretation of its success is, however, still subject of debate. Some proposed that thermalization is achieved at the level of hadrons through multiple collisions [6, 7]. Others [8], including the author, advocate that hadronization itself gives rise to an equilibrated system of hadrons (modulo the extra strangeness suppression in elementary collisions, not fully understood). In Hagedorn's words *hadrons are born in equilibrium* because it would be otherwise impossible to explain the statistical distributions observed in elementary collisions where multiplicities are such that freeze-out coincides with hadron, and, chiefly, the apparent full chemical equilibrium of strange particle abundances.

If hadronization itself gives rise to an equilibrated population of hadrons, can the later processes, if any, affect it? This is an important question for the understanding of the hadron production mechanism in both elementary and relativistic heavy ion collisions. We can sketch - in a very ideal fashion - the major steps in the process leading to the freeze-out in relativistic heavy ion collisions as points along a time-oriented axis as in figure 3. In fact, the freeze-out processes are continuous, species dependent, geometry dependent and overlapping. Nevertheless, the diagram of figure 3 is an useful tool for the purpose of illustration.

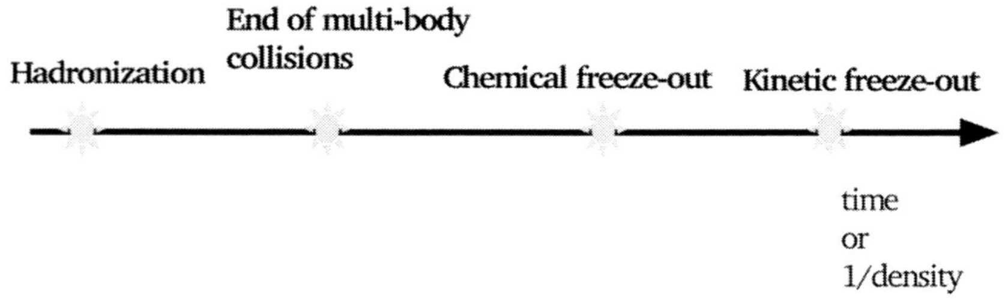

Figure 3. Sketch of the time sequence of processes in an expanding strongly interacting hadronic system: after hadrons are formed, a strongly interacting stage follows where particles may undergo multi-body collisions. Thereafter, mostly binary collisions survive till chemical freeze-out and, later, kinetic freeez-out.

After hadronization, the system begins to expand, but it can stay a little in the strong coupling regime where it cannot be described as a kinetic system, according to previous discussion. At some point, this regime, where chemical equilibrium is presumably ketp, ceases and the era of binary collisions sets in, provided that the number of particles is $> \mathcal{O}(100)$. For some time, binary collisions can be inelastic (chemical freeze-out) thereafter only elastic till kinetic freeze-out occurs and the system finally decouples.

Is this picture borne out by the data? The existence of a stage where binary collisions prevail is confirmed by at least three major evidences:

(i) The suppression of short-lived resonances compared to long-lived particles in the statistical model fits [9, 10, 11], an effect which is not seen in elementary collisions [12]. It can be explained by the rescattering of the resonance decay products in the expanding medium, not possible for a small system.

(ii) The difference between the temperatures determined in multiplicity fits and those in the fits of transverse spectra [11]. This means that elastic interactions continue after chemical freeze-out. Also this effect is not seen in elementary collisions [12].

(iii) The dependence of kinetic freeze-out temperatures as a function of the impact parameter [11, 13]. Collisional decoupling depends on particle multiplicity, as we have seen, as well as geometry, and it occurs at a higher density if multiplicity is lower.

The existence of a kinetic stage where hadrons interact elastically is by now generally accepted. Do we have a similar evidence for the abundances? In other words, do hadrons chemically decouple after some stage of binary collisions? It should be stressed that, if such a stage exists, hadronic abundances should show deviations from an initial chemical equilibrium situation [14].

Until recently, there was no major clue of such a phenomenon as statistical model fits were in agreement with the data to a satisfactory degree of accuracy. However, new measurements of anti-proton yield at SPS energy [15] and at LHC [16] resulted in significantly lower values compared to the statistical model predictions [17, 18]. This result has been soon interpreted as an effect of post-hadronization inelastic rescattering, where anti-baryons are annihilated before chemical freeze-out increasing pion multiplicity [19, 20, 21]. We point out that such annihilation should occur in a stage where 2-body collisions prevail (see figure 3) and Monte-Carlo simulations such as UrQMD [22] only include binary collisions. Howeever, as has been mentioned, the separation of N-ary to binary collisions stages in fact cannot be sharp and indeed it was pointed out [23] that the back-reaction of N pions colliding to regenerate antibaryons could continue for a longer time.

As a matter of fact, the reanalysis of the data assuming equilibrium at hadronization and subsequent post-hadronization inelastic rescattering stage using UrQMD predictions as correction factors gives a better fit to the data and a hadronization (or latest chemical equilibrium point) in agreement with the extrapolated QCD lattice line (see figure 4) [24]. Even though UrQMD does not include $N > 2$-nary collisions, this is a strong indication that inelastic processes following hadronization do play a role and that chemical freeze-out does *not* coincide with hadronization. Still, taking these processes properly into account, it is possible to reconstruct the original hadronization conditions.

5. Conclusions

There is little doubt that QGP has been produced in relativistic heavy ion collisions at high energy. The experimental programmes from SPS to LHC energies have allowed a verification of several QCD predictions for the phases of the strongly interacting matter. There are several features of the phase diagram which are still to be verified, such as the existence of a critical point, which will be the physics case of a next generation of experiments taking place at JINR, CERN and GSI. From a theoretical viewpoint, QGP is, to my knowledge, the only instance

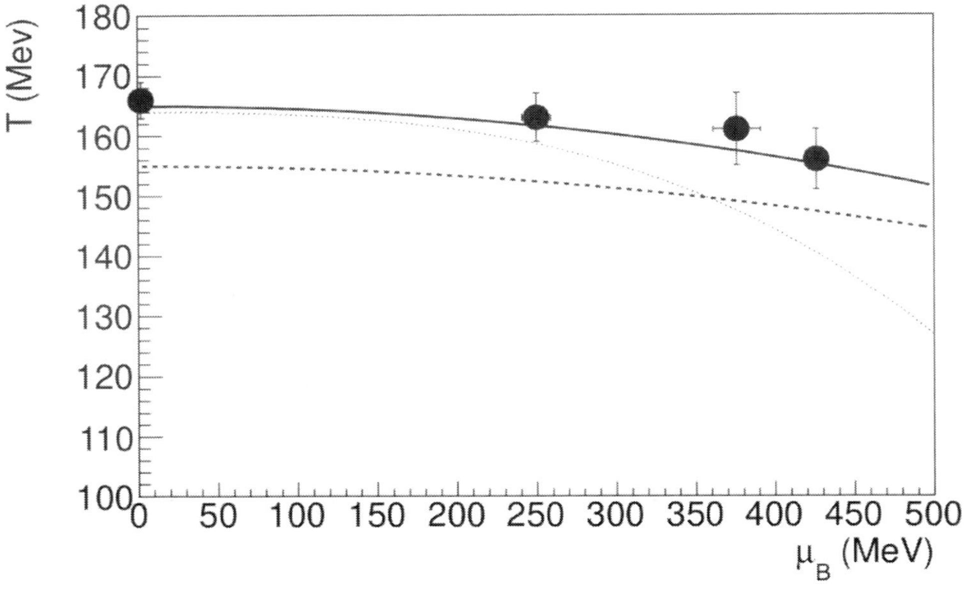

Figure 4. Reconstructed hadronization points in the (T, μ_B) plane compared with lattice QCD extrapolations (solid and dashed lines) and the chemical freeze-out line (dotted curve) (from Ref. [24]).

of fully relativistic matter (in the sense of locally equilibrated interacting system) created in a terrestrial laboratory, and this provided an unprecedented opportunity to develop, test and extend to the relativistic domain the theoretical tools utilized for the usual matter.

Acknowledgments

I would like to thank the organizers for their invitation to participate in a nice conference in a very enjoyable place.

References

[1] Aoki Y, Endrodi G, Fodor Z, Katz S D, and Szabo K K 2006 *Nature* **443** 675
[2] Heinz U, talk given at the workshop *Relativistic Aspects of Nuclear Physics* 2013, Rio de Janeiro, September 2013
[3] Casalderrey-Solana J, Liu H, Mateos D, Rajagopal K, and Wiedemann U A `arXiv:1101.0618 [hep-th]`; Adams A, Carr L D, Schäfer T, Steinberg P and Thomas J E 2012 *New J. Phys.* **14** 115009
[4] Hosoya A, Sakagami M, and Takao M 1984 *Annals Phys.* **154** 229
[5] Becattini F, *An Introduction to the Statistical Hadronization Model* `arXiv:0901.3643 [hep-ph]`; Becattini F and Fries R, in *Relativistic heavy ion physics*, Landolt-Börnstein 1-23 `arXiv:0907.1031 [nucl-th]` and references therein
[6] Braun-Munzinger P, Stachel J, and Wetterich C 2004 *Phys. Lett. B* **596** 61
[7] Noronha-Hostler J, Beitel M, Greiner C, and Shovkovy I 2010 *Phys. Rev. C* **81** 054909
[8] Becattini F and Heinz U W 1997 *Z. Phys. C* **76** 269; Heinz U 1999 *Nucl. Phys. A* **661** 140; Stock R 1999 *Phys. Lett. B* **456** 277; Castorina P, Kharzeev D, and Satz H 2007 *Eur. Phys. J. C* **52** 187
[9] Becattini F, Gazdzicki M, Keranen A, Manninen J, and Stock R 2004 *Phys. Rev. C* **69** 024905
[10] Markert C 2005 *J. Phys. G* **31** S169
[11] Blume C 2012 *Acta Phys. Polon. B* **43** 577
[12] Becattini F and Passaleva G 2002 *Eur. Phys. J. C* **23** 551

Journal of Physics: Conference Series **527** (2014) 012012 doi:10.1088/1742-6596/527/1/012012

[13] Heinz U and Kestin G 2006 *PoS CPOD* **2006** 038
[14] Bass S A and Dumitru A (2000) *Phys. Rev. C* **61** 064909
[15] Anticic T *et al.* [NA49 Collaboration] 2011 *Phys. Rev. C* **83** 014901
[16] Abelev B *et al.* [ALICE Collaboration] 2013 *Phys. Rev. C* **88** 044910
[17] Becattini F, Manninen J, and Gazdzicki M 2006 *Phys. Rev. C* **73** 044905
[18] Andronic A, Braun-Munzinger P, and Stachel J `arXiv:0707.4076 [nucl-th]`
[19] Pan Y and Pratt S `arXiv:1210.1577 [nucl-th]`
[20] Steinheimer J, Aichelin J, and Bleicher M 2013 *Phys. Rev. Lett.* **110** 042501
[21] Becattini F, Bleicher M, Kollegger T, Mitrovski M, Schuster T, and Stock R 2012 *Phys. Rev. C* **85** 044921
[22] Petersen H *et al.* 2008 *Phys. Rev. C* **78** 044901 and references therein
[23] Rapp R and Shuryak E V 2001 *Phys. Rev. Lett.* **86** 2980
[24] Becattini F, Bleicher M, Kollegger T, Schuster T, Steinheimer J, and Stock R 2013 *Phys. Rev. Lett.* **111** 082302

Journal of Physics: Conference Series **527** (2014) 012013 doi:10.1088/1742-6596/527/1/012013

Quark-Gluon Plasma: from lattice simulations to experimental results

G Aarts, C Allton
Department of Physics, College of Science, Swansea University, Swansea, United Kingdom

A Kelly, J-I Skullerud
Department of Mathematical Physics, National University of Ireland Maynooth, Maynooth, County Kildare, Ireland

S Kim
Department of Physics, Sejong University, Seoul 143-747, Korea

T Harris, S M Ryan
School of Mathematics, Trinity College, Dublin 2, Ireland

M P Lombardo
INFN-Laboratori Nazionali di Frascati, I-00044, Frascati (RM) Italy

Abstract. Theoretical studies of quarkonia can elucidate some of the important properties of the quark–gluon plasma, the state of matter realised when the temperature exceeds $\mathcal{O}(150)$ MeV, currently probed by heavy-ion collisions experiments at BNL and the LHC. We report on our results of lattice studies of bottomonia for temperatures in the range $100\,\text{MeV} \lesssim T \lesssim 450\,\text{MeV}$, introducing and discussing the methodologies we have applied. Of particular interest is the analysis of the spectral functions, where Bayesian methods borrowed and adapted from nuclear and condensed matter physics have proven very successful.

1. The plasma of quarks and gluons

Heavy ions colliding at ultrarelativistic energies produce a tiny fireball of a plasma of quarks and gluons — the state of matter thought to have existed slightly after the Big Bang. This experimental program started at the SPS in the 1980s, continued at RHIC and it is now running at the LHC, where the experiments ALICE, ATLAS and CMS are collecting and analyzing data from the collisions of lead nuclei. The most recent runs reached temperatures of about 500 MeV — approximately 5×10^{12} K — well above the temperature of the crossover from ordinary matter to the plasma of quarks and gluons, estimated to be at about 155 MeV [1, 2].

The analysis of the transition from ordinary matter to the quark-gluon plasma, and the nature of the plasma itself — the spectral properties and the residual interactions — is a very active field of theoretical research [3]. We are concerned with temperatures well above those where

Journal of Physics: Conference Series **527** (2014) 012013 doi:10.1088/1742-6596/527/1/012013

the synthesis of the lightest nuclei takes place, so strong interactions alone suffice to describe the system. Hence, the appropriate microscopic description is the relativistic field theory of the strong interactions, Quantum Chromodynamics, QCD. We also know that α_s, the QCD coupling at the scale of the temperatures of interest, can be as large as 0.5 and even larger, so that perturbation theory, which will ultimately be valid for very high temperatures, does not work, at least quantitatively, in the region which we wish to explore here. We therefore need a non-perturbative method, and we choose numerical simulations of QCD discretized on a lattice.

This note is devoted to the presentation of lattice results obtained by the FASTSUM collaboration, in particular those appearing in Refs. [4–7]. It updates and expands previous reviews [8, 9].

1.1. Why bottomonium?

Why is bottomonium such an interesting probe of the medium? We will be concerned with phenomena occurring at, or above, the crossover to a chirally symmetric, deconfined phase. In general both chiral symmetry restoration and deconfinement affect the spectrum of the theory: chiral symmetry will be seen in the light sector, by the degeneracy of the chiral partners. The heavier quarks, however, will be blind to chiral symmetry: $m_u, m_d \ll m_s \simeq T_c \simeq \Lambda_{QCD} \ll m_c, m_b$. For instance a sizeable fraction of the strange mass is due to the breaking of chiral symmetry, and for charm and bottom this contribution is completely negligible — modifications of the spectrum of charmonia and bottomonia come entirely from the gauge dynamics. In very short summary, then, quarkonia are ideal probes of the gluodynamics. Since their size is small, however, the sensitivity to deconfinement is not immediate: the short-range component of the potential, which is responsible for their binding, and hence the fundamental bound states, might well survive in the plasma, while the excited states dissolve. We talk of sequential suppression of quarkonia, and the goal of our studies is to make quantitative these very qualitative considerations.

Charmonia — which are easier to produce experimentally — have been studied since early SPS days. We know by now [10] that the experimental results for charmonia are also sensitive to cold nuclear matter effects which reduce the primordial charmonium number significantly — for instance by about 60% at SPS. There are also competing temporal scales — thermalization of the plasma and formation time of the bound states, as well as the intriguing observation that charmonia production rates at SPS and RHIC are quite similar. The latter effect can be at least in part explained by taking into account a regeneration mechanism for charmonia (feed-down): the higher dissociation rate at RHIC is compensated by a richer feed-down. All these considerations make the study of charmonium suppression patterns extremely fascinating but also challenging.

Bottomonium production, on the other hand, is much less prone to regeneration effects, and as such is a more promising observable for the spectral analysis of the quark–gluon plasma [10]. Since bottomonium production requires the larger energies available at the LHC, such data have become available only very recently. In the last few years, results showing sequential Upsilon suppression in PbPb collision at LHC energies have appeared [11, 12], and we will comment on those at the end.

2. Lattice QCD, relativistic and non–relativistic

Lattice calculations are performed in a discretized Euclidean space-time, which introduces technical scales: the lattice spacing a and the lattice spatial size, L. Each characteristic physical scale l should obviously fulfil the constraint $a \ll l \ll L$. Accommodating quarks with vastly different masses then poses a computational challenge.

Our strategy is to treat as accurately as possible the light quarks: in all our studies gauge configurations with dynamical light Wilson-type quark flavours are produced on highly

Journal of Physics: Conference Series **527** (2014) 012013 doi:10.1088/1742-6596/527/1/012013

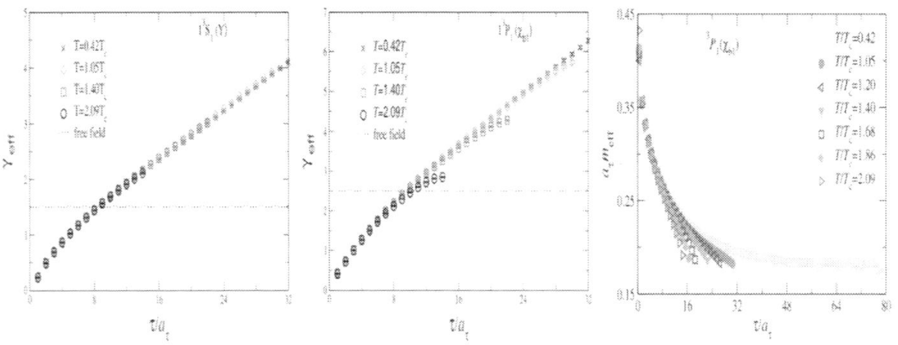

Figure 1. (Color online) Effective exponents $\gamma_{\rm eff}(\tau)$ for the Υ (left) and χ_{b1} (center), as a function of Euclidean time for various temperatures. The dotted line indicates the non-interacting result in the continuum, which is approached by the χ_{b1} results at the higher temperature $T = 2.1T_c$ [4]. Effective mass for the χ_{b1} (right), as a function of Euclidean time for various temperatures [5].

anisotropic lattices. Details of the lattice action and parameters can be found in Refs. [13, 14]. For the b quarks we use non-relativistic QCD (NRQCD). In our work we use a mean-field improved action with tree-level coefficients, which includes terms up to and including $\mathcal{O}(v^4)$, where v is the typical velocity of a bottom quark in bottomonium,$v_b^2 \simeq 0.1$. There is no (rest) mass term in the NRQCD action so one can dispense with the demanding constraint $a \ll 1/m_b$. In general, NRQCD relies on the separation of scales between the bottom quark and any other physical scale of the theory: in our work we study temperatures up to $2T_c \simeq 400$ MeV, hence $m_b \gg T$ and the application of NRQCD is fully justified.

3. Correlators in the plasma

In our studies we considered the S wave states Υ and η_b and the P wave states χ_{b0}, χ_{b1}, χ_{b2} and h_b. We found that the correlators in the different P wave channels behave in a very similar way, hence from now on for the P wave states we present results for the χ_{b1} channel only. The gauge configurations are generated with a dynamical first generation of quarks. More recently the strange quark has been added as well [15]: the general features of the results are unchanged, and a systematic, detailed analysis of the effect of a dynamical strange quark is in progress.

There is a further important simplification in NRQCD: to compute propagators we only need to solve an initial value problem, in contrast to the relativistic case where the same computation requires the inversion of a large sparse matrix. In the hadronic phase we expect that the large Euclidean time behaviour of the propagators is dominated by an exponential decay. In the QGP phase, one can use as guidance the behaviour found for free quarks in NRQCD [4, 5, 16], which yields the spectral functions

$$\rho_{\rm free}(\omega) \propto (\omega - \omega_0)^\alpha\, \Theta(\omega - \omega_0), \quad \text{where} \quad \alpha = \begin{cases} 1/2, & \text{S wave.} \\ 3/2, & \text{P wave.} \end{cases} \tag{1}$$

Here we supplemented [15] the free results by a threshold, ω_0, to account for the additive shift in the quarkonium energies which describes the residual interactions in the thermal medium. The correlation functions then have the following behaviour

$$G_{\rm free}(\tau) \propto \int d\omega\, e^{-\omega\tau}\rho(\omega) \propto \frac{e^{-\omega_0\tau}}{\tau^{\alpha+1}}. \tag{2}$$

Journal of Physics: Conference Series **527** (2014) 012013 doi:10.1088/1742-6596/527/1/012013

To visualize the temperature dependence and at the same time monitor the approach to a quasi-free behaviour we construct effective power plots [4], using the definition

$$\gamma_{\text{eff}}(\tau) = -\tau\frac{G'(\tau)}{G(\tau)} = -\tau\frac{G(\tau+a_\tau) - G(\tau-a_\tau)}{2a_\tau G(\tau)}, \tag{3}$$

where the prime denotes the (discretized) derivative. For a pure power law decay this yields a constant, $\gamma_{\text{eff}}(\tau) = \alpha + 1$. Taking into account the threshold ω_0, one finds $\gamma_{\text{eff}}(\tau) = \alpha + 1 + \omega_0\tau$. The results are shown in figure 1. We see that the Υ displays a very mild temperature dependence, while for the χ_{b1} the asymptotic behaviour of the effective power tends to flatten out with increasing temperature: this indicates that ω_0 — the slope — decreases with temperature. In the same panel also shown are the effective exponents in the continuum non-interacting limit ($\omega_0 = 0$). In the case of the χ_{b1}, we observe that the effective exponent — which can be read off the intercept of the asympotic straight line with the $\tau = 0$ axis — seems to approach the non-interacting result. Note that, even for $T \to \infty$ where $\omega_0 = 0$, the effective exponents on the lattice only become a straight line, $\gamma_{\text{eff}}(\tau) \to \alpha + 1$, at large enough τ due to the presence of lattice discretisation effects at smaller temporal separations.

A complementary description is offered by the effective masses,

$$a_\tau m_{\text{eff}}(\tau) = -\log[G(\tau)/G(\tau - a_\tau)], \tag{4}$$

which are shown in the rightmost panel of figure 1. When the correlator takes the form of a sum of exponentials, the ground state will show up as a plateau at large Euclidean times, provided that it is well separated from the excited states. This is indeed the case at the lowest temperature and leads to the zero-temperature spectrum discussed in Refs. [4,5]. Above T_c, we observe that the effective masses no longer follow the trend given by the correlator below T_c, but instead bend away from the low-temperature data. The results shown in figure 1 imply that the spectrum of the χ_{b1} has changed drastically. If isolated bound states persist, the ground state has to be much lighter and excited states cannot be well separated. A more natural explanation is that there is no exponential decay and bound states have melted, immediately above T_c. This interpretation is supported by the spectral function analysis presented next.

4. Spectral functions

Spectral functions play an important role in understanding how elementary excitations are modified in a thermal medium, from many-body physics — see the talk by Giuseppina Orlandini at this meeting [17] — to QCD, which is discussed here.

In general, the spectral decomposition of a (relativistic) zero-momentum Euclidean propagator $G(\tau)$ at finite temperature T is given by

$$G(\tau) = \int_{-\infty}^{\infty} \frac{d\omega}{2\pi} K(\tau,\omega)\rho(\omega), \tag{5}$$

where $\rho(\omega)$ is the spectral function and the kernel K is given by

$$K(\tau,\omega) = \frac{\cosh\left[\omega(\tau - 1/2T)\right]}{\sinh\left(\omega/2T\right)}. \tag{6}$$

In NRQCD the kinematical temperature dependence is always absent. This can be seen in a number of ways. Following Ref. [16], we write $\omega = 2M + \omega'$ and drop terms that are exponentially suppressed when $M \gg T$. The spectral relation (5) then reduces to its zero-temperature limit,

$$G(\tau) = \int_{-2M}^{\infty} \frac{d\omega'}{2\pi} \exp(-\omega'\tau)\rho(\omega'), \tag{7}$$

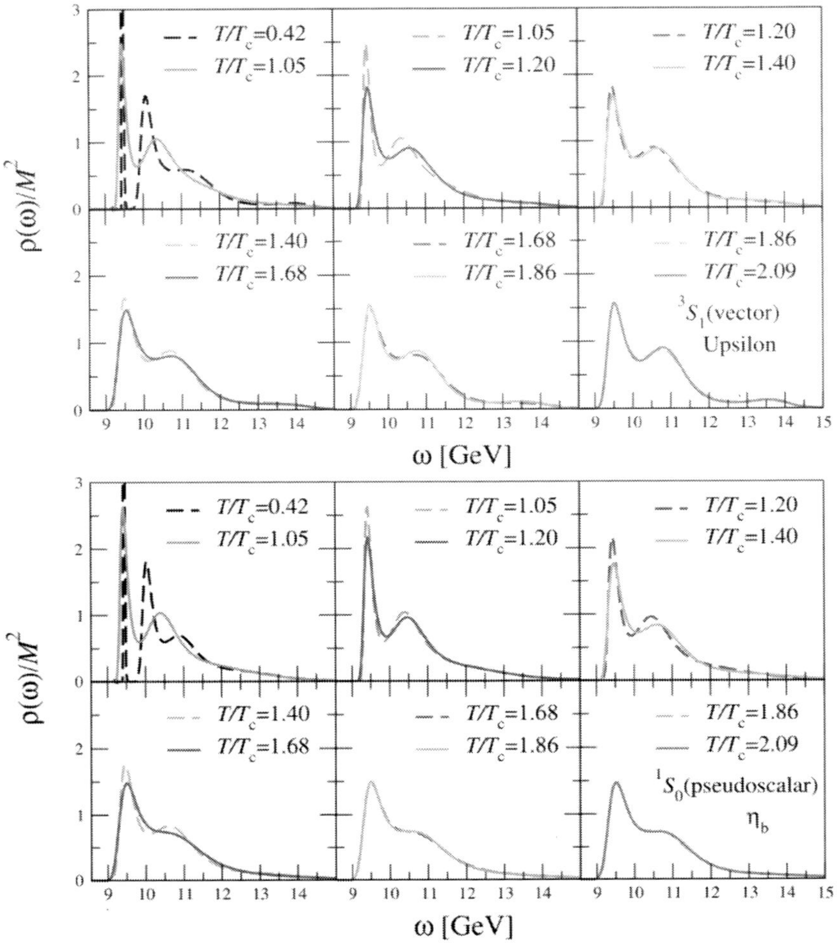

Figure 2. (Color online) Spectral functions $\rho(\omega)$, normalised with the heavy quark mass, in the vector (Υ) channel (upper panel) and in the pseudoscalar (η_b) channel (lower panel) for all temperature available. The subpanels are ordered from cold (top left) to hot (bottom right). Every subpanel contains two adjacent temperatures to facilitate the comparison [5].

even at nonzero temperature. Since the interesting physics takes place around the two-quark threshold, $\omega \sim 2M$, the region of interest is around $\omega' \sim 0$ and the lower limit becomes irrelevant. In summary, in the heavy-quark limit the spectral relation simplifies considerably, and temperature effects seen in the correlators are thus only due to changes in the light-quark–gluon system.

Despite these simplifications, the calculation of the NRQCD spectral functions using Euclidean propagators as an input remains a difficult, ill-defined problem. We will tackle it by using the Maximum Entropy Method (MEM) [18], which has proven successful in a variety of applications. We have carefully studied the systematics, including the dependence on the set of lattice data points in time, and on the default model $m(\omega)$ which enters in the parametrisation

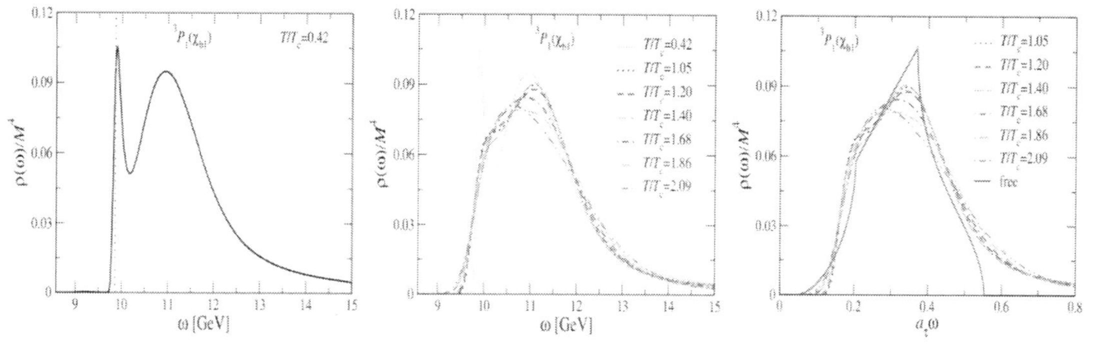

Figure 3. (Color online) Spectral function in the χ_{b1} channel at the lowest temperature (left) and at all temperatures (center). The dotted line on the left indicates the position of the ground state obtained with a standard exponential fit. Comparison with the free lattice spectral function above T_c (right) [7].

of the spectral function,

$$\rho(\omega) = m(\omega) \exp \sum_k c_k u_k(\omega), \tag{8}$$

where $u_k(\omega)$ are basis functions fixed by the kernel $K(\tau, \omega)$ and the number of time slices, while the coefficients c_k are to be determined by the MEM analysis [18]. We find that the results are insensitive to the choice of default model, provided that it is a smooth function of ω. It remains of interest, and it is an open avenue of research, to experiment with alternative prescriptions [19, 20], also on ensembles generated with different lattice actions [21].

The results for the spectral functions for the S wave states, Υ and η_b [5] can be seen in figure 2, which shows that as the temperature is increased the ground state peaks of both states remain visible. One caveat applies also here: the apparent width at zero temperature is most likely due to a lattice/MEM artefact, and this calls for an analysis of the discretization effects on the spectral function, which is one of our ongoing projects [22]. The peaks associated with the excited states become suppressed at higher temperature and are no longer discernible quickly above T_c.

Turning now to the P wave states, the result at the lowest temperature is given in figure 3 (left). The dotted vertical line indicates the mass of the lowest-energy state obtained with an exponential fit. We see from this that the narrow peak in the spectral function corresponds to the ground state. The second, wider structure is presumably a combination of excited states and lattice artefacts, see below. We note that we have not been able to extract the mass of the first excited state with an exponential fitting procedure. The spectral functions for all temperatures are shown in figure 3 (center). We find no evidence of a ground state peak for any of the temperatures above T_c. This is consistent with the interpretation of the correlator study presented above and supports the conclusion that the P wave bound states melt in the QGP.

In order to interpret the remaining structure, we compare it with the spectral function computed on the lattice, in the absence of interactions [5]. In figure 3, rightmost panel, we show the free lattice spectral functions, together with the spectral functions above T_c. In that plot we have adjusted the threshold to match our lattice results, and once this is done the lattice spectral function and the free one almost coincide. This lends further support to the conclusion drawn in Ref. [4] from an analysis of the correlators: the system in the P wave channels is approaching a system of noninteracting quarks, and the residual interactions can be described

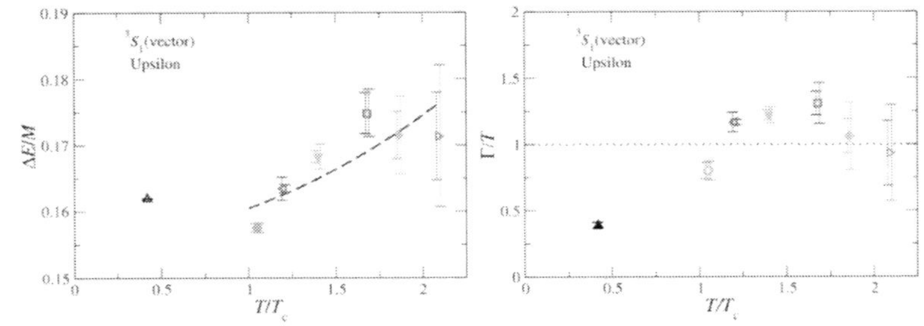

Figure 4. (Color online) Position of the ground state peak ΔE, normalised by the heavy quark mass (left), and the upper limit on the width of the ground state peak, normalised by the temperature (right), as a function of T/T_c in the vector (Υ) channel. Similar results have been obtained for the pseudoscalar (η_b) channel [5].

by a shift in the quarkonium energies, which does not affect the shape of spectral functions.

5. Comparison with analytic studies and momentum dependence

In this section we restrict our analysis to the S waves. From the computed spectral functions we can determine the mass (from the peak position) and (an upper bound on) the width of the ground state at each temperature.

In figure 4 we show the temperature dependence of the mass shift ΔE, normalised by the heavy quark mass and the temperature dependence of the width, normalised by the temperature.

We now contrast our results with analytic predictions derived assuming a weakly coupled plasma [20, 23–28]. According to Ref. [26], the thermal contribution to the width is given, at leading order in the weak coupling and large mass expansion, by

$$\frac{\Gamma}{T} = \frac{1156}{81}\alpha_s^3 \simeq 14.27\alpha_s^3, \tag{9}$$

i.e. the width increases linearly with the temperature. If we take as an estimate from our results that $\Gamma/T \sim 1$, we find that this corresponds to $\alpha_s \sim 0.4$, which is a reasonable result (it would be of interest to compute α_s directly on our configurations). In the same spirit the thermal mass shift is given in Ref. [26] by

$$\delta E_{\text{thermal}} = \frac{17\pi}{9}\alpha_s \frac{T^2}{M} \simeq 5.93\alpha_s \frac{T^2}{M}. \tag{10}$$

In these simulations we have $T_c \sim 220$ MeV, $M \sim 5$ GeV. Taking these values together with $\alpha_s \sim 0.4$ as determined above, Eq. (10) becomes

$$\frac{\delta E_{\text{thermal}}}{M} = 5.93\alpha_s \left(\frac{T_c}{M}\right)^2 \left(\frac{T}{T_c}\right)^2 \sim 0.0046 \left(\frac{T}{T_c}\right)^2. \tag{11}$$

In order to contrast our results with this analytical prediction, we have compared the temperature dependence of the peak positions to the simple expression

$$\frac{\Delta E}{M} = c + 0.0046 \left(\frac{T}{T_c}\right)^2, \tag{12}$$

where c is a free parameter. This is shown by the dashed line in figure 4 (left panel). The numerical results and the analytic ones are not inconsistent, within the large errors.

Table 1. Nonzero momenta [6]. Also indicated are the corresponding velocities $v = |\mathbf{p}|/M_S$ of the ground states in the vector (Υ) and pseudoscalar (η_b) channels, using the ground state masses determined previously [4], $M_\Upsilon = 9.460$ GeV and $M_{\eta_b} = 9.438$ GeV.

$\mathbf{n}$	(1,0,0)	(1,1,0)	(1,1,1)	(2,0,0)	(2,1,0)	(2,1,1)	(2,2,0)
$\|\mathbf{p}\|$ (GeV)	0.634	0.900	1.10	1.23	1.38	1.52	1.73
$v\ [\Upsilon(^3S_1)]$	0.0670	0.0951	0.116	0.130	0.146	0.161	0.183
$v\ [\eta_b(^1S_0)]$	0.0672	0.0954	0.117	0.130	0.146	0.161	0.183

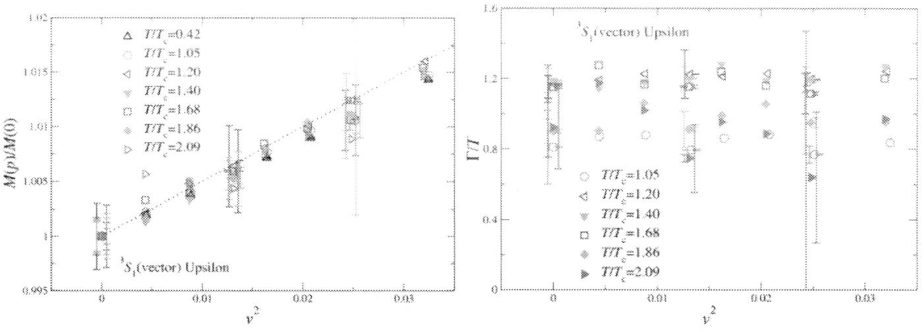

Figure 5. (Color online) Position of the ground state peak $M(\mathbf{p})/M(0)$ (left) and the upper limit on the width of the ground state peak, normalized by the temperature, Γ/T (right), as a function of the velocity squared (v^2) in the vector (Υ) channel Analogous results have been obtained for the the pseudoscalar (η_b) channel. The dotted line in the left figure represents $M(\mathbf{p})/M(0) = 1 + \frac{1}{2}v^2$ [6].

The analysis from effective theories predicts significant momentum effects at large momenta. Moreover current CMS results have been obtained at large momenta. There is therefore both phenomenological and experimental motivation to extend these studies to non-zero momenta [6]. The momenta and velocities that are accessible on the lattice are constrained by the discretization and the spatial lattice spacing. The lattice dispersion relation reads

$$a_s^2\mathbf{p}^2 = 4\sum_{i=1}^{3}\sin^2\frac{p_i}{2}, \qquad p_i = \frac{2\pi n_i}{N_s}, \qquad -\frac{N_s}{2} < n_i \le \frac{N_s}{2}. \tag{13}$$

To avoid lattice artefacts, only momenta with $n_i < N_s/4$ are used: we consider the combinations (and permutations thereof) given in Table 1. The largest momentum, using $\mathbf{n} = (2,2,0)$, is $|\mathbf{p}| \simeq 1.73$ GeV, corresponding to $v = |\mathbf{p}|/M_S \simeq 0.2$. Therefore, the range of velocities we consider is non-relativistic.

As in the zero-momentum case, we extract masses and widths from our correlators using MEM, and we display them in figure 5. We observe that the peak position increases linearly with v^2, as expected. Assuming the lowest-order, non-relativistic expression $M(\mathbf{p}) = M(0) + \mathbf{p}^2/2M(0)$, one finds

$$\frac{M(\mathbf{p})}{M(0)} = 1 + \frac{\mathbf{p}^2}{2M^2(0)} = 1 + \frac{1}{2}v^2, \tag{14}$$

which is indicated with the dotted lines in the left figures.

The dependence on the velocity can be compared with EFT predictions. In Ref. [28] a study of the velocity dependence was carried out in the context of QED, working in the rest frame of the bound state (i.e. the heat bath is moving). In order to compare with our setup, we consider the case in which the temperature is low enough for bound states to be present and where the velocities are non-relativistic. In that case, one finds [28], in the rest frame of the bound state and at leading order in the EFT expansion,

$$\frac{\Gamma_v}{\Gamma_0} = \frac{\sqrt{1-v^2}}{2v} \log\left(\frac{1+v}{1-v}\right), \tag{15}$$

where Γ_0 is the width at rest. Interpreting the width as an inverse lifetime, one can express this result in the rest frame of the heat bath by dividing with the Lorentz factor $\gamma = 1/\sqrt{1-v^2}$. An expansion for non-relativistic velocities then yields

$$\frac{\Gamma_v}{\Gamma_0} = 1 - \frac{2v^2}{3} + \mathcal{O}\left(v^4\right). \tag{16}$$

If we apply this result to our study of bottomonium, we find that the effect of the nonzero velocity shows up as a correction at the percent level (recall that $v^2 \lesssim 0.04$), which is beyond our level of precision but consistent with the observed v independence within errors. Similarly, additional thermal effects in the dispersion relation are currently beyond our level of precision. In summary, the observations in our low-momentum range are consistent with Ref. [28], and in order to observe the predicted non-trivial momentum dependence we need to explore larger momenta.

6. From lattice to experiments

We have presented our results for bottomonium in the quark-gluon plasma, for temperatures up to $2.1T_c$, at the threshold of the region currently explored by LHC heavy-ion experiments. Our analysis uses full relativistic dynamics for the light quarks, and a non-relativistic approach for the bottom quarks. The results are amenable to a successful comparison with effective models which we hope to further pursue in the future.

Our results demonstrate a pattern of suppression of bottomonia which apparently compares well with recent CMS results [11, 12], once we take into account that the temperatures reached in the collisions studied by CMS are similar to the ones of our lattices. However the systems under investigation are vastly different: Quantum Chromodynamics in thermal equilibrium in one case, and an expanding fireball, with an extremely complex experimental setup, in the other. For instance, our studies include processes between a b quark and thermal light quarks (and gluons), but do not include the thermal scattering of b quarks. Unravelling the details and the limitations of this comparison is a subject of active research, certainly beyond the scope of this note.

Acknowledgments

We thank Don Sinclair and Bugra Oktay for collaboration. This work was partly supported by the European Community under the FP7 programme HadronPhysics3. We acknowledge the support and infrastructure provided by the Trinity Centre for High Performance Computing and the IITAC project funded by the HEA under the Program for Research in Third Level Institutes (PRTLI) co-funded by the Irish Government and the European Union. The work of CA and GA is carried out as part of the UKQCD collaboration and the STFC funded DiRAC Facility. GA and CA are supported by STFC. GA is supported by the Royal Society, the Wolfson Foundation and the Leverhulme Trust. SK is grateful to STFC for a Visiting Researcher Grant and supported by the National Research Foundation of Korea grant funded by the Korea government (MEST)

No. 2012R1A1A2A04668255. SR is supported by the Research Executive Agency (REA) of the European Union under Grant Agreement number PITN-GA-2009-238353 (ITN STRONGnet) and the Science Foundation Ireland, grant no. 11-RFP.1-PHY-3201.

References

[1] Borsanyi S *et al.* [The Wuppertal-Budapest Collaboration] 2010 *JHEP* **1009** 073
[2] Bazavov A *et al.* [The HotQCD Collaboration] 2012 *Phys. Rev. D* **85** 054503
[3] Becattini F 2014 this volume
[4] Aarts G, Kim S, Lombardo M P, Oktay M S, Ryan S M, Sinclair D K, and Skullerud J I 2011 *Phys. Rev. Lett.* **106** 061602
[5] Aarts G, Allton C, Kim S, Lombardo M P, Oktay M B, Ryan S M, Sinclair D K, and Skullerud J I 2011 *JHEP* **1111** 103
[6] Aarts G, Allton C, Kim S, Lombardo M P, Oktay M B, Ryan S M, Sinclair D K, and Skullerud J I 2013 *JHEP* **1303** 084
[7] Aarts G, Allton C, Kim S, Lombardo M P, Ryan S M, and Skullerud J I 2013 *JHEP* **1312** 064
[8] Aarts G *et al.* 2013 *J. Phys. Conf. Ser.* **432** 012014
[9] Allton C *et al.* 2013 `arXiv:1310.5135`
[10] Rapp R and van Hees H 2010 R C Hwa, X-N Wang (eds) *Quark Gluon Plasma 4* (World Scientific) p 111
[11] Chatrchyan S *et al.* [The CMS Collaboration] 2011 *Phys. Rev. Lett.* **107** 052302
[12] Chatrchyan S *et al.* [The CMS Collaboration] 2012 *Phys. Rev. Lett.* **109** 222301
[13] Morrin R, Ó Cais A, Peardon M, Ryan S M, and Skullerud J I 2006 *Phys. Rev. D* **74** 014505
[14] Oktay M B and Skullerud J I 2010 `arXiv:1005.1209`
[15] Aarts G, Allton C, Harris T, Kim S, Lombardo M P, Ryan S M, and Skullerud J I 2014 `arXiv:1402.6210`
[16] Burnier Y, Laine M, and Vepsalainen M 2008 *JHEP* **0801** 043.
[17] Orlandini G 2014 private communication
[18] Asakawa M, Hatsuda T, and Nakahara Y 2001 *Prog. Part. Nucl. Phys.* **46** 459
[19] Rothkopf A 2013 *J. Comput. Phys.* **238** 106
[20] Burnier Y and Rothkopf A 2013 *Phys. Rev. Lett.* **111** 182003
[21] Kim S, Petreczky P, and Rothkopf A 2013 *PoS LATTICE* **2013** 169
[22] Harris T, Ryan S M, Aarts G, Allton C, Kim S, Lombardo M P, and Skullerud J I 2013 `arXiv:1311.3208`
[23] Laine M, Philipsen O, Romatschke P, and Tassler M 2007 *JHEP* **0703** 054
[24] Laine M 2007 *JHEP* **0705** 028
[25] Brambilla N, Ghiglieri J, Vairo A, and Petreczky P 2008 *Phys. Rev. D* **78** 014017
[26] Brambilla N, Escobedo M A, Ghiglieri J, Soto J, and Vairo A 2010 *JHEP* **1009** 038
[27] Brambilla N, Escobedo M A, Ghiglieri J, and Vairo A 2011 *JHEP* **1112** 116
[28] Escobedo M A, Soto J, and Mannarelli M 2011 *Phys. Rev. D* **84** 016008

Equations of state based on recent lattice QCD results at vanishing net-baryon density

P Alba[1], W Alberico[1], A Beraudo[2], M Bluhm[3], C Ratti[1]

[1] Dipartimento di Fisica, Università degli Studi di Torino & INFN, Sezione di Torino, via Giuria 1, I-10125 Torino, Italy

[2] Physics Department, Theory Unit, CERN, CH-1211 Genève 23, Switzerland

[3] Department of Physics, North Carolina State University, Raleigh, NC 27695, USA

E-mail: pgalba@to.infn.it

Abstract. We calculate various equations of state for QCD matter at vanishing net-baryon density starting from recent lattice QCD results at high temperatures, going to low temperatures using a hadron resonance gas (HRG) model to describe the confined phase. In order to have a better description of the experimental situation we use an implementation of the HRG model that allows the system to switch, after the chemical freeze-out, to a situation of partial chemical equilibrium. We explore the possibility of different freeze-out temperatures. Our results can be used in the hydrodynamic modeling of relativistic heavy-ion collisions at the LHC and at the highest RHIC beam energies.

1. Introduction

In relativistic heavy-ion collision experiments currently taking place at RHIC (Relativistic Heavy-Ion Collider) and at the LHC (Large Hadron Collider), the deconfined phase of strongly interacting matter, called Quark-Gluon Plasma (QGP), is transiently created. The QGP is formed immediately after the collision; the system then cools down and expands and matter goes through a phase transition from the QGP into a confined hadron gas. The transition is an analytic crossover, as unambiguously proven by first-principle simulations of QCD on the lattice [1]. The transition takes place at a pseudo-critical temperature T_c which has recently been estimated in lattice QCD simulations as well [2, 3].

Relativistic hydrodynamics (see e.g. the reviews in Refs. [4, 5]) has been very successful in modelling the collective dynamics of the matter created in the laboratory, starting from a stage immediately after thermalization until the kinetic freeze-out of final state hadrons. The evolution of the system is driven by the conservation equations for energy, momentum and for the additionally conserved charges (net-baryon number N_B, net-electric charge N_Q and net-strangeness N_S), under the assumption of local thermal equilibrium. The Equation of State (EoS) is an essential ingredient for hydrodynamic simulations: it provides a relation between energy density ϵ, pressure p and the densities n_B, n_Q and n_S of the conserved charges. Assuming the conservation of entropy, one needs to know the EoS only along adiabatic paths. In particular we will follow the path $n_B/s = 0$, i.e. we consider the situation of a vanishing n_B. In addition, to get closer to the situation realized in a heavy-ion collision, we impose $n_S = 0$, while the relationship between n_Q and n_B is fixed by the Z/A of the colliding nuclei.

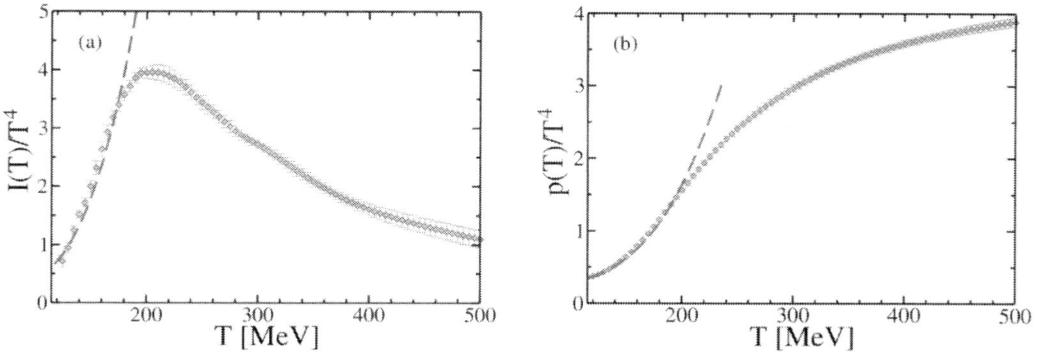

Figure 1. (Color online) Scaled interaction measure $I(T)/T^4$ (panel (a)) and pressure $p(T)/T^4$ (panel (b)) as functions of the temperature T. The symbols correspond to the lattice QCD results from [10]; the dashed curves are the results of the HRG model in thermodynamic equilibrium.

Lattice gauge theory simulations for the QCD equation of state reach nowadays unprecedented levels of accuracy. A basic quantity for the EoS is the interaction measure $I = \epsilon - 3p$, which has been calculated in [6, 7] and in [8, 9, 10]. The numerical results for $I(T)/T^4$ in [6, 7] show significant differences from those in [9, 10] in the transition region. We opt for utilizing in our work the most recent, continuum-extrapolated lattice QCD data from the WB-collaboration [10], corresponding to a system of 2+1 quark flavors with physical quark masses.

We build up a baseline QCD equation of state for $n_B = 0$, by combining a suitable parametrization of these lattice QCD results with a hadron resonance gas (HRG) model in thermodynamic equilibrium. We implement partial chemical equilibrium (i.e. a non-equilibrium situation) in the hadronic phase, in order to properly account for the actual chemical composition in the confined phase. This issue cannot be addressed within equilibrium lattice QCD thermodynamics, but it is of fundamental importance in order to reproduce the experimentally observed flow and p_T-spectra, as well as the correct particle ratios [11]. Because of the present uncertainty in its exact value [12, 13], we study various values for the chemical freeze-out temperature T_{ch}, below which the HRG is in partial chemical equilibrium. Having such QCD equations of state at hand will allow a more controlled determination of the QGP transport properties, as for example the shear (and bulk) viscosity coefficients, and even more important to check which modification to the thermodynamics would imply a variation of 15 MeV in T_{ch}.

This proceeding contribution is organized as follows: in section 2, we show how to combine the lattice QCD results with an HRG model in thermodynamic equilibrium. Section 3 deals with the inclusion of partial chemical equilibrium in the hadronic phase. In section 4, we discuss the obtained QCD equations of state and discuss their main features. For more details we refer the reader to Ref. [14].

2. Lattice QCD-based EoS

Continuum-extrapolated lattice QCD results for thermodynamic quantities in a system of 2+1 quark flavors with physical quark masses were presented in [10]. Their results for the scaled interaction measure $I(T)/T^4$ and for the scaled pressure $p(T)/T^4$ are shown in figure 1 panels (a) and (b), respectively. The other thermodynamic quantities can be obtained from $I(T)$ and $p(T)$ via $\epsilon(T) = 3p(T) + I(T)$ and $s(T) = (\epsilon(T) + p(T))/T$.

QCD thermodynamics in the hadronic phase can be understood in terms of the HRG model, which describes hadronic matter in thermodynamic equilibrium, see e.g. Refs. [15, 16]. The pressure of the model in the thermodynamic limit is given by

$$p(T,\{\mu_k\}) = \sum_k (-1)^{B_k+1} \frac{d_k T}{(2\pi)^3} \int d^3\vec{p}\ \ln\left[1 + (-1)^{B_k+1} e^{-(\sqrt{\vec{p}^2+m_k^2}-\mu_k)/T}\right], \tag{1}$$

where the sum is taken over all the hadronic (including resonances) states k included in the model. In equation (1), d_k and m_k are the degeneracy factor and the mass, and μ_k is the chemical potential of the hadron-species k. In chemical equilibrium, the latter reads $\mu_k = B_k\mu_B + Q_k\mu_Q + S_k\mu_S$, where B_k, Q_k and S_k are the respective quantum numbers of baryon charge, electric charge and strangeness, while μ_B, μ_Q and μ_S denote the chemical potentials associated with n_B, n_Q and n_S.

The particle number density of species k, $n_k = (\partial p/\partial \mu_k)_T$, is given by

$$n_k(T,\mu_k) = \frac{d_k}{(2\pi)^3} \int d^3\vec{p} \frac{1}{(-1)^{B_k+1} + e^{(\sqrt{\vec{p}^2+m_k^2}-\mu_k)/T}} \tag{2}$$

and the net-baryon density follows from $n_B = \sum_k B_k n_k$. Since the vanishing n_B is considered, all μ_k are set to zero in the chemical equilibrium case.

The HRG model used in this work contains states with mass up to 2 GeV as listed in the edition [17] of the Particle Data Book, as well as in the EoS-package provided in [18].

We construct an equation of state which serves as a baseline for the chemical equilibrium case by using our suitable parametrization of the lattice QCD results from [10] at high T and switching to the HRG model at low T around a temperature of 172 MeV. In order to avoid discontinuities in the thermodynamic quantities, which can generally arise in such an approach, we employ a straightforward interpolation procedure between the two parts in the interval 165 MeV $\leq T \leq$ 180 MeV. Such a procedure ensures that the pressure and its first and second derivatives with respect to T are continuous functions, like for example the speed of sound which shows a smooth behavior for all the temperatures in the considered range.

3. Hadron resonance gas in partial chemical equilibrium

A realistic description of a heavy ion collision assumes that the hadronic phase is not in complete chemical equilibrium, since the time scales for inelastic particle number changing processes are typically much larger than the lifetime of the hadronic stage [19]. This was first discussed in [20] and then considered in numerous works, see e.g. Refs. [11, 21, 22, 23, 24]. At the hadronization temperature T_c, hadronic matter is formed in chemical equilibrium; subsequently for temperatures below the chemical freeze-out temperature T_{ch} (with $T_{ch} \leq T_c$) the inelastic processes become suppressed, while the elastic interactions mediated by frequent strong resonance formation and decays (e.g. $\pi\pi \to \rho \to \pi\pi$, $K\pi \to K^* \to K\pi$, $p\pi \to \Delta \to p\pi$ etc.) continue to occur. As a consequence, the experimentally observed ratios of particle multiplicities of those species i, which are stable against strong decays within the lifetime of the system, are fixed at T_{ch}. Namely, for $T < T_{ch}$ their effective particle numbers $\bar{N}_i = N_i + \sum_r d_{r\to i} N_r$ are frozen, where N_i is the particle number of the stable hadron i, N_r the particle number of resonance r and $d_{r\to i}$ gives the average number of hadrons i produced in the decay of resonance r. The above sum is taken over all the resonances that decay into hadron i within the lifetime of the hadronic stage. Each stable particle species i acquires therefore an effective chemical potential $\mu_i(T)$. The chemical potentials of the resonances can the be written as a combination $\mu_r = \sum_i d_{r\to i}\, \mu_i$ of the ones of the stable particles.

The EoS is obtained as a highly-involved relation between p, ϵ and all charge densities: this is due to the conservation of the effective number $\bar{N}_i$ of each stable particle species i below T_{ch}.

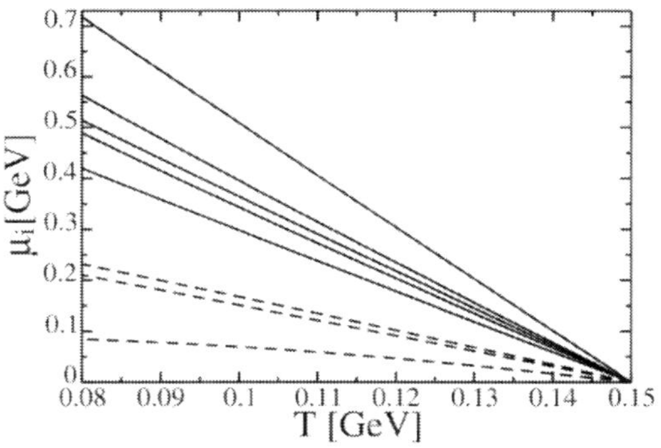

Figure 2. Temperature-dependence of the effective chemical potentials for selected hadronic states, for a chemical freeze-out temperature of $T_{ch} = 150$ MeV. The solid curves depict $\mu_i(T)$ for the baryons Ω^-, Ξ^-, Σ^-, Λ^0 and p from top to bottom, while the dashed curves show $\mu_i(T)$ for the mesons η, K^- and π^+ from top to bottom.

Assuming the conservation of entropy, the ratio between the effective particle number density and the entropy density $\bar{n}_i/s$ is fixed at T_{ch}. The following conditions, which imply that each $\bar{n}_i$ depends on all the effective chemical potentials $\mu_{i'}(T)$ (including $\mu_i(T)$), provide a practical tool to conserve all the $\bar{N}_i$ and to determine all the $\mu_i(T)$ for $T < T_{ch}$:

$$\frac{\bar{n}_i(T, \{\mu_{i'}(T)\})}{s(T, \{\mu_{i'}(T)\})} = \frac{\bar{n}_i(T_{ch}, \{0\})}{s(T_{ch}, \{0\})}. \tag{3}$$

Besides being needed for the determination of the EoS, the knowledge of all the $\mu_i(T)$ is necessary to determine the final state hadron abundances.

In this work, we consider as stable particle species the mesons π^0, π^+, π^-, K^+, K^-, K^0, $\overline{K}^0$ and η and the baryons p, n, Λ^0, Σ^+, Σ^0, Σ^-, Ξ^0, Ξ^- and Ω^- as well as their respective anti-baryons, i.e. in total 26 different states. For the chemical freeze-out temperature, we consider different values, namely $T_{ch}/\mathrm{MeV} = 145$, 150, 155 and 160. These are within the range of the T_c-values determined in lattice QCD [2, 3].

In figure 2, we show the behavior of the effective chemical potentials $\mu_i(T)$ as functions of the temperature, for some exemplary particle species in the case $T_{ch} = 150$ MeV. The $\mu_i(T)$ increase with decreasing T, as one would expect since the stable states become more populated by resonance decays at low temperatures. In the related published paper [14] we provide a parametrization of such chemical potentials, with a table of the parameters needed to obtain them.

4. Discussion and Conclusions

We obtained various equations of state by continuously combining our parametrization of the lattice QCD data [10] as a function of T with the HRG model either in full or in partial chemical equilibrium in the hadronic phase. The latter case has been studied for various T_{ch}-values. In hydrodynamic simulations, however, the EoS is used in the form $p(\epsilon, n_B)$, i.e. as a function of ϵ and n_B, as well as the results for the effective chemical potentials μ_i are required for the determination of the particle abundances. In figure 3, we show our results for the different equations of state $p(\epsilon)$ supplemented by the corresponding $T(\epsilon)$ for $n_B = 0$. In figure 3 we focus on the energy density regions, in which the confinement transition and the chemical freeze-out take place.

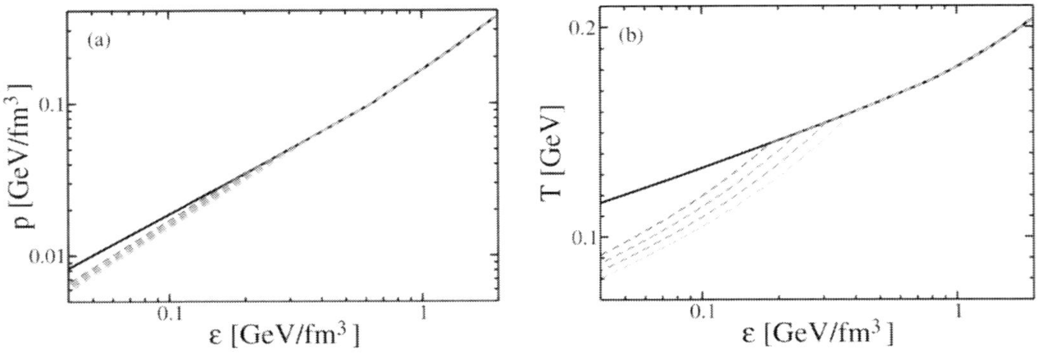

Figure 3. (Color online) (a) Equations of state $p(\epsilon)$ for $n_B = 0$ zoomed into the regions in ϵ, in which the confinement transition and the chemical freeze-out occur. The solid curve corresponds to the HRG model in chemical equilibrium. For the dashed curves, partial chemical equilibrium is included below ϵ_{ch}. The value of ϵ_{ch} depends on the value of the freeze-out temperature T_{ch}. We consider $T_{ch}/\text{MeV} = 145$, 150, 155 and 160 (from top to bottom in the figure, respectively). (b) Temperature vs. energy density $T(\epsilon)$ for the equations of state with chemical equilibrium (solid curve) and with partial chemical equilibrium (dashed curves) (labelling as in panel (a)).

The differences in $p(\epsilon)$ between chemical equilibrium (solid curve) and partial chemical equilibrium (dashed curves) in the hadronic phase turn out to be small, while the ϵ-dependence of T is significantly influenced, for $\epsilon < \epsilon_{ch}$, by the chemical freeze-out (see panel (b) in figure 3). Our results are collected in tabulated form and made available in [25]. For practical convenience, in [14] we also provide parameterizations of these numerical results as functions of ϵ.

The high precision in the provided parameterization is motivated by our goal to maintain thermodynamic consistency and continuity in the second derivatives up to a high numerical accuracy.

The temperature-dependence of the speed of sound c_s^2 obtained by numerical differentiation, is shown in figure 4 (solid curve) and compared to the lattice QCD results available from the WB-collaboration [9]. In the case of full chemical equilibrium, our curve agrees with the lattice QCD data within error-bars: in particular, we find a rather large $c_s^2(T)$ in the confinement transition region. This indicates that our EoS is rather stiff. The dashed curves correspond to the partial chemical equilibrium case. One observes a discontinuity in $c_s^2(T)$ at $T = T_{ch}$, which is characteristic for the chemical freeze-out. As expected, the behavior of $c_s^2(T)$ in the non-equilibrium situation is different from the trend seen in equilibrium lattice QCD thermodynamics.

In summary, we constructed QCD equations of state for vanishing net-baryon density based on recent continuum-extrapolated lattice QCD results in the physical quark mass limit at high T [10], continuously combined with a HRG model at low T. The latter was considered to be either in chemical or in partial chemical equilibrium. Focusing on the partial chemical equilibrium case, we studied different values for the chemical freeze-out temperature within its presently expected range [12].

In the present work, we focused ourselves to the $n_B = 0$ case. Therefore, our results can be used as an input in the hydrodynamic modeling of high-energy heavy-ion collisions at the LHC and at RHIC top beam energies and at mid-rapidity. We will address the equation of state in partial chemical equilibrium at finite density in a forthcoming publication.

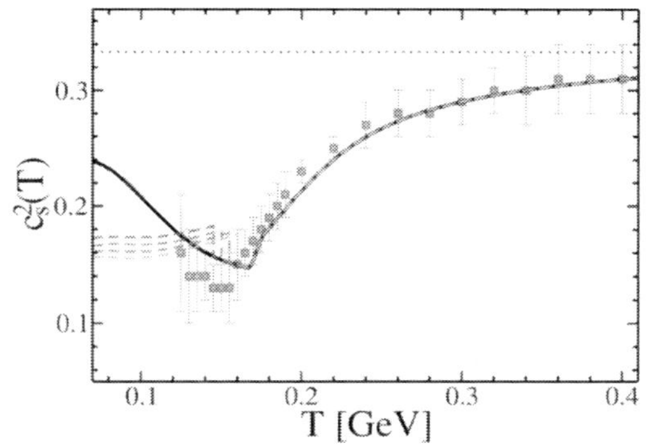

Figure 4. (Color online) Temperature-dependence of the squared speed of sound $c_s^2(T)$. The solid curve corresponds to the HRG model in chemical equilibrium. The symbols represent available equilibrium lattice QCD data from [9]. The dashed curves highlight $c_s^2(T)$ when partial chemical equilibrium is assumed in the hadronic phase. We consider T_{ch}/MeV= 145, 150, 155 and 160 (from top to bottom, respectively).

References

[1] Aoki Y, Endrodi G, Fodor Z, Katz S D, and Szabo K K 2006 *Nature* **443** 675
[2] Borsanyi S *et al.* 2010 *J. High Energy Phys.* **1009** 073
[3] Bazavov A *et al.* 2012 *Phys. Rev. D* **85** 054503
[4] Kolb P F and Heinz U in *Quark Gluon Plasma 3*, edited by Hwa R C and Wang X N 2004 *World Scientific, Singapore* p. 634
[5] Gale C, Jeon S, and Schenke B 2013 *Int. J. of Mod. Phys. A* **28** 1340011
[6] Bazavov A *et al.* 2009 *Phys. Rev. D* **80** 014504
[7] Cheng M *et al.* 2010 *Phys. Rev. D* **81** 054504
[8] Aoki Y, Fodor Z, Katz S D, and Szabo K K 2006 *J. High Energy Phys.* **0601** 089
[9] Borsanyi S *et al.* 2010 *J. High Energy Phys.* **1011** 077
[10] Borsanyi S *et al.* 2011 *Proc. of Sci.* **LATTICE2011** 201
[11] Hirano T and Tsuda K 2002 *Phys. Rev. C* **66** 054905
[12] Borsanyi S, Fodor Z, Katz S D, Krieg S, Ratti C, and Szabo K K 2013 *Phys. Rev. Lett.* **111** 062005
[13] Borsanyi S, Fodor Z, Katz S D, Krieg S, Ratti C, and Szabo K K 2014 `arXiv:1403.4576`
[14] Bluhm M *et al.* 2013 `arXiv:1306.6188`
[15] Karsch F, Redlich K, and Tawfik A 2003 *Phys. Lett. B* **571** 67
[16] Tawfik A 2005 *Phys. Rev. D* **71** 054502
[17] Eidelman S *et al.* [Particle Data Group] 2004 *Phys. Lett. B* **592** 1
[18] Huovinen P and Petreczky P 2010 *Nuc. Phys. A* **837** 26
[19] Teaney D 2002 `arXiv:nucl-th/0204023`
[20] Bebie H, Gerber P, Goity J L, and Leutwyler H 1992 *Nuc. Phys. B* **378** 95
[21] Rapp R 2002 *Phys. Rev. C* **66** 017901
[22] Kolb P F and Rapp R 2003 *Phys. Rev. C* **67** 044903
[23] Huovinen P 2008 *Eur. Phys. J. A* **37** 121
[24] Del Zanna L *et al.* 2013 `arXiv:1305.7052`
[25] `http://personalpages.to.infn.it/~ratti/EoS/Equation_of_State/Home.html`

Relativistic distribution function for particles with spin at local thermodynamical equilibrium

E Grossi[1]

[1] Università di Firenze and INFN Sezione di Firenze, Florence Italy

E-mail: grossi@fi.infn.it

Abstract. We present an extension of relativistic single-particle distribution function for weakly interacting particles at local thermodynamical equilibrium including spin degrees of freedom, for massive spin 1/2 particles. We infer, on the basis of the global equilibrium case, that at local thermodynamical equilibrium particles acquire a net polarization proportional to the vorticity of the inverse temperature four-vector field. The obtained formula for polarization also implies that a steady gradient of temperature entails a polarization orthogonal to particle momentum. The single-particle distribution function in momentum space extends the so-called Cooper-Frye formula to particles with spin 1/2 and allows to predict their polarization, and particularly of Lambda hyperons, in relativistic heavy ion collisions at the freeze-out.

1. Introduction

The single-particle distribution function is the main quantity in kinetic theory and its form at local thermodynamical equilibrium for relativistic, weakly interacting, gases is well known. For spinless particles, it is simply the Bose-Einstein distribution with x-dependent values of temperature and chemical potential (which can be defined as Bose-Jüttner distribution):

$$f(x,p) = \frac{1}{e^{\beta(x)\cdot p - \xi(x)} - 1} \tag{1}$$

where $\beta = \frac{1}{T_0} u$ is the inverse temperature four-vector, T_0 being the proper temperature measured by a comoving thermometer with the four-velocity u and $\xi = \mu_0/T_0$ is the ratio between the proper chemical potential μ_0 and T_0. The above formula has, as straightforward consequence, the invariant momentum spectrum at local thermodynamical equilibrium, the so-called Cooper-Frye formula [1]:

$$\varepsilon \frac{dN}{d^3p} = \int_\Sigma d\Sigma_\mu p^\mu f(x,p) \tag{2}$$

where Σ is a space-like 3-dimensional hypersurface. This formula is widespreadly used in e.g. relativistic heavy ion collisions to calculate hadronic spectra at the end of the hydrodynamical stage.

The distribution (1), multiplied by a degeneracy factor $(2S + 1)$, is also used for particles with spin (for fermions replacing the -1 with $+1$ in the denominator) being understood that f means the *total* particle density in phase space, i.e. summed over polarization states. However, in general, particles may not evenly populate the various polarization states and one may then

Journal of Physics: Conference Series **527** (2014) 012015 doi:10.1088/1742-6596/527/1/012015

wonder what is the appropriate extension of (1) in this case. Indeed, in this work, we will answer this question and provide a generalization of (1) and (2) including the spin degrees of freedom. We will argue that a non-even population of the polarization states arises when the inverse temperature four-vector field has a non-vanishing antisymmetric part of its gradient and calculate the polarization vector for massive spin 1/2 particles. Phenomenologically, this extension may have several interesting applications. For instance, it would make possible to predict the value of particle polarization in relativistic heavy ion collisions [2, 3, 4, 5, 6, 7] at the hydrodynamical decoupling, provided that local thermodynamical equilibrium applies to spin degrees of freedom as well. The detailed analysis and calculation can be found in [8].

2. Single particle distribution function with spin

At global thermodynamical equilibrium with finite angular momentum density the density operator is well known [9, 10] and reads:

$$\widehat{\rho} = \frac{1}{Z}\exp[-\widehat{H}/T + \mu\widehat{Q}/T + \boldsymbol{\omega}\cdot\widehat{\mathbf{J}}/T]\mathsf{P}_V \tag{3}$$

where $\boldsymbol{\omega}$ is a constant fixed vector whose physical meaning is that of an angular velocity and T is the global temperature, that is the temperature of a thermostat in contact with the system or that measured by a thermometer at rest with respect to the external inertial observer. The P_V operator is the projector operator onto localized states [11] which is needed in order to avoid the relativistic singularity at $r = c/\omega$. For this distribution, it has been shown the single-particle distribution function of an ideal relativistic Boltzmann gas of particles with spin S can be obtained from just statistical mechanics arguments [11]:

$$f(x,p)_{rs} = \mathrm{e}^{\xi}\,\mathrm{e}^{-\beta\cdot p}\;\frac{1}{2}\left(D^S([p]^{-1}\mathsf{R}_{\hat{\omega}}(i\omega/T)[p]) + D^S([p]^{\dagger}\mathsf{R}_{\hat{\omega}}(i\omega/T)[p]^{\dagger-1})\right)_{rs} \tag{4}$$

where $\beta = \frac{1}{T}(1, \boldsymbol{\omega}\times\mathbf{x})$ is the inverse temperature four-vector; λ is the fugacity; $[p]$ is the SL(2,C) matrix corresponding to the Lorentz transformation taking the time unit vector $\hat{t}$ into $\hat{p}$ (so-called standard transformation); D^S stands for the $(S,0)$ irreducible representation of SL(2,C); R is the SL(2,C) corresponding of a rotation, which is calculated for an imaginary angle $i\omega/T$.

For $S = 1/2$ case this formula can be express using the Dirac spinors $u(p)$ and $v(p)$, thanks to the Weyl's representation, with the normalization $\bar{u}(p)_r u(p)_s = 2m\delta_{rs}$ and with $C = i\sigma_2$ ($\boldsymbol{\sigma}$ being Pauli matrices) [12, 13] read:

$$\begin{pmatrix} u_+(p) \\ u_-(p)\end{pmatrix} = \sqrt{m}\begin{pmatrix} D^S([p]) \\ D^S([p]^{\dagger-1})\end{pmatrix} \qquad \begin{pmatrix} v_+(p) \\ v_-(p)\end{pmatrix} = \sqrt{m}\begin{pmatrix} D^S([p]C^{-1}) \\ D^S([p]^{\dagger-1}C)\end{pmatrix} \tag{5}$$

and $D^{1/2}(\mathsf{R}_{\hat{\omega}}(i\omega/T)) = \exp[(-\omega/T)\sigma_3/2]$ one can rewrite the (4), for spin 1/2 particles in the Boltzmann limit as:

$$f_{rs}(x,p) = \mathrm{e}^{\xi}\,\mathrm{e}^{-\beta\cdot p}\;\frac{1}{2m}\bar{u}_r(p)\exp[(\omega/T)\,\Sigma_z]u_s(p)\;, \tag{6}$$

being

$$\Sigma_z = \frac{1}{2}\begin{pmatrix} \sigma_3 & 0 \\ 0 & \sigma_3 \end{pmatrix}\;, \tag{7}$$

and similarly for antiparticles. The extension to Fermi-Dirac statistics of this formula is not a straightforward calculation like in the non-rotating case, hence we make an *ansatz* about this extension which reproduces the usual Fermi-Dirac distribution in the non-rotating case and, at the same time, has the (4) for $S = 1/2$ case as its Boltzmann limit.

$$\begin{aligned} f(x,p)_{rs} &= \frac{1}{2m}\bar{u}_r(p)\left(\exp[\beta\cdot p - \xi]\exp\left[-\frac{1}{2}\varpi_{\mu\nu}\Sigma^{\mu\nu}\right] + I\right)^{-1} u_s(p) \\ \bar{f}(x,p)_{rs} &= -\frac{1}{2m}\bar{v}_s(p)\left(\exp[\beta\cdot p + \xi]\exp\left[\frac{1}{2}\varpi_{\mu\nu}\Sigma^{\mu\nu}\right] + I\right)^{-1} v_r(p) \end{aligned} \tag{8}$$

whereas ϖ is the covariant form for the angular velocity:

$$\varpi_{\mu\nu} = (\omega/T)(\delta^1_\mu\delta^2_\nu - \delta^1_\nu\delta^2_\mu) = \sqrt{\beta^2}\Omega_{\mu\nu} \tag{9}$$

and $\Omega_{\mu\nu}$ turns out to be the acceleration tensor of the Frenet-Serret tetrad of the β field lines [11]. The above second equality holds for a rigid velocity field only [11, 14].

3. Single particle distribution function at local thermodinamical equilibrium

In quantum relativistic statistical mechanics, local thermodinamical equilibrium density reads:

$$\widehat{\rho}_{\rm LE} = \frac{1}{Z_{\rm LE}}\exp\left[-\int {\rm d}^3{\rm x}\ \beta_\nu(x)\widehat{T}^{0\nu}(x) - \frac{1}{2}\varpi_{\mu\nu}(x)\widehat{\mathcal{S}}^{0,\mu\nu}(x) - \xi(x)\widehat{\jmath}^0(x)\right] \tag{10}$$

and it is obtained by maximizing entropy $S = -{\rm tr}[\widehat{\rho}\log\widehat{\rho}]$ with the constraints of given local values of mean energy-momentum, angular momentum and charge density [15, 16]. As entropy is not conserved in nonequilibrium situation, the above operator breaks covariance (there cannot be invariant spatial integrals of non-conserved currents) and it is time dependent. This operator is used in derivations of the relativistic Kubo formulae of transport coefficients [17] and, in comparison with usual formulations, it has an additional term involving the spin tensor, obtained in Ref. [18]. In principle, all quantities at local thermodynamical equilibrium in quantum relativistic statistical mechanics should be calculated using (10) as density operator, including the covariant Wigner function of the Dirac field. However, the full calculation is quite complicated and goes beyond the scope of this work. At the lowest order of approximation, however, we know that the single particle distribution function must yield the same formal expression at global thermodynamical equilibrium with space-time dependent intensive thermodynamics functions, that is space-time dependent β, ϖ and ξ. Hence, the single-particle distribution functions (4) for $S = 1/2$ case become:

$$\begin{aligned} f(x,p)_{rs} &= \frac{1}{2m}\bar{u}_r(p)\left(\exp[\beta(x)\cdot p - \xi(x)]\exp\left[-\frac{1}{2}\varpi(x)_{\mu\nu}\Sigma^{\mu\nu}\right] + I\right)^{-1} u_s(p) \\ \bar{f}(x,p)_{rs} &= -\frac{1}{2m}\bar{v}_s(p)\left(\exp[\beta(x)\cdot p + \xi(x)]\exp\left[\frac{1}{2}\varpi(x)_{\mu\nu}\Sigma^{\mu\nu}\right] + I\right)^{-1} v_r(p) \end{aligned} \tag{11}$$

where β is the inverse temperature four-vector, ξ is the ratio between comoving chemical potential and temperature, $\Sigma^{\mu\nu} = (i/4)[\gamma^\mu, \gamma^\nu]$ are the generators of Lorentz transformations of 4-components spinors, $u(p)$ and $v(p)$ are the spinors solutions of the free Dirac equation.

Using this distribution functions we obtain the free Dirac part of the canonical stress-energy tensor as:

$$T^{\mu\nu}(x) = \int \frac{{\rm d}^3{\rm p}}{\varepsilon}\ p^\mu p^\nu \sum_r (f_{rr} + \bar{f}_{rr}) \tag{12}$$

and the current:

$$j^\mu(x) = \frac{1}{2}\int \frac{{\rm d}^3{\rm p}}{\varepsilon}\ p^\mu \sum_r \left[{\rm tr}\, f_{rr} - {\rm tr}\, \bar{f}_{rr}\right] \tag{13}$$

Journal of Physics: Conference Series **527** (2014) 012015 doi:10.1088/1742-6596/527/1/012015

both involving the traces of $f_{rs}, \bar{f}_{rs}$. If ϖ, which is adimensional in natural units, is small enough such that one can write the expansion:

$$\left(\mathrm{e}^{\beta(x)\cdot p\mp\xi(x)}\exp\left[\mp\frac{1}{2}\varpi(x):\Sigma\right]+I\right)^{-1}=\sum_{n=1}^{\infty}(-1)^{n+1}\mathrm{e}^{-n\beta(x)\cdot p\pm n\xi(x)}\exp\left[\pm\frac{n}{2}\varpi(x):\Sigma\right] \quad (14)$$

where : is a shorthand for the 2 tensor contraction, i.e. $\varpi(x):\Sigma=\varpi^{\mu\nu}(x)\Sigma_{\mu\nu}$, then it is possible to obtain an approximate expression of those traces for $\varpi_{\mu\nu}\ll 1$. Note that at full rotational equilibrium this condition amounts to require $\hbar\omega/KT\ll 1$ (natural units purposely restored) which is a normally fulfilled condition. Hence:

$$\begin{aligned}\mathrm{tr}\left(\exp\left[\pm\frac{n}{2}\varpi(x):\Sigma\right]\right) &\simeq \mathrm{tr}\left(I\pm\frac{n}{2}\varpi(x):\Sigma+\frac{n^2}{4}\varpi(x):\Sigma\,\varpi(x):\Sigma\right)\\ &= 4+\frac{n^2}{4}\varpi(x)^{\lambda\rho}\varpi(x)^{\sigma\tau}\,\mathrm{tr}\left(\Sigma_{\lambda\rho}\Sigma_{\sigma\tau}\right) \quad (15)\end{aligned}$$

where the tracelessness of Σ matrices has been used. By using known formulae for the traces of γ matrices it can be shown that:

$$\mathrm{tr}\left(\Sigma_{\lambda\rho}\Sigma_{\sigma\tau}\right)=g_{\lambda\sigma}g_{\rho\tau}-g_{\lambda\tau}g_{\rho\sigma}$$

whence the equation (15) becomes:

$$\mathrm{tr}\left(\exp\left[\pm\frac{n}{2}\varpi(x):\Sigma\right]\right)\simeq 4+\frac{n^2}{2}\varpi(x):\varpi(x)\,. \quad (16)$$

Therefore, we have:

$$\begin{aligned}\sum_r f_{rr} &\simeq \sum_{n=1}^{\infty}(-1)^{n+1}\mathrm{e}^{-n\beta(x)\cdot p+n\xi(x)}\left(4+\frac{n^2}{2}\varpi(x):\varpi(x)\right)\\ &= 4n_F+\frac{1}{2}n_F(1-n_F)(1-2n_F)\varpi(x):\varpi(x) \quad (17)\end{aligned}$$

where:

$$n_F=\frac{1}{\mathrm{e}^{\beta(x)\cdot p-\xi(x)}+1}$$

and similarly for $\sum_r \bar{f}_{rr}$ with the replacement $\xi\to-\xi$. One can then plug the equation (17) and its corresponding for $\sum_r \bar{f}_{rr}$ into equations (12) and (13) to obtain, e.g., for the charge density:

$$j^0(x)=2\int \mathrm{d}^3\mathrm{p}(n_F-\bar{n}_F)+\varpi(x):\varpi(x)\frac{1}{4}\int \mathrm{d}^3\mathrm{p}[n_F(1-n_F)(1-2n_F)-\bar{n}_F(1-\bar{n}_F)(1-2\bar{n}_F)]\,.$$

For $\varpi=0$ one recovers the usual expression; it is worth noting that the lowest order correction to charge density and stress-energy tensor is quadratic in ϖ, i.e. in $\hbar\omega/KT$ at equilibrium, then it's a small contribution.

While the general physical meaning of the fields β and ξ can be easily inferred from the equilibrium limit (β is the local inverse temperature four-vector field and ξ the ratio between the comoving chemical potential $\mu_0(x)$ and the local comoving temperature $T_0(x)=1/\sqrt{\beta^2}$), $\varpi_{\mu\nu}(x)$'s expression cannot be uniquely obtained from the equilibrium distribution. The reason of this ambiguity is that at rotational equilibrium, the tensor ϖ is:

$$\varpi_{\mu\nu}=\sqrt{\beta^2}\Omega_{\mu\nu} \quad (18)$$

where $\Omega_{\mu\nu}$ is the acceleration tensor of the Frenet-Serret tetrad of the β field lines [11] and [14]:

$$\varpi_{\mu\nu} = -\frac{1}{2}(\partial_\mu \beta_\nu - \partial_\nu \beta_\mu) \tag{19}$$

that is the relativistic generalization of angular velocity of fluid. In a nonequilibrium situation, the right hand sides of equations (19) and (18) differ and it is not obvious which one applies, perhaps neither. However, if the system is not too far from equilibrium, ϖ cannot be too distant from the right hand side of (19). Particularly, the difference must be of the 2nd order in the gradients of the β field (for instance: $(\partial \cdot \beta)(\partial_\mu \beta_\nu - \partial_\nu \beta_\mu)$ or second order gradients like $\partial_\mu \partial_\nu \beta^\lambda$) which vanish at equilibrium [14]. For the lowest-order formulation of relativistic hydrodynamics, the expression (19) is sufficient to determine the expression of spin-related quantities.

4. Polarization

The main consequence of the distribution functions (11) is that spin 1/2 particles get polarized at local thermodynamical equilibrium. The polarization four-vector, for a particle with mass m and four-momentum p is defined as:

$$\Pi_\mu = -\frac{1}{2}\epsilon_{\mu\rho\sigma\tau} S^{\rho\sigma} \frac{p^\tau}{m} \tag{20}$$

where $S^{\rho\sigma}$ is the mean value of the *total* angular momentum operator of the single particle. However the Levi-Civita tensor makes the orbital part of the total angular momentum irrelevant, so we are left with the spin-tensor density in the phase space $\mathcal{S}^{0,\rho\tau}(x,p)$ contribution only:

$$\langle \Pi_\mu(x,p) \rangle = -\frac{1}{2} \frac{1}{\sum_r f_{rr}} \epsilon_{\mu\rho\sigma\tau} \frac{\mathrm{d}\mathcal{S}^{0,\rho\sigma}(x,p)}{\mathrm{d}^3 p} \frac{p^\tau}{m} . \tag{21}$$

The *canonical* spin tensor can be computed using (11):

$$\mathcal{S}^{\lambda,\mu\nu}(x) = \frac{1}{2}\int \frac{\mathrm{d}^3\mathrm{p}}{2\varepsilon} \sum_{rs} \left(f_{rs}(x,p)\bar{u}_s(p)\{\gamma^\lambda, \Sigma^{\mu\nu}\} u_r(p) \right) - \left(\bar{f}_{rs}(x,p)\bar{v}_r(p)\{\gamma^\lambda, \Sigma^{\mu\nu}\} v_s(p) \right) . \tag{22}$$

We can write the spin tensor as:

$$\mathcal{S}^{\lambda,\mu\nu} = \frac{1}{2}\int \frac{\mathrm{d}^3\mathrm{p}}{\varepsilon} \left(p^\lambda \Theta^{\mu\nu} + p^\nu \Theta^{\lambda\mu} + p^\mu \Theta^{\nu\lambda} + p^\lambda \bar{\Theta}^{\mu\nu} + p^\nu \bar{\Theta}^{\lambda\mu} + p^\mu \bar{\Theta}^{\nu\lambda} \right) \tag{23}$$

where

$$\Theta^{\mu\nu} \equiv \sum_r \left(f_{rs}\bar{u}_s(p)\Sigma^{\mu\nu} u_r(p) \right) \qquad \bar{\Theta}^{\mu\nu} \equiv -\operatorname{tr}\left(\bar{f}_{rs}\bar{v}_r(p)\Sigma^{\mu\nu} u_s(p) \right) .$$

Altogether, the form (23) of the canonical spin tensor only depends on the fact that the distribution function $f_{rs}\bar{f}_{rs}$ are a superposition of an even number of γ matrices. The full antisymmetry of the indices is now apparent, although it was already contained in the operator definition ensuing from the properties of γ matrices. Thanks to trace cyclicity, Θ can be written as a derivative with respect to the ϖ tensor:

$$\begin{aligned}
\Theta^{\mu\nu} &= \sum_{n=1}^{\infty} (-1)^{n+1} \mathrm{e}^{-n\beta(x)\cdot p + n\xi(x)} \operatorname{tr}\left(\exp\left[\frac{n}{2}\varpi(x) : \Sigma\right] \Sigma^{\mu\nu} \right) \\
&= \sum_{n=1}^{\infty} (-1)^{n+1} \mathrm{e}^{-n\beta(x)\cdot p + n\xi(x)} \frac{1}{n}\frac{\partial}{\partial \varpi_{\mu\nu}} \operatorname{tr}\left(\exp\left[\frac{n}{2}\varpi(x) : \Sigma\right] \right) \\
&= \frac{\partial}{\partial \varpi_{\mu\nu}} \operatorname{tr}\left(\log\left\{ I + \mathrm{e}^{-\beta(x)\cdot p + \xi(x)} \exp\left[\frac{1}{2}\varpi(x) : \Sigma\right] \right\} \right)
\end{aligned} \tag{24}$$

where the (14) has been used. Then, using the approximation (16), the equation (24) becomes:

$$\begin{aligned}\Theta^{\mu\nu} &\simeq \sum_{n=1}^{\infty}(-1)^{n+1}\mathrm{e}^{-n\beta(x)\cdot p+n\xi(x)}\frac{1}{n}\frac{\partial}{\partial\varpi_{\mu\nu}}\left(4+\frac{n^2}{2}\varpi(x):\varpi(x)\right)\\ &= \sum_{n=1}^{\infty}(-1)^{n+1}\mathrm{e}^{-n\beta(x)\cdot p+n\xi(x)}n\varpi(x)^{\mu\nu} = n_F(1-n_F)\varpi(x)^{\mu\nu}\ . \end{aligned} \tag{25}$$

Likewise, it can be shown that:

$$\bar{\Theta}^{\mu\nu} \simeq \bar{n}_F(1-\bar{n}_F)\varpi(x)^{\mu\nu}\ . \tag{26}$$

Not far from equilibrium, for small value of $\varpi^{\mu\nu}$, the polarization vector becomes:

$$\langle\Pi_\mu(x,p)\rangle = -\frac{1}{4\sum_r f_{rr}}\epsilon_{\mu\rho\sigma\tau}\frac{1}{\varepsilon}\left(p^0\Theta^{\rho\sigma}+p^\sigma\Theta^{0\rho}+p^\rho\Theta^{\sigma 0}\right)\frac{p^\tau}{m} = -\frac{1}{2\sum_r f_{rr}}\epsilon_{\mu\rho\sigma\tau}\,\Theta^{\rho\sigma}\frac{p^\tau}{m} \tag{27}$$

and at the lowest orther in ϖ becomes:

$$\langle\Pi_\mu(x,p)\rangle \simeq \frac{1}{16}\epsilon_{\mu\rho\sigma\tau}\,(1-n_F)\left(\partial^\rho\beta^\sigma-\partial^\sigma\beta^\rho\right)\frac{p^\tau}{m} = \frac{1}{8}\epsilon_{\mu\rho\sigma\tau}\,(1-n_F)\partial^\rho\beta^\sigma\frac{p^\tau}{m} \tag{28}$$

where $n_F = 1/(\exp(\beta(x)\cdot p - \xi(x))+1)$ is the Fermi-Dirac distribution function.

The above formula has the remarkable consequence that quasi-free particles get polarized not only in a vorticous flow (what was pointed out in previous works [19]), but also in a steady temperature gradient without velocity flow, i.e. when $\nabla\beta^0 \neq 0$.In the non-relativistic the predicted polarization reads:

$$\Pi = (\Pi^0, \mathbf{\Pi}) = (1-n_F)\frac{\hbar p}{8mKT^2}(0, \nabla T \times \hat{\mathbf{p}})$$

which is usually tiny but could be relevant in some extreme situations.

It may be of interest, e.g. for relativistic heavy ion collisions, to calculate the space-integrated mean polarization vector. For a three-dimensional spacelike hypersurface Σ, one has:

$$\langle\Pi_\mu(p)\rangle \equiv \frac{\displaystyle\int \mathrm{d}\Sigma_\lambda\,\frac{p^\lambda}{\varepsilon}(-1/2)\epsilon_{\mu\rho\sigma\tau}\frac{\mathrm{d}\mathcal{S}^{0,\rho\sigma}}{\mathrm{d}^3\mathrm{p}}\frac{p^\tau}{m}}{\displaystyle\int \mathrm{d}\Sigma_\lambda\,\frac{p^\lambda}{\varepsilon}\sum_r f_{rr}(x,p)} \simeq \frac{1}{8}\epsilon_{\mu\rho\sigma\tau}\frac{p^\tau}{m}\frac{\displaystyle\int \mathrm{d}\Sigma_\lambda\,p^\lambda\,n_F(1-n_F)\partial^\rho\beta^\sigma}{\displaystyle\int \mathrm{d}\Sigma_\lambda\,p^\lambda n_F}\ . \tag{29}$$

For antiparticles, one gets the same formula, with $\bar{n}_F = 1/(\exp(\beta(x)\cdot p+\xi(x))+1)$ replacing n_F. Here an important comment is in order: the fact that local thermodynamical equilibrium implies the same orientation for the polarization vector of particles and antiparticles (unlike e.g. in the electromagnetic field) is a general outcome and does not depend on the introduced approximations. It stems from the fact that the spin tensor, as well as the angular momentum, is a charge-conjugation even operator, or, more simply stated, that thermal and mechanical effects do not "see" the internal charge of the particles.

This formula (29) may be used to predict the polarization of particles, in particular Λ hyperons, produced at the freeze-out (primary particles) in a relativistic heavy-ion collision, after the hydrodynamical evolution.

In a heavy ion collision the polarization could exist also in plasma phase if it were described as weakly interacting particles, since is an effect of the local equilibrium. Nevertheless it's not the case for QGP and it's unknown the effect of strong coupling on the polarization. However at the freeze-out it's assumed, to compute the spectra that the fluid instantaneously produce free streaming hadrons distributed with a free particle-distribution function, then we can think that also happened with the spin's degrees of freedom. In a more realistic model of the particle production the interaction between hadrons turn of slowly and this polarization effect could be modified.

The polarization of the primary particles emitted is proportional to the thermal vorticity (19) at freeze-out hypersurface and could give an observable contribution [20].

The observation of this effect would be a confirmation of the achievement of the local thermodynamical equilibrium also to spin degrees of freedom and would also indicate that vorticity and circulation may persist in the hydrodinamical evolution up to freeze-out.

References

[1] Cooper F and Frye G 1974 *Phys. Rev. D* **10** 186
[2] Baznat M, Gudima K, Sorin A, and Teryaev O 2013 *Phys. Rev. C* **88** 061901
[3] Huang X G, Huovinen P, and Wang X N 2011 *Phys. Rev. C* **84** 054910
[4] Gao J H, Chen S W, Deng W T, Liang Z T, Wang Q, and Wang X N 2008 *Phys. Rev. C* **77** 044902
[5] Gao J H, Liang Z T, Pu S, Wang Q, and Wang X N 2012 *Phys. Rev. Lett.* **109** 232301
[6] Betz B, Gyulassy M, and Torrieri G 2007 *Phys. Rev. C* **76** 044901
[7] Csernai L, Magas V K, and Wang D J 2013 *Phys. Rev. C* **87** 034906
[8] Becattini F, Chandra V, Del Zanna L, and Grossi E 2013 *Annals Phys.* **338** 32
[9] Landau L and Lifshitz L 1980 *Statistical Physics, Part 1* (Oxford, England: Pergamon Press)
[10] Vilenkin A 1980 *Phys. Rev. D* **21** 2260
[11] Becattini F and Tinti L 2010 *Annals Phys.* **325** 1566
[12] Moussa P and Stora R 1966 *Angular Analysis of Elementary Particle Reactions*, in Proceedings of the 1966 International School on Elementary Particles, Hercegnovi (Gordon and Breach, New York, London)
[13] Becattini F and Ferroni L 2007 *Eur. Phys. J. C* **52** 597
[14] Becattini F 2012 *Phys. Rev. Lett.* **108** 244502
[15] Zubarev D N 1971 *Non Equilibrium Statistical Thermodynamics* (Nauk, Moscow english translation: New York, Consultant Bureau, 1974)
[16] Zubarev D N , Morozov V G, and Röpke G 1997 *Statistical Mechanics of Nonequilibrium Processes. Volume 2: Relaxation and Hydrodynamic Processes* (Akademie Verlag, Berlin)
[17] Hosoya A, Sakagami M, and Takao M 1984 *Annals Phys.* **154**, 229
[18] Becattini F and Tinti L 2011 *Phys. Rev. D* **84** 025013
[19] Becattini F and Piccinini F 2008 *Annals Phys.* **323** 2452
[20] Becattini F, Csernai L, and Wang D J 2013 *Phys. Rev. C* **88** 034905

Journal of Physics: Conference Series **527** (2014) 012016 doi:10.1088/1742-6596/527/1/012016

Transport coefficients of Quark-Gluon Plasma in a Kinetic Theory approach

A Puglisi[1,2], **S Plumari**[1,2], **F Scardina**[1,2], **V Greco**[1,2]

[1] Department of Physics and Astronomy, University of Catania, Via S. Sofia 64, I-95125 Catania (Italy)

[2] Laboratorio Nazionale del Sud, INFN-LNS, Via S. Sofia 63, I-95125 Catania (Italy)

E-mail: puglisia@lns.infn.it

Abstract. One of the main results of heavy ions collision at relativistic energy experiments is the very small shear viscosity to entropy density ratio of the Quark-Gluon Plasma, close to the conjectured lower bound $\eta/s = 1/4\pi$ for systems in the infinite coupling limit. Transport coefficients like shear viscosity are responsible of non-equilibrium properties of a system: Green-Kubo relations give us an exact expression to compute these coefficients. We computed shear viscosity numerically using Green-Kubo relation in the framework of Kinetic Theory solving the relativistic transport Boltzmann equation in a finite box with periodic boundary conditions. We investigated different cases of particles, for one component system (gluon matter), interacting via isotropic or anisotropic cross-section in the range of temperature of interest for HIC. Green-Kubo results are in agreement with Chapman-Enskog approximation while Relaxation Time approximation can underestimates the viscosity of a factor 2. Another transport coefficient of interest is the electric conductivity σ_{el} which determines the response of QGP to the electromagnetic fields present in the early stage of the collision. We study the σ_{el} dependence on microscopic details of interaction and we find also in this case that Relaxation Time Approximation is a good approximation only for isotropic cross-section.

1. Introduction

High energy Heavy Ion Collisions (HIC) represent the only way to have experimental access in the high temperature and small baryon density region of the nuclear matter phase diagram. Relativistic Heavy Ion Collider (RHIC) at BNL and Large Hadron Collider (LHC) at CERN have produced a new state of matter, the Quark-Gluon plasma, in which quarks and gluons are not confined. The azimuthal asymmetry in momentum space [1], the ellptic flow v_2, has shown that the QGP has a very small shear viscosity to entropy density ratio η/s [2], very close to the conjectured lower-bound limit $\eta/s = 1/4\pi$ [3] for a strongly interacting system in the limit of infinite coupling.

Calculation of transport coefficients is of great interest to characterize the non-equilibrium dynamics and dissipative effects of QGP created in HIC.

While shear and bulk viscosity are studied enough, only very recently electric conductivity σ_{el} has captured large attention due to the strong electric field created in the collision region [4, 5]. Electric conductivity represents the response of the system to the applied electric field: this coefficients is responsible for the production of an electric current generated by the charged particles (quarks) in the early stage of the collision and for collective flow due to the presence of such fields [6].

The paper is organized as follows. In section 2, we discuss the method for calculating the shear viscosity η using the Green-Kubo relation in the framework of transport theory solving the relativistic Boltzmann equation and, after a brief overview of analytical approximation like Relaxation Time Approximation (RTA) and Chapman-Enskog (CE) scheme [7, 8], we show the comparison between our results using the Green- Kubo relation and these schemes of approximation for anisotropic cross-section and for a more realistic case of anisotropic and energy-dependent cross-section. The knowledge of the exact analytical formula for shear viscosity leads to the development of a transport approach at fixed η/s in order to have a comparison with experimental data in the same way of hydrodynamics models [9, 10, 11]. In section 3, we discuss how to compute the electric conductivity σ_{el} in the box for a system of quarks, antiquarks and gluons, and we show its dependence on anisotropy of the cross-section comparing numerical results with RTA.

2. Green-Kubo relation for shear viscosity

Green-Kubo relation gives an exact expression to compute transport coefficient like shear and bulk viscosity, electric and thermal conductivity [12]. Each coefficient can be expressed in terms of correlation function of a flux in thermal equilibrium. For shear viscosity Green-Kubo relation assumes the following form [8]

$$\eta = \frac{1}{T}\int_0^\infty dt \int_V d^3x \, \langle \pi^{xy}(\mathbf{x},t)\pi^{xy}(\mathbf{0},0)\rangle \tag{1}$$

where T is the temperature, π^{xy} is the xy component of the energy-momentum tensor, while $\langle \ldots \rangle$ denotes the ensamble average. In this work we compute numerically the correlation function $\langle \pi^{xy}(t)\pi^{xy}(0)\rangle$ solving the ultrarelativistic Boltzmann transport equation for a particle system in a static box of volume V with periodic boundary conditions at thermal equilibrium. The parton cascade developed [13] solves the Relativistic Boltzmann-Vlasov equation:

$$p^\mu \partial_\mu f(x,p) + M(x)\partial_\mu M(x)\partial_p^\mu f(x,p) = \mathcal{C}(x,p) \tag{2}$$

where $f(x,p)$ is the distribution function for on-shell particles with mass $M(x)$, $\mathcal{C}(x,p)$ is the collision integral that for a one-component system can be written as

$$\mathcal{C}(x,p) = \int_2 \int_{1'} \int_{2'} (f_{1'}f_{2'} - f_1 f_2)\,|\mathcal{M}_{1'2'\to 12}|\delta^4(p_1+p_2-p_1'p_2') \tag{3}$$

with $\int_j = \int d^3p_j/(2\pi)^3 2E_j$, $\mathcal{M}$ denotes the transition matrix for the elastic process linked to the differential cross-section $|\mathcal{M}|^2 = 16\pi s(s-4M^2)d\sigma/dt$.

In this work we solve numerically the collision integral to evaluate shear viscosity using Green-Kubo relation and compare this one with the results of the RTA and CE approximations. The dynamic of the system is simulated via Monte Carlo methods based on the stochastic interpretation of the transition [14] according to which collision probability of two particles in a cell of volume ΔV_{cell} and time-step Δt_{step} is

$$P = v_{rel}\frac{\sigma_{tot}}{N_{test}}\frac{\Delta t}{\Delta V_{cell}} \tag{4}$$

where σ_{tot} is the total cross-section and $v_{rel} = \sqrt{s(s-4M^2)}/2E_1E_2$ is the relative velocity of incoming particles. The number of test-particles N_{test} is chosen to reproduce known quantities, such energy density ϵ and pressure P, and reduces statistical fluctuations.

The shear component of the energy-momentum tensor is given by

$$\pi^{xy}(x,t) = T^{xy}(x,t) = \int \frac{d^3p}{(2\pi)^3}\frac{p^x p^y}{E} f(x,p;t). \tag{5}$$

It is useful to write the time correlation function $\langle \pi^{xy}(t)\pi^{xy}(0)\rangle$ as follows

$$\langle \pi^{xy}(t)\pi^{xy}(0)\rangle = \left\langle \lim_{T_{max}\to\infty} \frac{1}{T_{max}} \int_0^{T_{max}} dt'\, \pi^{xy}(t+t')\pi^{xy}(t') \right\rangle \tag{6}$$

where T_{max} is the maximum time chosen in simulation, $N_{T_{max}} = T_{max}/\Delta t$ is the maximum number of time steps, $i\Delta t = t$ while $\langle \ldots \rangle$ denotes the average over numerically generated events.

In figure 1 is shown an example of time fluctuations of $\pi^{xy}(t)$ for a system in thermal equilibrium: the average $\langle \pi^{xy}\rangle = 0$ red dashed line. In figure 2 we show the correlation function $\langle \pi^{xy}(t)\pi^{xy}(0)\rangle$ as a function of time for different values of total cross-section σ_{tot}. In these calculations the particles are distributed uniformly in coordinate space while in momentum space according to thermal distribution function $f(p) \sim e^{-E/T}$ ($T = 0.4$ GeV). In the particular case of figure 2 the particles are massless and interact via an isotropic energy-independent differential cross-section. As we can see, $\langle \pi^{xy}(t)\pi^{xy}(0)\rangle$ is a decreasing exponential function $\langle \pi^{xy}(t)\pi^{xy}(0)\rangle = \langle \pi^{xy}(0)\pi^{xy}(0)\rangle e^{-t/\tau}$ so Green-Kubo relation equation (1) simplifies to

$$\eta = \frac{V}{T}\langle \pi^{xy}(0)\pi^{xy}(0)\rangle \tau \tag{7}$$

where τ is determined performing a fit on the temporal range in which the time correlation function is a decreasing exponential [13, 15, 16, 17].

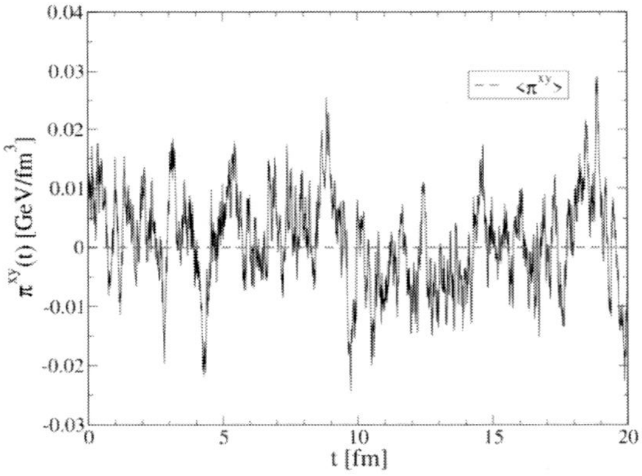

Figure 1. (Color online) xy component of energy momentum tensor as a fucntion of time at thermal equilibrium. Average value of $\langle \pi^{xy}\rangle = 0$ at thermal equilibrium (red dashed line).

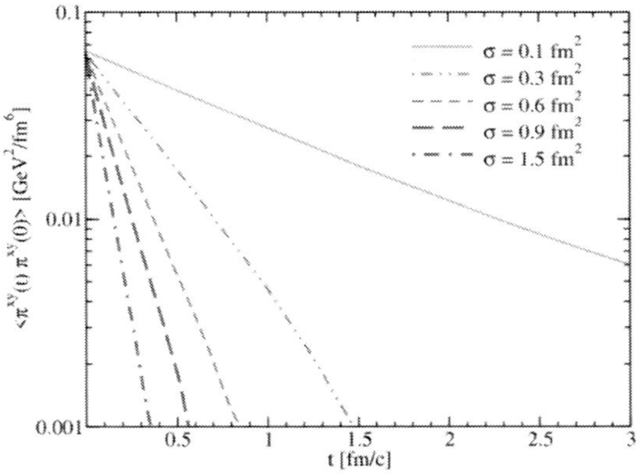

Figure 2. (Color online) Correlation function $\langle \pi^{xy}(t)\pi^{xy}(0)\rangle$ as a function of time for different value of total cross-section σ_{tot}. These results are for massless gluons interacting via isotropic cross-section; temperature is fixed to $T = 0.4$ GeV.

2.1. Shear viscosity for anisotropic scatterings

In this section we compare Green-Kubo results for shear viscosity η, for a system of massless gluon interacting with an anisotropic cross-section, with analytical expressions already present in Literature: RTA and CE scheme.

We choose a typical pQCD-inspired cross-section with the infrared singularity regularized by a Debye thermal mass m_D:

$$\frac{d\sigma}{dt} = \frac{9\pi\alpha_s^2}{2}\frac{1}{(t-m_D^2)^2}\left(1+\frac{m_D^2}{s}\right) \tag{8}$$

where s, t are the Mandelstam variables. This kind of cross-section is typically used in parton cascade approaches [14, 18, 19, 20, 21, 22]. The total cross-section is $\sigma_{tot} = 9\pi\alpha_s^2/m_D^2$, which is energy and temperature independent. In equation (8) m_D is a parameter that regulates the anisotropy of the scattering: we change it to modify the anisotropy but keeping constant the total cross-section.

The transport cross-section is:

$$\sigma_{tr}(s) = \int dt \, \frac{d\sigma}{dt} \sin^2\Theta = \sigma_{tot} h(a) \tag{9}$$

where $h(a) = 4a(1+a)[(2a+1)\log(1+1/a) - 2]$ and $a = m_D^2/s$. For $m_D \to \infty$ the function $h(a) \to 2/3$ and we recover the isotropic limit $\sigma_{tr} = (2/3)\sigma_{tot}$.

The difference between the two schemes of approximation, RTA and CE, resides in the different way of approximating the collision integral (3). The starting hypothesis of the RTA is that the collision integral can be approximated by $\mathcal{C}[f] = -(f - f^{eq})/\tau$, where τ is the so called relaxation time, and one can obtain the following relation:

$$\eta_{RTA} = \frac{4}{15}\rho\langle p\rangle\tau = \frac{4}{15}\frac{\langle p\rangle}{\sigma_{tot}} \qquad \eta^*_{RTA} = \frac{4}{15}\rho\langle p\rangle\tau_{tr} = \frac{4}{15}\frac{\langle p\rangle}{\langle h(a)v_{rel}\rangle\sigma_{tot}} \tag{10}$$

where $\langle p\rangle$ is the average momentum, $\tau^{-1} = \langle\rho\sigma v_{rel}\rangle$ is the relaxation time while $\tau_{tr}^{-1} = \langle\rho\sigma_{tr}v_{rel}\rangle$ is the transport relaxation time . The modified RTA η^*_{RTA} differs from the simple one only in considering the transport cross-section σ_{tr} instead of the total one.

On the other hand, the CE approach is based on a variational approach to the collision integral [23, 13] and one can obtain the following formula:

$$\eta_{CE} = \frac{4}{15}\rho\langle p\rangle\tau_\eta = \frac{4}{15}\frac{\langle p\rangle}{g(a)\sigma_{tot}} \quad , \quad g(a) = \frac{1}{50}\int dy\, y^6 \left[(y^2+\frac{1}{3})K_3(2y) - yK_2(2y)\right] h\left(\frac{a^2}{y^2}\right) \tag{11}$$

with K_n being the Bessel functions and $a = m_D/2T$. Note that equation (11) is valid only for massless particles (see [13] for the general formula). The main difference between the two schemes is that in CE the shear viscosity depends in a natural way on the transport cross-section.

In figure 3 we show shear viscosity η as a function of Debye mass m_D, that represent the anisotropy of the cross-section, for three different values of temperature $T = 0.3, 0.4, 0.5$ GeV. Circles are Green-Kubo results and they are not affected by any kind of approximations. On the left panel lines represent the modified RTA equation (10) while on the right CE equation (11): as we can see, the modified RTA can understimate η of about a factor 2 for $m_D \leq 1.5$ GeV however it is in agreement with the isotropic limit; CE approximation is able to account for the correct value of η for all the range of temperature and anisotropy explored. This means that the correct formula describing the dependence of shear viscosity on microscopic details is just the CE approximation.

2.2. Shear viscosity of a gluon plasma

In this section we investigate the more realistic case of shear viscosity of a gluon plasma. We consider massless gluons in thermal equilibrium interacting via two-body collisions corresponding to direct u and t channels:

$$\frac{d\sigma^{gg\to gg}}{dq^2} = 9\pi\alpha_s^2\frac{1}{(q^2+m_D^2)^2}, \quad \alpha_s(T) = \frac{4\pi}{11\log\left(\frac{2\pi T}{\Lambda}\right)^2}, \quad \Lambda = 200\,MeV \tag{12}$$

where $m_D = T\sqrt{4\pi\alpha_s}$ is Debye mass according to Hard-Thermal-Loop (HTL) calculations, and α_s is the pQCD running coupling constant.

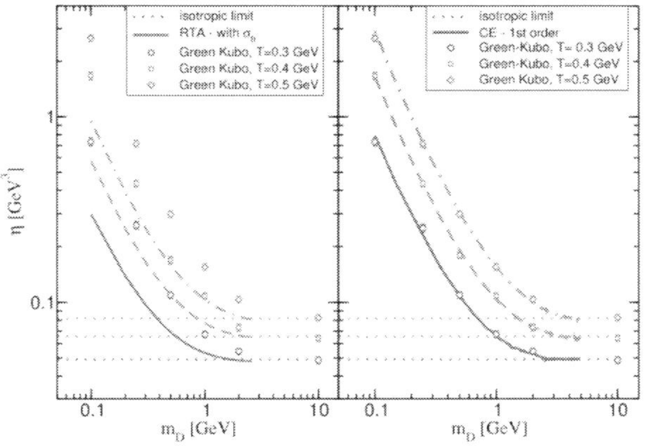

Figure 3. (Color online) Shear viscosity η as a function of the Debye mass m_D for three different value of temperature $T = 0.3$ GeV (blue solid line), $T = 0.4$ GeV (red dashed line) and $T = 0.5$ GeV (green dash-dotted line). Dotted lines represent the isotropic limit.

The total cross-section in this scheme is energy and temperature dependent:

$$\sigma_{tot} = \frac{9\pi\alpha_s^2}{m_D^2}\frac{s}{s+m_D^2}. \tag{13}$$

For this realistic case we have computed the η/s as a function of temperature in the range $0.2 \leq T \leq 1.0$ GeV using Green-Kubo formula. Entropy density s/T^3 is constant as for free gas. In figure 4 Green-Kubo results (dark circles) are compared with the analytical expression given by: CE (orange dashed line), modified RTA (dark dash-dotted line), simple RTA (green dotted line) and next-to-leading-log order (NLL) (red line) [24]. Ref. [24] is a perturbative calculation which includes also the s channel and all interference terms and considers a gluon propagator with both longitudinal and transverse components. We find again that CE approximation is in excellent agreement with Green-Kubo results while the modified RTA significantly understimates η/s by about a 20% at $T \sim 0.2$ GeV. We find also a good agreement for $T \geq 0.4$ GeV between Green-Kubo, CE and NLL from [24]: this means that the u and t channels are the dominant contribution to shear viscosity η.

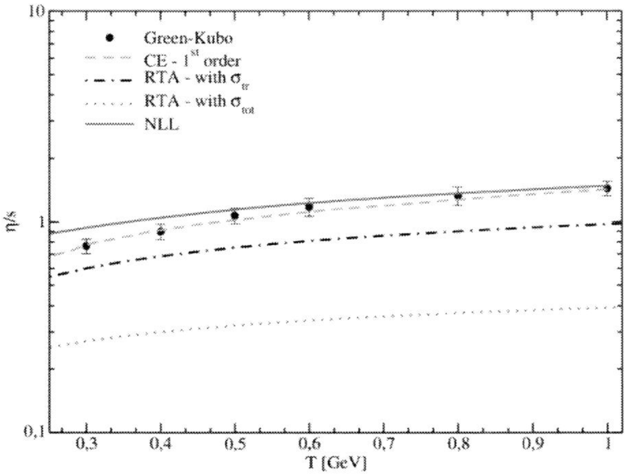

Figure 4. (Color online) Shear viscosity to entropy densisty ratio η/s of a gluon plasma interacting through the differential cross-section in equation (8) as a function of temperature T. Green dotted line is the simple RTA, dark dot-dashed line is the modified RTA, orange dashed line is CE while red line represents results of [24].

We note that the agreement between CE and Green-Kubo supplies a relatively simple analytical expression that can allow one to develop a kinetic transport theory at fixed $\eta/s(T)$ with much higher accuracy with respect to first attempt in RTA [18]: the idea is to change locally the total cross-section in equation (11) keeping constant the value η/s[9, 10, 11].

3. Electric conductivity

Computing transport coefficients of the QGP is mandatory to study non-equilibrium and dissipative effects during the space-time evolution of the fireball created in Heavy Ion Collisions experiments at ultrarelativistic energy. Another transport coefficients of interest is the electric conductivity σ_{el} of the QGP: it represents the response of the system to an applied electric field.

In this section we compute the electric conductivity for a system of massive particles (quark, antiquark and gluons) in a box in thermal equilibrium with an external electric field E. To simulate a constant electric field it is sufficient to modify the equation of motion of each particle as follows [4]:

$$\frac{d}{dt} p_z^i = q_i e E_z \tag{14}$$

where q_j is the charge of the particles and we have chosen the electric field along the $z-$direction. The presence of such electric field will develop an electric current whose density is given by:

$$j_z(t) = \frac{1}{V} \sum_i \frac{e q_i p_z^i(t)}{m_i} \tag{15}$$

where the sum runs over particles of mass m and V is the volume of the system.

In figure 5 we show electric density current $j_z(t)$ as a function of time for different value of the applied electric field: the current saturates towards the equilibrium value proportional to eE. Electric conductivity can be computed as the ratio between the equilibrium current j_z^{eq} and the applied electric field:

$$\sigma_{el} = \frac{j_z^{eq}}{E_z}. \tag{16}$$

In figure 6 we plot electric conductivity to temperature ratio as a function of the applied electric field: as we can see, σ_{el}/T is a constant in the range of eE explored, in agreement with other results obtained in [4].

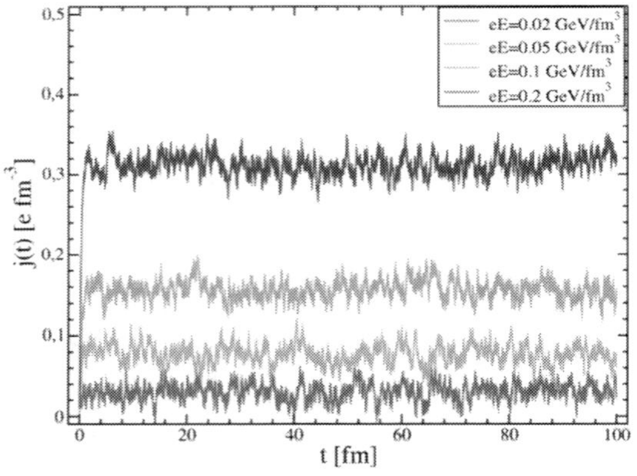

Figure 5. (Color online) Electric current j_z as a function of time for different values of the applied electric field.

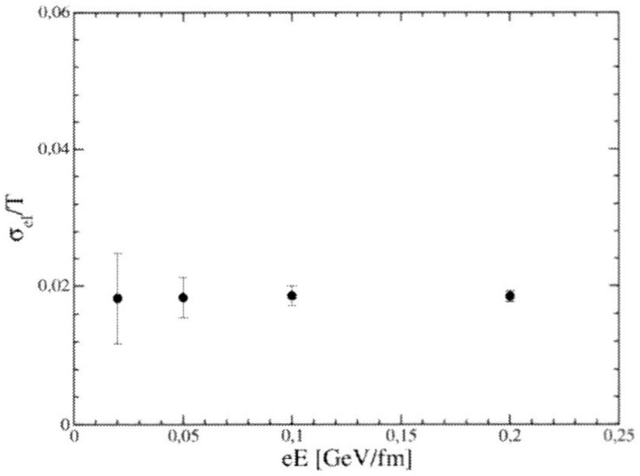

Figure 6. Electric conductivity σ_{el} over temperature as a function of the applied electric field.

3.1. Electric conductivity: anisotropic case

In this section our aim is to study the dependence of σ_{el} on microscopic details, in particular on anisotropy of the cross-section, so we consider a quark-gluon plasma interacting via anisotropic cross-section equation (8): in this simulations we fixed $\sigma_{tot} = 10$ mb, $T = 0.3$ GeV, $eE = 0.06$ GeV/fm and $m_q = m_g = 0.4$ GeV, typical masses of quasiparticle model for $T = 2.0\,T_c$ [25].

The electric conductivity is expected to be described by the well known Drude model in kinetic theory according to:

$$\sigma_{el} = \sum_j \frac{e^2 q_j^2 \rho_j \tau_j}{m_j} \tag{17}$$

where the sum runs over charged species, ρ is the particle density, τ is the collision time. For a multi-component system, interacting via the same cross-section, one can generalize the transport relaxation time as follows:

$$\tau_{tr,i}^{-1} = \sum_j \langle \rho_j \sigma_{ij} v_{rel}^{ij} \rangle = \sigma_{tot} \langle v_{rel} h(a) \rangle (\rho_q + \rho_{\bar{q}} + \rho_g) \tag{18}$$

where j runs over the species, σ_{ij} is the cross-section for a ij collision and $\rho_{q,\bar{q},g}$ is the particle density.

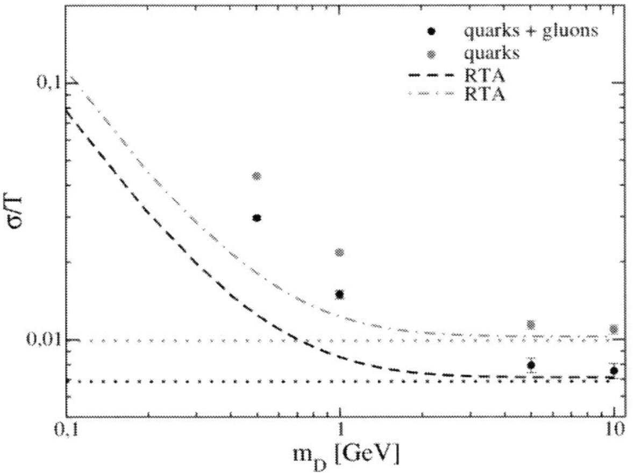

Figure 7. (Color online) Electric conductivity as a function of Debye mass m_D. Red circles are obtained considering only quarks while dark circles quarks and gluons. Dashed lines represent RTA for the two system. Red and dark dotted lines are the isotropic limit for the two systems.

In figure 7 we show numerical results of electric conductivity as a function of Debye mass m_D for a system of only quarks (red circles) and a system of quarks and gluons (dark circles). Dashed lines represent RTA while dotted lines are the isotropic limit that are reached for $m_D \to \infty$. The presence of gluons decreases the value of electric conductivity because they have not electric charge but contribute to the collision integral so they represent a resistance. As we can see from figure 7 RTA is a good approximation only in the isotropic limit while for $m_D \leq 1$ GeV it understimates the results of about a factor 2: the deviation between Gree-Kubo and RTA appears to be similar but even larger than the one seen also for the shear viscosity.

4. Conclusions

Recent developments in the physics of strong interactions at finite T are focusing many efforts on the calculation of transport coefficients being interested in the non-equilibrium dynamics of the QGP created in HIC: shear viscosity η is linked directly to the physical observable of elliptic flow $v_2(p_T)$. There are several approximation schemes that relate the viscosity η to the microscopic details such as cross-section. The validity of such approximations was not yet investigated. We developed a method to solve numerically the Green-Kubo formula for shear viscosity for a relativistic Boltzmann gas. Our objective was to compare the CE and RTAs to Green-Kubo results that in principle should give the correct results.

We have found that, for the realistic case of massless gluons interacting via pQCD-like cross-section, CE approximation is in agreement with Green-Kubo results at the level of 4% [13]: this agreement supplies an analytical expression that can allow one to develop kinetic transport theory at fixed viscosity with very good precision while RTA in general understimates the viscosity even by a factor 2.

We have studied the microscopic dependence of electric conductivity σ_{el}: investigating the properties of such coefficient is very interesting in order to quantify the response of the QGP to the initial electromagnetic fields present in the central region during the collions. We have found that RTA understimates the correct values of electric conductivity σ_{el} of about 50% fo anisotropic cross-section ($m_D \leq 2$ GeV) while it is a good approximation in the isotropic limit $m_D \to \infty$.

In the next future it would be interesting to apply the methods for the calculation of η/s and σ_{el} in order to explore quasiparticle models, where till now only RTA have been considered [25].

Acknowledgments

V Greco and F Scardina acknowledge ERC-StG2010 under the QGPDyn Grant.

References

[1] Romatschke P and Romatschke U 2007 *Phys. Rev. Lett.* **99** 172301
[2] Song H and Heinz U W 2008 *Phys. Rev. C* **78** 024902
[3] Kovtun P K, Son D T, and Starinets A O 2005 *Phys. Rev. Lett.* **94** 111601
[4] Cassing W, Linnyk O, Steinert T, and Ozvenchuk V 2013 *Phys. Rev. Lett.* **110** 182301
[5] Marty R, Bratkovskaya E, Cassing W, Aichelin J, and Berrehrah H 2013 *Phys. Rev. C* **88** 045204
[6] Tuchin K 2013 *Adv. High Energy Phys.* **2013** 490495
[7] Reif F 1965 *Fundamentals of Statistical and Thermal Physics* (McGraw-Hill, New York)
[8] de Groot S R, van Leeuwen W A, and Weert C 1980 *Relativistic Kinetic Theory, Principles and Applications* (North-Holland, Amsterdam)
[9] Plumari S, Greco V, and Csernai L P 2013 arXiv:1304.6566
[10] Ruggieri M, Scardina F, Plumari S, and Greco V 2013 *Phys. Lett. B* **727** 177
[11] Ruggieri M, Scardina F, Plumari S, and Greco V 2013 arXiv:1312.6060
[12] Green M S 1954 *J. Chem. Phys.* **22** 398; Kubo R 1957 *J. Phys. Soc. Jpn.* **12** 570
[13] Plumari S, Puglisi A, Scardina F, and Greco V 2012 *Phys. Rev. C* **86** 054902
[14] Xu Z and Greiner C 2005 *Phys. Rev. C* **71** 064901
[15] Demir N and Bass S A 2009 *Phys. Rev. Lett.* **102** 172302
[16] Muronga A 2004 *Phys. Rev. C* **69** 044901
[17] Wesp C, El A, Reining F, Xu Z, Bouras I, and Greiner C 2011 *Phys. Rev. C* **84** 054911
[18] Ferini G, Colonna M, Di Toro M, and Greco V 2009 *Phys. Lett. B* **670** 325
[19] Zhang B, Gyulassy M, and Ko C M 1999 *Phys. Lett. B* **455** 45
[20] Molnar D and Gyulassy M 2002 *Nucl. Phys. A* **697** 495
[21] Greco V, Colonna M, Di Toro M, and Ferini G 2009 *Prog. Part. Nucl. Phys.* **62** 562
[22] Plumari S, Baran V, Di Toro M, Ferini G, and Greco V 2010 *Phys. Lett. B* **689** 18
[23] Wiranata A and Prakash M 2012 *Phys. Rev. C* **85** 054908
[24] Arnold P, Moore G D, and Yaffe L G 2003 *J. High Energy Phys.* **05** 051
[25] Plumari S, Alberico W M, Greco V, and Ratti C 2011 *Phys. Rev. D* **84** 094004

Journal of Physics: Conference Series **527** (2014) 012017 doi:10.1088/1742-6596/527/1/012017

Overview of hydrodynamic modeling of heavy ion collision with ECHO-QGP

Valentina Rolando

Università di Ferrara, INFN-Sezione di Ferrara

E-mail: rolando@fe.infn.it

Abstract. ECHO-QGP is a numerical code, aimed to reproduce the dynamics of matter produced in heavy ions collisions space following its evolution through the phase transition from QGP to HG, using viscous hydrodynamic principles. Even though ECHO-QGP is being used for nuclear physics applications, its beginnings were designed for astrophysics purposes, treating general relativistic magneto-hydrodynamics. For this reason it praises second-order treatment of causal relativistic viscosity effects in every general relativity metrics and will be extended to include evolution of the electromagnetic fields coupled to the plasma.

1. Introduction

The idea of applying ideal hydrodynamics and its to strongly interacting matter goes back to 1953, when Landau for the first time presented it in [1]. Only few years ago though, with the first RHIC data [2], this approach has been proved to predict quantitatively experimental observables [3]. The validity of hydrodynamics has been later confirmed with the LHC data [4,5], showing evidences of QGP formation, early thermalisation and presence of viscosity but with very small ratio η/s. In the low p_T region, hydrodynamics well reproduces transverse momentum spectra of the emitted hadrons in central and semicentral collisions, and the anisotropy in momentum space generated at very early stages of the collision, detectable through the elliptic flow v_2. Nevertheless it still fails in describing other features such as two-particle correlation.

In this context, we present ECHO-QGP which is a numerical code for the description of heavy ion collisions in a (3+1)-D space and, as an outcome of our collaboration, foreseen to reproduce the above described features.

In the first section of the present work we will explain the formalism and the basics hydrodynamic equations used in ECHO-QGP and in the following we will describe the accessory theories used as setup in ECHO-QGP. In the third we explain the last stage of the simulation: the conversion of hydrodynamic fields in particles. In the last section we draw our conclusions and propose future outlooks. The present work does not include all the formal derivation of equations, the algorithmic details, and the description of the many tests ECHO-QGP has undergone, which can be found in [6–8].

2. Relativistic hydrodynamics

In order to describe the space time evolution of a fluid, we need a set of equations composed by the conservation law of the momentum-energy tensor $T^{\mu\nu}$, together with a continuity equation for each conserved current N^μ:

$$d_\mu N^\mu = 0, \tag{1}$$
$$d_\mu T^{\mu\nu} = 0. \tag{2}$$

For a perfect fluid this set of (five) equations contains just six unknowns: charge density n, energy density e, pressure P, and the fluid four velocity u^μ. The last equation needed to close up this set is a complete equation of state (EOS).

For the case of viscous hydrodynamics, not only the energy-momentum tensor becomes dependent on the viscous fields, but we must take into account the evolution of the shear and bulk part of the viscous-stress tensor (respectively $\pi^{\mu\nu}$ and Π). The shear viscous tensor is defined as $\pi^{\mu\nu} = [\frac{1}{2}(\Delta^\mu_\alpha \Delta^\nu_\beta + \Delta^\mu_\beta \Delta^\nu_\alpha) - \frac{1}{3}\Delta^{\mu\nu}\Delta_{\alpha\beta}]T^{\alpha\beta}$, and it satisfies the orthogonality ($\pi^{\mu\nu}u_\nu = 0$) and traceless ($\pi^\mu_\mu = 0$) conditions. In the most general case the expressions for the energy momentum tensor and conserved current are:

$$N^\mu = nu^\mu + V^\mu, \tag{3}$$
$$T^{\mu\nu} = eu^\mu u^\nu + (P + \Pi)\Delta^{\mu\nu} + \pi^{\mu\nu} + w^\mu u^\nu + w^\nu u^\mu, \tag{4}$$

where we included the isotropic pressure ($P + \Pi = \frac{1}{3}\Delta_{\mu\nu}T^{\mu\nu}$); the energy-momentum flow orthogonal to u^μ ($w^\mu = -\Delta^\mu_\alpha T^{\alpha\beta}u_\beta$) and the particles diffusion flux ($V^\mu = \Delta^\mu_\alpha N^\alpha$).

Since for this work we left out the case of finite baryon density, it is very convenient (and commonly adopted) the Landau frame, which allows to neglect both the continuity equation and set $w^\mu = 0$.

All the conservation laws are split in the direction parallel and orthogonal to u^μ, so we can decompose the covariant derivative of the fluid velocity *irreducible tensorial parts* as

$$d_\mu u_\nu = \sigma_{\mu\nu} + \omega_{\mu\nu} - u_\mu Du_\nu + \tfrac{1}{3}\Delta_{\mu\nu}\theta, \tag{5}$$

where we define the (transverse, traceless, and symmetric) *shear tensor* $\sigma_{\mu\nu}$, the (transverse, traceless, and antisymmetric) *vorticity tensor* $\omega_{\mu\nu}$, and the *expansion scalar* θ. Using the definition (5) and after a bit of manipulation, the conservation laws respectively for energy and momentum, bulk and shear viscous parts of stress tensor (including terms up to second-order in the velocity gradients), can be expressed as

$$De = -(e + P + \Pi)\theta - \pi^{\mu\nu}\sigma_{\mu\nu} \tag{6}$$
$$D\Pi = -\frac{1}{\tau_\Pi}(\Pi + \zeta\theta) - \tfrac{4}{3}\Pi\theta, \tag{7}$$
$$(e + P + \Pi)Du_\nu + \nabla_\nu(P + \Pi) + \Delta^\beta_\nu \nabla_\alpha \pi^\alpha_\beta + Du^\mu\,\pi_{\mu\nu} = 0, \tag{8}$$
$$D\pi^{\mu\nu} = -\frac{1}{\tau_\pi}(\pi^{\mu\nu} + 2\eta\sigma^{\mu\nu}) - \tfrac{4}{3}\pi^{\mu\nu}\theta + \mathcal{I}^{\mu\nu}_1 + \mathcal{I}^{\mu\nu}_2, \tag{9}$$

where the source term $\mathcal{I}^{\mu\nu}_1 = (\pi^{\lambda\mu}u^\nu + \pi^{\lambda\nu}u^\mu)Du_\lambda$ comes from the orthogonal projection, while $\mathcal{I}^{\mu\nu}_2 = -\lambda(\pi^{\mu\lambda}\omega^\nu_\lambda + \pi^{\nu\lambda}\omega^\mu_\lambda)$ is the vorticity contribution term.

Once the evolution equations are explicitly displayed, we need of complementary theories to fix the values of the viscosity coefficients (η, ζ), the shear and bulk relaxation times (τ_π and τ_Π) and the other second-order transport parameter ($\lambda \equiv \lambda_2/\eta$). In our analysis we exploited the $\mathcal{N} = \triangle$ Super-symmetric Yang Mills theory [10] for the the parameter λ_2 which is still unknown in a non-perturbative domain in QCD. Given RHIC results [11] on elliptic flow measurements fitted with Glauber type initial conditions [12], we use a value for η/s variable within the range $0.08 \leq \eta/s \leq 0.16$, while concerning the bulk viscosity, we used the relation $\zeta = 2\eta\,(\frac{1}{3} - c_s^2)$, referring to the study of strongly coupled gauge theories [13]. For the shear and bulk relaxation times values τ_π and τ_Π, we follow the same choice made by other groups [14, 15]: $\tau_\pi = \tau_\Pi = \frac{3\eta}{sT}$ which are extracted from kinetic theory [16, 17] and from AdS-CFT [10, 18]. The vorticity contribution $\omega_{\mu\nu}$ in the present analysis will be mostly ignored by fixing $\lambda = 0$.

2.1. Implementation in ECHO-QGP

Our aim in this section is to write all the conservation laws in a conservative form suitable for a numerical code, and in particular for ECHO-QGP. In order to do so, we must isolate the time dependency, and reduce our set of equations in matricial form:

$$\partial_0 \mathcal{U} + \partial_k \mathcal{F}^k = \mathcal{S} \tag{10}$$

This expression must be suitable for every general relativity diagonal metrics. We can eventually write our set of 12 equations in the matricial form (of equation 10) by manipulating equations (3)-(9) and obtaining:

$$\textit{conservative variables} \quad \mathcal{U} = |g|^{\frac{1}{2}} \begin{pmatrix} N \equiv N^0 \\ S_i \equiv T^0_i \\ E \equiv -T^0_0 \\ N\Pi \\ N\pi^{ij} \end{pmatrix} ; \quad \textit{fluxes} \quad \mathcal{F}^k = |g|^{\frac{1}{2}} \begin{pmatrix} N^k \\ T^k_i \\ -T^k_0 \\ N^k\Pi \\ N^k\pi^{ij} \end{pmatrix} ; \tag{11}$$

$$\textit{source terms} \quad \mathcal{S} = |g|^{\frac{1}{2}} \begin{pmatrix} 0 \\ \frac{1}{2}T^{\mu\nu}\partial_i g_{\mu\nu} \\ -\frac{1}{2}T^{\mu\nu}\partial_0 g_{\mu\nu} \\ n[-\frac{1}{\tau_\pi}(\Pi + \zeta\theta) - \frac{4}{3}\Pi\theta] \\ n[-\frac{1}{\tau_\pi}(\pi^{ij} + 2\eta\sigma^{ij}) - \frac{4}{3}\pi^{ij}\theta + \mathcal{I}^{ij}_0 + \mathcal{I}^{ij}_1 + \mathcal{I}^{ij}_2] \end{pmatrix} \tag{12}$$

The algorithm in ECHO-QGP, which takes care of solving this set of equations can be summarized in three stages: to retrieve primitive variables; to compute fluxes; to evolve the time step. The retrieving of the primitive variables is performed applying the orthogonality conditions for the shear viscous stress tensor, and subsequently inverting the equations (10)-(12).

The fluxes are computed by recovering the primitive variables at first, obtaining left and right states at cell interfaces, and then for each component and at each intercell, upwind fluxes are worked out (HLL two-state formula). Since the velocities are both primitive variables and necessary to extrapolate the fluxes, these two tasks are concatenated one into another, to obtain the final set of primitive variables at the current time step. The evolution equations are updated in time via a second or third order Runge-Kutta time-stepping routine.

ECHO-QGP has been widely tested, and proved agreement to analytic solutions, semi-analytic solutions and other publicly available numerical codes (See for further details the agreement between ECHO-QGP [6] and the tests in [19–26]).

2.2. Initial conditions

The parameters listed in this section, together with the shape of the initial distribution for energy density or entropy density, fluid velocities and viscous-stress tensor, are the choices we made to perform the current work, but we would like to underline that any set of parameters or functional form can easily be implemented by the user in ECHO-QGP.

The current version of the ECHO-QGP, uses both optical or Monte Carlo Glauber initial conditions.

For the first choice, ECHO-QGP provides an initial profile for the energy density distribution (or alternatively for the entropy density), which in the transverse plane can be written as

$$e(\tau_0, \mathbf{x}; b) = e_0 \left[(1-\alpha)\frac{n_{\text{part}}(\mathbf{x}; b)}{n_{\text{part}}(\mathbf{0}; 0)} + \alpha \frac{n_{\text{coll}}(\mathbf{x}; b)}{n_{\text{coll}}(\mathbf{0}; 0)} \right], \tag{13}$$

where the soft $n_{\rm part}$ and the hard $n_{\rm coll}$ contributions are parametrized by the parameter α running from 0 to 1. In this scheme, e_0 is $e_0 = e(\tau_0, \mathbf{x} = 0; b = 0)$; and the density of participants and of binary collisions of the colliding nuclei A and B, can be written in the transverse plane as

$$n_{\rm coll}(\mathbf{x}; b) = AB\, \sigma_{\rm in}^{NN}\, \widehat{T}_A(\mathbf{x}+\mathbf{b}/2)\widehat{T}_B(\mathbf{x}-\mathbf{b}/2) \qquad n_{\rm part}(\mathbf{x}; b) \equiv n^A_{\rm part}(\mathbf{x}; b) + n^B_{\rm part}(\mathbf{x}; b)$$
$$n^{A(B)}_{\rm part}(\mathbf{x}; b) = A(B)\, \widehat{T}_{A(B)}(\mathbf{x} \overset{+}{\underset{(-)}{}} \mathbf{b}/2)\left\{1 - [1 - \widehat{T}_{B(A)}(\mathbf{x} \overset{-}{\underset{(+)}{}} \mathbf{b}/2)\sigma_{\rm in}^{NN}]^{B(A)}\right\}$$

We denoted the inelastic nucleon-nucleon cross-section as $\sigma_{\rm in}^{NN}$, and $\widehat{T}_{A/B}(\mathbf{x})$ is the usual nuclear thickness function [27]. To extend the initialization to the (3+1)-D space we used a profile given by a function that factorizes the dependency along the longitudinal and transverse plane, following the indications in [28, 29]. Our choice for the energy density profile is a gaussian curve, with σ_η width, flattened in the central region $|\eta_s| \leq \Delta_\eta$. Given a beam rapidity Y_b, the factorization of the initial energy density profile reads:

$$e(\tau_0, \mathbf{x}, \eta_s; b) = \tilde{e}_0\, \theta(Y_b - |\eta_s|)\, f^{\rm pp}(\eta_s) F(\mathbf{x}; b) \tag{14}$$
$$F(\mathbf{x}; b) = \left[\alpha n_{\rm coll}(\mathbf{x}; b) + (1-\alpha)\left(\frac{Y_b - \eta_s}{Y_b} n^A_{\rm part}(\mathbf{x}; b) + \frac{Y_b + \eta_s}{Y_b} n^B_{\rm part}(\mathbf{x}; b)\right)\right] \tag{15}$$
$$f^{\rm pp}(\eta_s) = \exp\left[-\theta(|\eta_s| - \Delta\eta)\frac{(|\eta_s| - \Delta\eta)^2}{\sigma_\eta^2}\right]. \tag{16}$$

The Glauber Monte Carlo routine instead, allows to perform event-by-event simulations. It shapes the energy density profile sampling the positions of nucleons of each nucleus from a Woods-Saxon distribution and then it reallocates them into the respective center-of-mass frame. The impact parameter is sampled in the range $b \in [0, b_{\rm max}]$, with respect to the distribution $dP = 2\pi b db$. The criterion of collision between two nucleons, i (from nucleus A) and j (from nucleus B), is $(x_i - x_j)^2 + (y_i - y_j)^2 < \sigma_{\rm NN}/\pi$, the event is accepted if at least one binary nucleon-nucleon collision happened. As in the optical Glauber model, we combine the soft and the hard components to the collision, but this time every participant nucleon or collision is a source of energy density:

$$e(\tau_0, \mathbf{x}) = \frac{K}{2\pi\sigma}\left\{(1-\alpha)\sum_{i=1}^{N_{\rm part}} \exp\left[-\frac{(\mathbf{x} - \mathbf{x}_i^{\rm part})^2}{2\sigma^2}\right] + \alpha \sum_{i=1}^{N_{\rm coll}} \exp\left[-\frac{(\mathbf{x} - \mathbf{x}_i^{\rm coll})^2}{2\sigma^2}\right]\right\}. \tag{17}$$

where the parameter σ plays the role of the variance of gaussian stretching of the source.

The dependence along the longitudinal direction can be inserted *a posteriori* as in the optical-Glauber initialization of equation (16).

At $\tau = \tau_0$ we set, as initial condition, zero transverse flow velocities and a longitudinal flow given by the Bjorken's solution ($Y = \eta_s$, Y being the fluid rapidity). Finally, note that even if $u^\mu \equiv (1, 0, 0, 0)$ in the LRF at τ_0, in Bjorken coordinates the derivatives appearing in θ and $\sigma^{\mu\nu}$ in Ref. [6] do not all vanish, thus Π and the components of $\pi^{\mu\nu}$ must be initialized somehow. We use the first-order expressions $\Pi = -\zeta\theta$ and $\pi^{\mu\nu} = -2\sigma^{\mu\nu}$, thus $\Pi = -\zeta/\tau$, and $2\pi^{xx} = 2\pi^{yy} = -\tau^2\pi^{\eta\eta} = \frac{4}{3}\eta/\tau$, at $\tau = \tau_0$, with all other components set to zero.

2.3. Equation of state

The code is already designed to handle any form for $P = \mathcal{P}(e)$. Even if it is not a complete EOS, it is the most convenient choice of parametrization of the EOS in case of vanishing baryon density. We refer as EOS-I to the ideal, ultrarelativistic, non-interacting QGP with 3 light flavors ($g = 37$) which has been included in ECHO-QGP in its analytical form:

$$P = e/3 = \frac{g\pi^2}{90}T^4, \quad c_s^2 = \frac{1}{3}$$

Journal of Physics: Conference Series **527** (2014) 012017 doi:10.1088/1742-6596/527/1/012017

ECHO-QGP contains other, more realistic, QCD EOS's too; in particular a EOS by Laine and Schröder [30], arising from a weak-coupling QCD calculation with realistic quark masses and already used in UVH2+1 by Luzum and Romatschke [31, 32], and a second one obtained by matching a Hadron-Resonance-Gas EOS at low temperature with the continuum-extrapolated lattice-QCD results by the Budapest-Wuppertal collaboration [33, 34].

3. Decoupling

In order to compare to actual data we need to produce quantities related to the experiments, which are not naturally arising from the hydrodynamics evolution itself. The variety of techniques to generate particles from the hydro stage is wide, and has a long history. One of the currently most used technique of decoupling comes from Ref. [35, 36], which basically exploit the fact that, since the particle mean free paths strongly depend on the temperature of the medium, it can be assumed that below a certain temperature T_{freeze} (or equivalently "outside" a constant-temperature hypersurface) particles stop interacting and propagate as free streaming particles. Once determined the oriented hypersurface of constant temperature Σ, the total emission of primary particles flowing through it becomes the sum of all its points, that behave like sources. Referring to the particle species as i, the Cooper-Frye formula reads

$$E\frac{d^3N_i}{dp^3} = \frac{d^3N_i}{dyp_Tdp_Td\phi} = \frac{g_i}{(2\pi)^3}\int_\Sigma \frac{-p^\mu d^3\Sigma_\mu}{\exp\left(-\frac{u^\mu p_\mu+\mu_i}{T_{\rm freeze}}\right)\pm 1} \tag{18}$$

From the numerical point of view, the tough task for this recipe is to correctly find the hypersurface. There are refined techniques to do so, like CORNELIUS [37] or MUSIC's [38], but as a starting point, we choose to perform the simplest way of reconstructing it. This choice translates in a "pixelized" hypersurface, meaning that it can be orthogonal to only one direction at time:

$$d^3\Sigma_\mu = \begin{pmatrix} dV_{\perp\tau} \\ dV_{\perp x} \\ dV_{\perp y} \\ dV_{\perp\eta} \end{pmatrix} = \begin{pmatrix} \tau\Delta x\Delta y\Delta\eta_s\, s^\tau \\ \tau\Delta y\Delta\eta_s\Delta\tau\, s^x \\ \tau\Delta\eta_s\Delta\tau\Delta x\, s^y \\ \frac{1}{\tau}\Delta\tau\Delta x\Delta y\, s^\eta \end{pmatrix} \qquad s^\mu \equiv -\mathrm{sign}\left(\frac{\partial T}{\partial x^\mu}\right) \tag{19}$$

where the aim of the vector s^μ is to ensure that the hypersurface is oriented towards the colder region. We scan the whole space-time grid of the hydrodynamics evolution and select those adjacent cells (say A an B) where the temperature changes in such a way that: $(T_A - T_{\rm freeze})(T_{\rm freeze} - T_B) > 0$. An example of how the hypersurface comes across is displayed in figure 1. Here, even using a modestly refined technique, the hypersurface is very smooth in the case of optical Glauber initial conditions. Using this discretization, the Cooper-Frye formula (18), reduces to:

$$\frac{d^3N_i}{p_{\rm T}dp_{\rm T}dyd\phi} = \frac{g_i}{(2\pi)^3}\sum_{T_{\rm freeze}} \frac{m_{\rm T}\cosh(\eta_s-{\rm y})\,dV^{\perp\tau} + p_{\rm T}(\cos\phi\, dV^{\perp x} + \sin\phi\, dV^{\perp y}) - m_{\rm T}\sinh(\eta_s-{\rm y})\,\tau dV^{\perp\eta_s}}{\exp\left[\left(u^\tau m_{\rm T}\cosh(\eta_s-{\rm y}) - p_{\rm T}(u^x\cos\phi - u^y\sin\phi) + \tau u^\eta m_{\rm T}\sinh(\eta_s-{\rm y}) - \mu_i\right)/T\right]\pm 1} \tag{20}$$

where we explicitly wrote, in Bjorken coordinates the particle momentum and transverse mass:

$$p^\mu = (m_{\rm T}\cosh({\rm y}-\eta); p_{\rm T}\cos\phi, p_{\rm T}\sin\phi, m_{\rm T}\sinh({\rm y}-\eta))\,, \qquad m_{\rm T} = (p_{\rm T}^2+m^2)^{1/2} \tag{21}$$

Despite this algorithm is still at a much less refined stage, producing only primary spectra and using a rough reconstruction of the hypersurface, it has been tested against a well-known code such as AZHYDRO [39, 40], and proved a good agreement.

Anyway, the corrections to the primary particle spectra related to the decay of unstable particles have been shown to be significant and they must be included to reproduce the experimental data [39, 41]. Moreover, recent studies showed how the best way to reproduce data is to combine hydrodynamics and transport model [28, 37, 42–49]: in these works the Cooper-Frye prescription is not used to perform the freeze-out directly, but it is just exploited as a way of changing the physical description, leaving the hadronization task to the transport models.

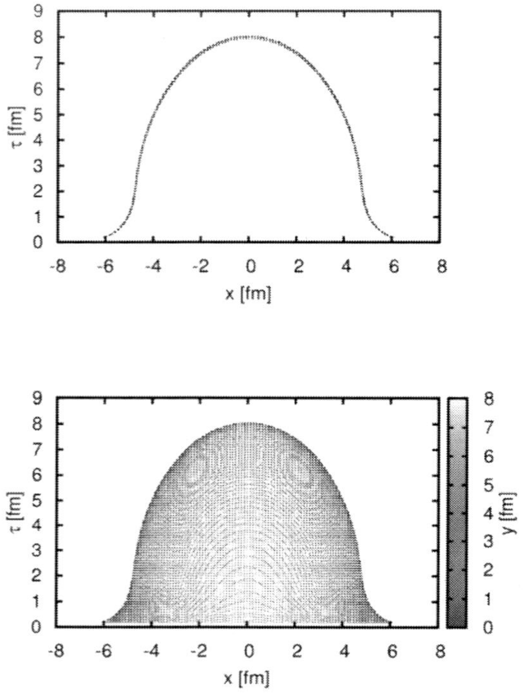

Figure 1. (Color online) Hypersurface shape in Bjorken coordinates, in the upper panel for zero value of the y coordinate, in the lower one with a map coloring for posistive y coordinates.

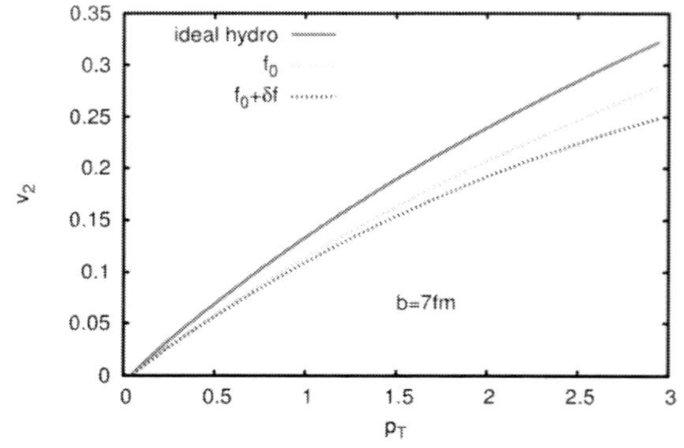

Figure 2. (Color online) Elliptic flow suppression due to the inclusion of viscosity only in the hydro stage (green line, f_0) and in the decoupling $f_0 + \delta f$). We quantitatively agree with other groups on the overall contribution of the inclusion of viscosity (about 20% at $p_T \simeq 1.5$ GeV).

In order to extend the Cooper-Frye prescription to the presence of viscosity, the particle distribution must be corrected by a small deviation $f(x,p) = f_0(x,p) + \delta f(x,p)$. For the present work we neglect in δf the contribution of the bulk viscosity, whose formalization is still controversial, but we include the shear viscosity contribution as in [50, 51], letting it be:

$$\delta f(x,p) = f_0(1 \pm f_0)\frac{p^\alpha p^\beta \pi_{\alpha\beta}}{2T^2(e+p)} \tag{22}$$

As shown in fig. (2), the inclusion of the viscous contribution nicely reproduces the elliptic flow suppression of about 20% at 1.5 GeV transverse momentum in agreement with the results in [52–56] but the issue on why some groups find the most part of the suppression due to the hydro evolution and others to the decoupling process is still to be understood.

4. Summary and outlook

Although its late birth, ECHO-QGP is answering to all the diagnostics at best, and in the next future our collaboration aim to achieve and overdue the state of the art in all its features. In the perspective of distributing the code, we must point out that at the current stage it allows many customization, including different metrics, the possibility of completely switching off the viscosity (not only set very small parameters), a quite wide range of initial condition possibilities and the inclusion of user-defined equation of state. We are currently working on the implementation of the event-by-event decoupling, including a more refined technique of detecting the hyper-surface, and the inclusion of non vanishing initial fluid four velocity profile. As a further item to address, we would like to include in ECHO-QGP the possibility of dealing with a finite-density EOS and finally, to recover the possibility of evolving also electromagnetic fields, which would characterize uniquely ECHO-QGP among over hydrodynamical codes.

References

[1] Landau L D 1953 *Akad. Nauk Ser. Fiz.* **17** 51
[2] Adams J *et al.* (The STAR Collaboration) 2005 *Nucl. Phys. A* **757** 102; Adcox K *et al.* (The PHENIX Collaboration) 2005 *Nucl. Phys. A* **757** 184; Back B *et al.* (The PHOBOS Collaboration) 2005 *Nucl. Phys. A* **757** 28; Arsence A *et al.* (The BRAHMS Collaboration) 2005 *Nucl. Phys. A* **757** 1
[3] Kolb P F and Heinz U W 2003 arXiv:nucl-th/0305084
[4] Krzewicki M (The ALICE Collaboration) 2011 *QM proceedings* (arXiv:nucl-ex/1107.0080v1); Aamodt K *et al.* (The ALICE Collaboration) 2010 *Phys. Rev. Lett.* **105** 252302; Aamodt K *et al.* (The ALICE Collaboration) 2011 *Phys. Rev. Lett.* **106**, 032301
[5] Hirano T, Huovinen P, and Nara Y 2011 *Phys. Rev. C* **84** 011901; Chen X F, Hirano T, Wang E, Wang X N, and Zhang H 2011 *Phys. Rev. C* **84** 034902
[6] Del Zanna L, Chandra V, Inghirami G, Rolando V, Beraudo A, De Pace A, Pagliara G, Drago A, and Becattini F 2013 *Eur. J. Phys. C* **73** 2524
[7] Del Zanna L, Zanotti O, Bucciantini N, and Londrillo P 2007 *Astron. Astrophys.* **473** 11
[8] Del Zanna L, Bucciantini N, and Londrillo P 2003 *Astron. Astrophys.* **400** 397
[9] Romatschke P 2007 *Eur. J. Phys. C* **52** 203
[10] Romatschke P 2010 *Class. Quantum Grav.* **27** 025006; Baier R *et al.* 2008 *JHEP* **0804** 100
[11] Alver B *et al.* (The PHOBOS Collaboration) 2008 *Int. J. Mod. Phys. E* **16** 331
[12] Kolb P F *et al.* 2001 *Nucl. Phys. A* **696** 197
[13] Kanitscheider I and Skenderis K 2009 *JHEP* **04** 062
[14] Song H and Heinz U W 2010 *Phys. Rev. C* **81** 024905
[15] Bozek P 2012 *Phys. Rev. C* **85** 034901
[16] Israel W 1976 *Ann. Phys.* (N.Y.) **100** 310; Israel W and Stewart JM 1979 *Ann. Phys.* (N.Y.) **118** 349
[17] Baier R, Romatschke P, and Wiedemann U A 2006 *Phys. Rev. C* **73** 064903
[18] Natsuume M and Okamura T 2008 *Phys. Rev. D* **77** 066014 [*Erratum-ibid D* **78** 089902]; Bhattacharyya S, Hubeny V S, Minwalla S, and Rangamani M 2008 *JHEP* **0802** 045
[19] Takamoto M and Inutsuka S 2011 *Jour. Comp. Phys.* **230** 7002
[20] Bucciantini N and Del Zanna L 2013 *MNRAS* **428** 71
[21] Molnàr E, Niemi H, and Rischke D H 2010 *Eur. Phys. J. C* **65** 615
[22] Muronga A 2002 *Phys. Rev. Lett.* **88** 062302 [*Erratum-ibid* **89** 159901]; Muronga A 2004 *Phys. Rev. C* **69** 034903
[23] Patra B K, Agotiya V, and Chandra V (2010) *Eur. Phys. J. C* **67** 465
[24] Baier R, Romatschke P, Son T D, Starinets A O, and Stephanov M A 2008 *JHEP* **0804** 100
[25] Baym G *et al.* 1983 *Nucl. Phys. A* **407** 541
[26] Gubser S S 2010 *Phys. Rev. D* **82** 085027
[27] De Vries H, De Jager C W, and De Vries C 1987 *At. Data Nucl. Data Tables* **36** 495
[28] Hirano T, Heinz U W, Kharzeev D, Lacey R, and Nara Y 2006 *Phys. Lett. B* **636** 299
[29] Adil A and Gyulassy M 2005 *Phys. Rev. C* **72** 034907
[30] Laine M and Schröder Y 2006 *Phys. Rev. D* **73** 085009
[31] Luzum M and Romatschke P, 2008 *Phys. Rev. C* **78** 034915 [*Erratum-ibid C* **79** 039903]
[32] Romatschke P and Romatschke U 2007 *Phys. Rev. Lett.* **99** 172301
[33] Borsanyi S *et al.* 2010 *JHEP* **1011** 077
[34] Bluhm M, Alba P, Alberico W, Beraudo A, and Ratti C 2013 arXiv:1306.6188

[35] Milekhin A 1958 *Zh. Eks. Teor. Fiz.* **35** 1185
[36] Cooper F and Frye G 1974 *Phys. Rev. D* **10** 186
[37] Huovinen P and Petersen H 2012 *Eur. Phys. J. A* **48** 171
[38] Ryu S, Jeon S, Gale C, Schenke B, and Young C 2013 *Nucl. Phys. A* **904-905** 389c
[39] Kolb P F, Sollfrank J, and Heinz U W 2000 *Phys. Rev. C* **62** 054909
[40] Kolb P F and Rapp R 2003 *Phys. Rev. C* **67** 044903
[41] Sollfrank J, Koch P, and Heinz U W 1991 *Z. Phys. C* **52** 593
[42] Bass S A and Dumitru A 2000 *Phys. Rev. C* **61** 064909
[43] Teaney D, Lauret J, and Shuryak E V 2001 `arXiv:nucl-th/0110037`
[44] Nonaka C and Bass S A 2007 *Phys. Rev. C* **75** 014902
[45] Petersen H, Steinheimer J, Burau G, Bleicher M, and Stocker H 2008 *Phys. Rev. C* **78** 044901
[46] Werner K, Karpenko I, Pierog T, Bleicher M, and Mikhailov K 2010 *Phys. Rev. C* **82** 044904
[47] Song H, Bass S A, Heinz U W, Hirano T, and Shen C 2011 *Phys. Rev. Lett.* **106** 192301
[48] Karpenko I, Sinyukov Y, and Werner K 2013 *Phys. Rev. C* **87** 024914
[49] Hirano T, Huovinen P, Murase K, and Nara Y 2013 *Prog. Part. Nucl. Phys.* **70** 108
[50] Romatschke P 2010 *Int. J. Mod. Phys. E* **19** 1
[51] Huovinen P 2013 *Int. J. Mod. Phys. E* **22** 1330029
[52] Dusling K and Teaney D 2008 *Phys. Rev. C* **77** 034905
[53] Romatschke P and Romatschke U 2007 *Phys. Rev. Lett.* **99** 172301
[54] Song H and Heinz U W 2008 *Phys. Rev. C* **78** 024902
[55] Jeon S and Kapusta J I 2001 *Phys. Rev. C* **63** 011901
[56] Molnar D and Huovinen P 2008 *J. Phys. G* **35** 104125

Journal of Physics: Conference Series **527** (2014) 012018 doi:10.1088/1742-6596/527/1/012018

Elliptic flow and shear viscosity of the shattered color glass condensate

Marco Ruggieri

Department of Physics and Astronomy, University of Catania, Via S. Sofia 64, I-95125 Catania

E-mail: marco.ruggieri@lns.infn.it

Francesco Scardina, Salvatore Plumari, Vincenzo Greco

Department of Physics and Astronomy, University of Catania, Via S. Sofia 64, I-95125 Catania
INFN-Laboratori Nazionali del Sud, Via S. Sofia 62, I-95123 Catania, Italy

Abstract. In this contribution, we report on our results about the computation of the elliptic flow of the quark-gluon-plasma produced in relativistic heavy ion collisions, simulating the expansion of the fireball by solving the relativistic Boltzmann equation for the parton distribution function tuned at a fixed shear viscosity to entropy density ratio η/s. We emphasize the role of saturation in the initial gluon spectrum modelling the shattering of the color glass condensate, causing the initial distribution to be out of equilibrium. We find that the saturation reduces the efficiency in building-up the elliptic flow, even if the thermalization process is quite fast $\tau_{therm} \approx 0.8\,\mathrm{fm/c}$, and the pressure isotropization even faster $\tau_{isotr} \approx 0.3\,\mathrm{fm/c}$. The impact of the initial non-equilibrium manifests for non-central collisions and can modify the estimate of the viscosity respect to the assumption of full thermalization in p_T-space.

1. Introduction

In the last decade it has been reached a general consensus that ultra-relativistic heavy-ion collisions (uRHICs) at the Relativistic Heavy-Ion Collider (RHIC) and the Large Hadron Collider (LHC) create a hot and dense strongly interacting quark and gluon plasma (QGP) [1, 2, 3, 4]. A main discovery has been that the QGP has a very small shear viscosity to entropy density ratio, η/s, which is more than one order of magnitude smaller than the one of water [5, 6], and close to the lower bound of $1/4\pi$ conjectured for systems at infinite strong coupling [7]. A key observable to reach such a conclusion is the elliptic flow [8], $v_2 = \langle (p_x^2 - p_y^2)/(p_x^2 + p_y^2) \rangle$. In fact, the expansion of the created matter generates a large anisotropy of the emitted particles that can be primarily measured by v_2. The origin of v_2 is the initial spatial eccentricity, $\epsilon_x = \langle y^2 - x^2 \rangle / \langle x^2 + y^2 \rangle$ of the overlap region in non-central collisions, which is responsible for different pressure gradients in the transverse plane thus favoring flow preferably along the x direction rather than y direction. The observed large v_2 is considered a signal of a very small η/s because if η/s was large then viscosity would damp the flow, hence reducing anisotropy in momentum space. Calculations agree in indicating an average η/s of the QGP lying in the range $4\pi\eta/s \approx 1 - 3$ [9, 10, 11, 12, 13, 14, 15, 16, 17, 18, 19, 20].

Along with the existence of a deconfined QGP matter and the understanding of its properties, the uRHIC program offers the opportunity to verify the picture in which the two colliding

Journal of Physics: Conference Series **527** (2014) 012018 doi:10.1088/1742-6596/527/1/012018

nuclei are described as two sheets of Color Glass Condensate (CGC) [21]. The CGC would be generated by the very high density of the gluon distribution function at low x (parton longitudinal momentum fraction), which triggers a saturation of the distribution for p_T below a saturation scale, Q_s. The determination of the shear viscosity η/s of the QGP and the search for the CGC are related: in fact the uncertainty on the initial condition translates into an uncertainty on the theoretical estimate of η/s [11, 22, 23, 24].

In this contribution we report our results [19, 25] about the computation of the elliptic flow of the quark-gluon-plasma produced in relativistic heavy ion collisions, simulating the expansion of the fireball by solving the relativistic Boltzmann equation for the parton distribution function tuned at a fixed shear viscosity to entropy density ratio η/s [17, 18, 25, 26, 27]. The advantage of using kinetic theory is that starting from a one-body phase space distribution function $f(x,p)$, and not from the energy-momentum tensor $T^{\mu\nu}$, it is straightforward to initialize simulations from a non-equilibrium distribution function, while hydrodynamics relies on the gradient expansion of the stress tensor and therefore is applicable only if initial deviations from local equilibrium are small.

In this study we will consider several kinds of initial conditions, two of them related to the shattering of the CGC which takes place after the collision. For describing the latter we make use of the KLN model [28, 29, 30, 31, 32], which has been largely employed to study the dynamics of HIC and the viscosity of the QGP within hydrodynamical simulations [11, 24, 22, 31, 33, 34, 35]. The uncertainty in the initial condition translates into an uncertainty on η/s of at a least a factor of two as estimated by mean of several viscous hydrodynamical approaches [24, 22, 11, 23]. We point out that the implementation of the shattered CGC initialization in hydrodynamics takes into account only the different space distribution respect to a geometric Glauber model, discarding the key and more peculiar feature of the damping of the distribution for p_T below the Q_s saturation scale. We have found by mean of kinetic theory that this has a significant impact on the build-up of v_2, hence on the determination of η/s.

In our study we neglect the initial time evolution of the chromo-electric and chromo-magnetic fields (the glasma) produced immediately after the collision. According to the commonly accepted picture of heavy ion collisions, our approach should be justified as soon as the initial strong fields decay into particle quanta. Characteristic time for the glasma decay is of the order of $\tau_0 \approx 1/Q_s$, which turns out to be of the order of a fraction of fm/c [36, 37, 38]. From this time on the field contribution to the pressure is less important, and the dynamics is dominated by the scatterings among partons, which justifies the use of kinetic theory to describe the sequent evolution of the system. The advantage to use kinetic theory from τ_0 rather than hydro is that the nonequilibrium initial distribution is not problematic as kinetic theory is built to study the evolution of a generic $f(x,p)$ distribution function.

2. Initial conditions for the simulations

In this section we specify the initial conditions we implement in our simulations. The first one is based on the Glauber model, see [39] for a review, with an $\boldsymbol{x}$-space distribution given by the a standard mixture $0.85N_{part} + 0.15N_{coll}$ and a $\boldsymbol{p}$-space thermalized spectrum in the transverse plane at a time $\tau_0 = 0.6\,\text{fm/c}$ for the RHIC runs and $\tau_0 = 0.3$ fm/c for the LHC runs. Maximum temperature is assumed $T_0 = 0.34$ GeV for the Au-Au collisions at RHIC energy and $T_0 = 0.51$ GeV for the Pb-Pb collisions at LHC energy. We also assume $y = \eta$ at $\tau = \tau_0$, where η corresponds to space-time rapidity; we also assume boost invariance in the longitudinal direction at $\tau = \tau_0$ which implies independence of the initial distribution on y. Following the nomenclature introduced in [19, 25] we will refer to this case as Th-Glauber.

The other two kinds of initial conditions we use in our simulations [19, 25] are based on a model of gluon production for the shattered CGC named factorized-KLN (fKLN in the following) [28, 30, 31], in which the initial spectrum is assumed to be

$$\frac{dN}{dyd^2\boldsymbol{x}_\perp d^2\boldsymbol{p}_T} = \kappa\frac{p_Ap_B}{p_T^2}\int^{p_T} d^2\boldsymbol{k}_T\alpha_S(Q^2)\phi_A(x_1,k_T^2;\boldsymbol{x}_\perp)\phi_B(x_2,(\boldsymbol{p}_T-\boldsymbol{k}_T)^2;\boldsymbol{x}_\perp)\ ; \qquad (1)$$

the coordinate space distribution is obtained by integration of the above equation over $\boldsymbol{p}_T$. Here $(\boldsymbol{p}_T, y)$ correspond to transverse momentum and momentum rapidity of the produced gluons respectively, and $x_{1,2} = p_T\exp(\pm y)/\sqrt{s}$. In equation (1) $p_{A,B}$ denote the probability to find one nucleon at a given transverse coordinate, $p_A(\boldsymbol{x}_\perp) = 1 - [1 - \sigma_{in}T_A(\boldsymbol{x}_\perp)/A]^A$ where σ_{in} is the inelastic cross section and T_A corresponds to the usual thickness function of the Glauber model. The overall constant κ in equation (1) is fixed in order to reproduce the experimental multiplicity in the most central collisions.

The specific form of the unintegrated gluon distribution function for gluons $\phi_{A,B}$ is not important here; it is enough to remind that it embeds saturation in the sense that $\phi_{A,B}$ as a function of p_T is constant for $p_T < Q_s$, where Q_s corresponds to the saturation scale. Here we show results obtained within a parameter set such that $\langle Q_s\rangle_{y=0} \approx 1$ GeV for the most central collisions at longitudinal momentum fraction $x = 0.01$.

Having described the fKLN model we are now ready to define the other two kinds of initializations we use in simulations. Firstly we assume that both the initial p_T-spectrum and the coordinate space distribution are given by equation (1) and its integral over $\boldsymbol{p}_T$, which corresponds to the proper implementation of the fKLN model. We call this initialization as fKLN initialization. This kind of initial condition is not implemented in hydro simulations because the initial strong deviation from equilibrium cannot be studied within viscous hydrodynamics. In the fKLN initialization we assume $\tau_0 = 0.2$ fm/c because no assumption on equilibration needs to be introduced; this time can be interpreted as the time needed for the strong initial fields to decay to a parton liquid, as explained in the Introduction. On the other hand, we also consider another initial condition which is close to the one usually implemented in hydro simulations, in which we assume the coordinate space distribution given by the KLN model and a p_T-thermalized spectrum. In this way we assume some thermalization occured in a short time, leaving at the same time the distribution in coordinate space unaffected. We call this initialization as Th-fKLN initialization. For the latter we assume the same initialization times used for the Th-Glauber model. In the two cases we assume $y = \eta$ at $\tau = \tau_0$.

3. Kinetic theory at fixed η/s

For the dynamical evolution of the initial conditions we employ transport theory as a base of a simulation code of the fireball expansion created in relativistic heavy-ion collision [17, 26, 18, 27]; we assume the system can be described by a gluon distribution function $f(\boldsymbol{x},\boldsymbol{p},t)$ which evolves according to the Relativistic Boltzmann Transport (RBT) equation:

$$p_\mu\partial^\mu f = \int d\Gamma_2 d\Gamma_{1'}d\Gamma_{2'}(f_{1'}f_{2'} - ff_2)|\mathcal{M}|^2\delta^4(p+p_2-p_{1'}-p_{2'})\ , \qquad (2)$$

with $d^3\boldsymbol{p}_k = 2E_k(2\pi)^3d\Gamma_k$, and $\mathcal{M}$ corresponds to the transition amplitude. At variance with the standard use of transport theory we have developed an approach that, instead of focusing on specific microscopic calculations of the scattering matrix, fixes the total cross section cell by cell in the grids in order to have a wished value of η/s. By means of this procedure we translate transport theory to hydrodynamical language since we use the Boltzmann equation to simulate the dynamical evolution of a fluid which dissipates with a specified shear viscosity, in analogy to what is done within hydrodynamical simulations. Hence this approach offers a bridge between transport theory and hydrodynamics.

The advantage of the kinetic theory approach at fixed η/s, compared to hydro simulations, is twofold: the initial condition does not need to be close to thermal equilibrium, because we do not perform a gradient expansion of the energy-momentum tensor. Moreover we do not need to specify an ansatz for the deviations from equilibrium. This approach to kinetic theory has been also considered in [40] where it has been shown that transport theory at fixed η/s reproduces the results of viscous hydrodynamics for one-body observable like $T^{\mu\nu}$ or entropy density also in the limit in which the system is not in the dilute regime. This is not surprising because looking at the Boltzmann collision integral in terms of viscosity allows the analytical derivation of second order viscous hydrodynamics [41, 42, 43, 44]. Despite the fact we consider only the $2 \leftrightarrow 2$ processes to compute the collision integral, once the system is close to the hydro regime with many collisions happening in a short time range, the detail of the single scattering is lost and what matters is only the viscosity of the fluid and not the specific microscopic process producing such viscosity. We have checked we work in a regime where the specific process is not relevant by changing the microscopic two-body scattering matrix from anisotropic to isotropic and adjusting the total cross section to keep the same η/s. We have found that leaving unchanged all the other parameters, the elliptic flow is not affected by this change for $p_T \le 2.5$ GeV [45].

Once η/s is fixed, we compute the total cross section in each cell of the coordinate space of our grid. The Chapmann-Enskog approximation supplies the relation between η/s an cross section with quite good approximation, in agreement with the results obtained using the Green Kubo formula [27, 46]:

$$\sigma_{tot} = \frac{1}{15}\frac{\langle p\rangle}{\rho\, g(a)}\frac{1}{\eta/s} = \frac{1}{15}\langle p\rangle\, \tau_\eta \ , \tag{3}$$

which is valid for a generic differential cross section $d\sigma/dt \sim 1/(t-m_D^2)^2$ [46]. In the above equation $a = T/m_D$, with m_D the screening mass regulating the angular dependence of the cross section, while $g(a)$ corresponds to a function whose analytic form can be found in [19, 25]. In the regime were viscous hydrodynamic applies (not too large η/s and p_T) the specific microscopic detail of the cross section is irrelevant and our approach is an effective way to employ transport theory to simulate a fluid at a given η/s.

The η/s we use in our study [25] is as follows: in the plasma phase it is a constant, whose numerical value is fixed case by case; we then implement a kinetic freezout by assuming that η/s increases smoothly in a temperature range from the plasma phase to a hadron phase values which is fixed by referring to the estimates in [47, 48, 49]. In this way we take into account scatterings in the hadron phase as well, which however give a very tiny contribution to the collective flow because of the damping due to the larger viscosity.

4. Thermalization

In the upper left panel of figure 1 we collect the initial spectra, $dN/2\pi p_T dp_T$, integrated over the momentum rapidity window $|y| < 0.5$, for the case of a Au-Au collision $\sqrt{s} = 200A$ GeV for the fKLN and Th-Glauber initial conditions at their respective initial times τ_0 and at the final time $\tau = 8.2$ fm/c. For the case of the Th-fKLN, we find that the spectra are the same as in the case of the Th-Glauber, therefore we do not plot them in the figure. We have also shown by the dashed line the spectrum at $\tau = 1.2$ fm/c for the case of the fKLN initial condition. The results in figure 1 are obtained with $4\pi\eta/s = 1$. We notice that initially the fKLN spectrum is quite far from a thermalized spectrum; nevertheless the spectrum thermalizes in the transverse plane within 1 fm/c, since its p_T dependence becomes exponential with a slope very similar to the Th-Glauber. In [25] we have also checked thermalization is achieved in the full momentum space within $\tau_{therm} \approx 0.8$ fm/c by computing the time evolution of the quantity $T^* \equiv T\tau^{1/3}$, with $T = E/3N$ representing the temperature in the case of a thermalized system. In the case of 1D dissipation-less expansion T^* saturates to a constant as a function of time, which we find

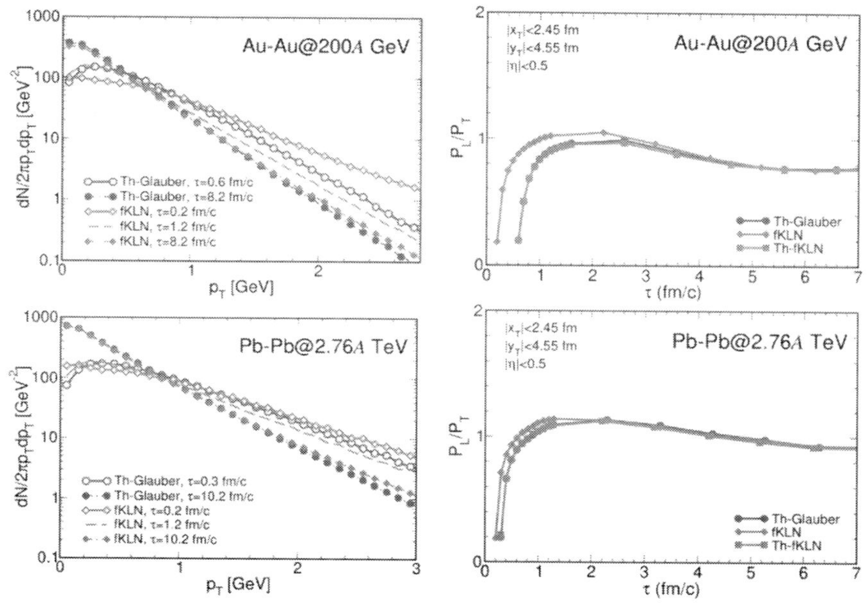

Figure 1. (Color online) Left panels: time evolution of spectra, for collisions at RHIC (upper left) and LHC (lower left) energies. Right panels: time evolution of P_L/P_T. In all panels $b = 7.5$ fm and $4\pi\eta/s = 1$. The spectra for the Th-fKLN case are not shown because we do not find visible deviations from the Th-Glauber initialization. Adapted from Ref. [25].

to happen for the fKLN case within τ_{therm}. Results for the LHC runs are shown on the lower left panel of figure 1.

On the upper right panel of figure 1, which refers to the case of Au-Au collision, we plot the ratio P_L/P_T where P_L, P_T correspond to the longitudinal and transverse pressure respectively. In the initial stage of the collision there is a strong anisotropy due to the fact that $P_L \ll P_T$; however the system efficiently removes the anisotropy in the case of small η/s, and in fact we find that the time required to get isotropic pressure $P_L \approx P_T$ is $\tau_{iso} \approx 0.3$ fm/c. In [25] we show that in case η/s is quite large and comparable with the perturbative QCD estimates, the system is not efficient in removing the initial anisotropy, which is quite natural because the large viscosity damps the flow which instead is necessary to transfer momentum and equilibrate the pressures. Results for the LHC runs are shown on the lower right panel of figure 1.

5. Elliptic flow

In the left panel of figure 2, we collect our results for the differential elliptic flow for the case of a Au-Au collision at RHIC energy with $b = 7.5$ fm. In the right panel we plot the same quantity for the case of Pb-Pb collisions at LHC energy and same impact parameter. To guide the eye, in the figure we also plot experimental data for v_2 in the relevant centrality class [50]. Since no hadronization process is yet included in our approach, the comparison with the charged hadrons v_2 has to be taken with care, even if it indicates that the azimuthal asymmetries generated by RBT are in the correct range.

Firstly we focus on the thermalized initializations, Th-Glauber and Th-fKLN. In the case of Au-Au collisions at RHIC energy, the v_2 obtained by Th-Glauber with $4\pi\eta/s = 1$ is obtained by Th-fKLN with $4\pi\eta/s \approx 2$, because momentum distribution is the same but initial eccentricity of

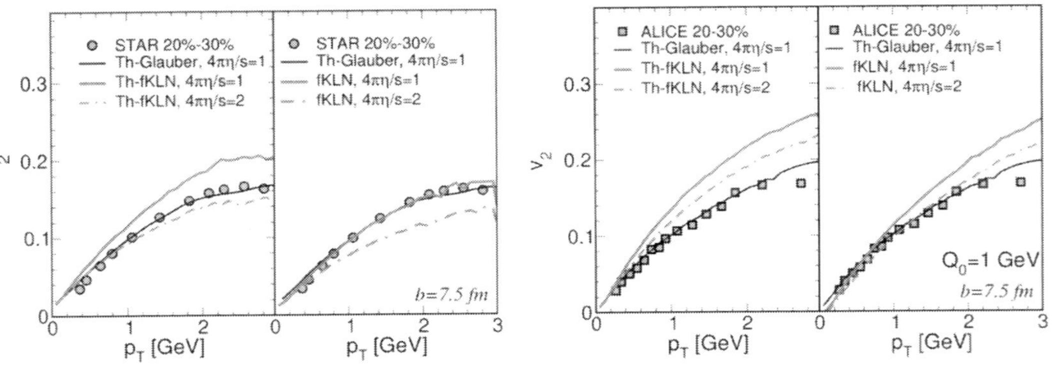

Figure 2. (Color online) Elliptic flow $v_2(p_T)$ at midrapidity $|y| < 0.5$ for different initial conditions and η/s as in the legend. Calculations in the left panel refer to Au-Au collisions at $\sqrt{s} = 200A$ GeV, while in the right panel we plot results for Pb-Pb collisions at $\sqrt{s} = 2.76A$ TeV. Adapted from Ref. [25].

Th-fKLN is larger than the one of the Th-Glauber initial condition, requiring a larger viscosity in the former case to damp the flow. These results are in agreement with the ones obtained from viscous hydrodynamics [11, 23, 24], showing the solidity and consistency of our transport approach at fixed η/s.

Next we discuss the result for the fKLN model, represented by the green lines in figure 2, with the proper distribution function implemented in both the $\boldsymbol{x}$ and $\boldsymbol{p}$ spaces. We find that fKLN with a $4\pi\eta/s = 1$ gives a $v_2(p_T)$ quite similar to the Th-Glauber, even if the initial eccentricity in this case is larger. For fKLN with $4\pi\eta/s = 2$ the differential elliptic flow would be too small. In other words the initial out-of-equilibrium fKLN distribution reduces the efficiency in converting ϵ_x into v_2. Our interpretation is that the initial large eccentricity of the fKLN configuration is compensated by the key feature of an almost saturated initial distribution in $\boldsymbol{p}$-space below the saturation scale Q_s and probably by the softer tail at $p_T > Q_s$. We obtain similar results at the LHC energy. In this case the Th-fKLN overestimate the Th-Glauber v_2 at both $4\pi\eta/s = 1$ and 2. When the full fKLN is implemented we can see that the damping effect already discussed at RHIC is such that with an $4\pi\eta/s = 2$ one is closer to the Th-Glauber with $4\pi\eta/s = 1$.

The effect we find on v_2 is more important for peripheral collisions. In fact in figure 3 we plot v_2 at $p_T = 2$ GeV for different initializations, as a function of the number of participants. This figure permits to visualize and summarize the dependence of $v_2(p_T)$ on the centrality class, comparing the impact of the initial distribution on the final v_2. We notice that the discrepancy between Th-fKLN and fKLN initializations becomes less relevant for more central collisions, implying that the effect of the initial momentum distribution is not negligible if one considers non-central collisions. We also note that for central collisions at RHIC Th-Glauber and Th-fKLN for $4\pi\eta/s = 1$ predict the same v_2 and the effect is of KLN generating larger v_2 disappears. This is seen also in viscous hydro simulation and it is a further confirmation that our approach converge to viscous hydro when the same thermal initial conditions are employed.

6. Conclusions

In this contribution we have reported our results [19, 25] on thermalization and building-up of the elliptic flow for fireballs produced in relativistic heavy ion collisions both at RHIC and LHC energies. We have put emphasis on the role of a nonequilibrium initial condition on the generation of the collective flow, when the KLN initialization is properly implemented in

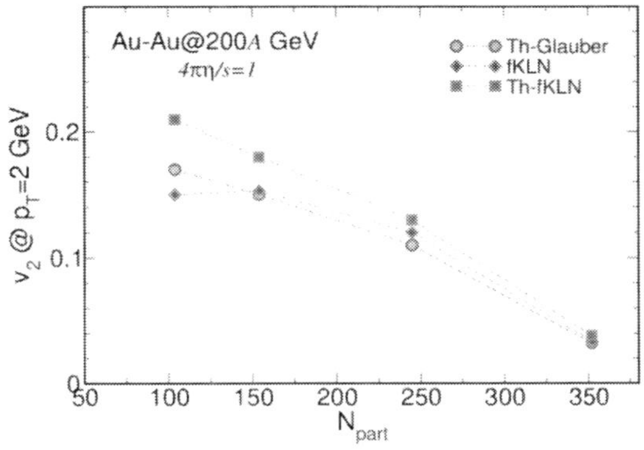

Figure 3. (Color online) Elliptic flow $v_2(p_T)$ at midrapidity $|y| < 0.5$ and at $p_T = 2$ GeV for different initial conditions, computed at $4\pi\eta/s = 1$. All the calculations refer to Au-Au collisions at $\sqrt{s} = 200$ GeV. Adapted from Ref. [25].

coordinate and momentum space. Our study is based on kinetic theory at fixed η/s. Fixing the shear viscosity over entropy density ratio permits to forget about microscopic processes which give that specific value of the ratio itself, and to focus on the effect of shear viscosity on thermalization, isotropization and building of the collective flows.

For what concerns thermalization, in both RHIC and LHC runs we have found that thermalization times are $\tau_{therm} \approx 0.8$-1 fm/c. We have then focused our attention on the elliptic flow production when the initial distribution has a saturation scale built in it. We have found that the amount of elliptic flow produced in heavy ion collisions depends not only on the pressure gradients and the η/s of the system, but also on the initial distribution in momentum space. In particular, an initial condition characterized by a momentum distribution with a saturation scale generates smaller v_2 respect to the thermal one. Assuming the fKLN distribution as the one arising from the shattering of the CGC, the effect of the initial nonequilibrium distribution affects the estimate of η/s of about a factor of two. However we have also that this effect is maximal for semi-peripheral collisions, becoming quite small for very central collisions.

In order to make more precise comparison with experimental data we are currently implementing also fluctuating initial conditions which will allow to extend the present study to all the v_n harmonics relevant in HIC's. This will allow to see if fKLN can account for the measured v_3 or the non-equilibrium damps such harmonics even more than what seen in hydrodynamics. In [51] it is found that harmonics up to the fifth order can be reproduced by combining CYM early-time with hydro late-evolution evolutions; however in the calculations of [51] a small deviation from equilibrium has to be assumed in order to use viscous hydrodynamical equations. Therefore it will be interesting to compute the higher order harmonics combining the CYM initial spectrum with the dynamics embedded in the kinetic equations where the assumption of sudden thermalization can be relaxed, which will be the subject of future studies.

Acknowledgments

V G and F S acknowledge the ERC-STG funding under the QGPDyn grant.

References

[1] Adams J *et al.* [The STAR Collaboration] 2005 *Nucl. Phys. A* **757** 102; Adcox K *et al.* [The PHENIX Collaboration] 2005 *Nucl. Phys. A* **757** 184
[2] Aamodt K *et al.* [The ALICE Collaboration] 2010 *Phys. Rev. Lett.* **105** 252302

[3] Jacak B and Muller B 2012 *Science* **337** 310
[4] Fries R *et al.* 2008 *Ann. Rev. Nucl. Part. Sci.* **58** 177
[5] Csernai L *et al.* 2006 *Phys. Rev. Lett.* **97** 152303
[6] Lacey R *et al.* 2007 *Phys. Rev. Lett.* **98** 092301
[7] Kovtun D *et al.* 2005 *Phys. Rev. Lett.* **94** 111601
[8] Ollitrault J Y 1992 *Phys. Rev. D* **46** 229
[9] Romatschke P and Romatschke U 2007 *Phys. Rev. Lett.* **99** 172301
[10] Song H and Heinz W W 2008 *Phys. Rev. C* **78** 024902
[11] Song H *et al.* 2011 *Phys. Rev. C* **83** 054910
[12] Schenke B *et al.* 2010 *Phys. Rev. C* **82** 014903
[13] Niemi *et al.* 2011 *Phys. Rev. Lett.* **106** 212302
[14] Xu Zand Greiner C 2009 *Phys. Rev. C* **79** 014904
[15] Xu Z, Greiner C, and Stocker H 2008 *Phys. Rev. Lett.* **101** 082302
[16] Bratkovskaya E *et al.*2011 *Nucl. Phys. A* **856** 162
[17] Ferini G, Colonna M, Di Toro M, and Greco V 2009 *Phys. Lett. B* **670** 325
[18] Plumari S and Greco V 2012 *AIP Conf. Proc.* **1422** 56
[19] Ruggieri M *et al.* 2013 *Phys. Lett. B* **727** 177
[20] Uphoff J *et al.* 2014 `arXiv:1401.1364`
[21] McLerran L and Venugopalan R 1994 *Phys. Rev. D* **49** 2233; 1994 *Phys. Rev. D* **49** 3352; 1994 *Phys. Rev. D* **50** 2225
[22] Alver B *et al.*2010 *Phys. Rev. C* **82** 034913
[23] Adare A *et al.* [The PHENIX Collaboration] 2011 *Phys. Rev. Lett.* **107** 252301
[24] Luzum M and Romatschke P 2008 *Phys. Rev. C* **78** 034915
[25] Ruggieri M et al. 2013 `arXiv:1312.6060`
[26] Plumari S *et al.* 2010 *Phys. Lett. B* **689** 18
[27] Plumari S *et al.* 2013 *J. Phys. Conf. Ser.* **420** 012029
[28] Kharzeev D, Levin E, and Nardi M 2005 *Nucl. Phys. A* **747** 609; Kharzeev D, Levin E, and McLerran L 2003 *Phys. Lett. B* **561** 93; Kharzeev D and Nardi M 2001 *Phys. Lett. B* **507** 121; Kharzeev D and Levin E 2001 *Phys. Lett. B* **523** 79; Hirano T and Nara Y 2004 *Nucl. Phys. A* **743** 305
[29] Hirano T *et al.* 2006 *Phys. Lett. B* **636** 299
[30] Drescher W and Nara U 2007 *Phys. Rev. C* **75** 034905; 2007 *Phys. Rev. C* **76** 041903
[31] Hirano T and Nara Y 2009 *Phys. Rev. C* **79** 064904
[32] Drescher H *et al.* 2006 *Phys. Rev. C* **74** 044905
[33] Shen C *et al.* 2013`arXiv:1308.2440`
[34] Shen C*et al.* 2013 `arXiv:1308.2111`
[35] Song H *et al.* 2013 `arXiv:1311.0157`
[36] Ryblewski R and Florkowski W 2013 *Phys. Rev. D* **88** 034028
[37] Gelis F and Epelbaum T 2013 `arXiv:1307.2214`
[38] Fukushima K 2013 `arXiv:1307.1046`
[39] 2007 Miller M L *et al.Ann. Rev. Nucl. Part. Sci.* **57** 205
[40] Huovinen P and Molnar D 2009 *Phys. Rev. C* **79** 014906
[41] Grad H 1949 *Commun. Pure Appl. Math.* **2** 331; 1949 *Commun. Pure Appl. Math.* **2** 325
[42] Denicol G *et al.* 2012 *Eur. Phys. J. A* **11** 170g
[43] Tsumura K and Kunihiro T 2013 `arXiv:1311.7059`
[44] Bazow D *et al.* 2013`arXiv:1311.6720`
[45] Greco V 2013 *Anisotropic Flow from a Kinetic Theory Approach* talk given at INPC 2013, 2-6 June 2013, Florence (Italy)
[46] Plumari S *et al.* 2012 *Phys. Rev. C* **86** 054902
[47] Prakash M *et al.* 1993*Phys. Rep.* **227** 321
[48] Chen *et al.* 2007 *Phys. Rev. D* **76** 114011
[49] Demir N and Bass S 2009 *Phys. Rev. Lett.* **102** 172302
[50] Adams J *et al.* [The STAR Collaboration] 2005 *Phys. Rev. C* **72** 014904; Dobrin A 2011 *J. Phys. G* **38** 124170
[51] Gale C*et al.* 2013 *Phys. Rev. Lett.* **110** 012302
[52] Song H and Heinz U 2008 *Phys. Lett. B* **658** 279

Journal of Physics: Conference Series **527** (2014) 012019 doi:10.1088/1742-6596/527/1/012019

Thermodynamic instabilities in dense asymmetric nuclear matter and in compact stars

A Lavagno[1], **A Drago**[2], **G Pagliara**[2], **D Pigato**[1]

[1]Department of Applied Science and Technology, Politecnico di Torino and INFN, Sezione di Torino, Torino, Italy

[2]Dip. di Fisica e Scienze della Terra dell'Università di Ferrara and INFN Sez. di Ferrara, Ferrara, Italy

Abstract. We investigate the presence of thermodynamic instabilities in compressed asymmetric baryonic matter, reachable in high energy heavy ion collisions, and in the cold β-stable compact stars. To this end we study the relativistic nuclear equation of state with the inclusion of Δ-isobars and require the global conservation of baryon and electric charge numbers. Similarly to the low density nuclear liquid-gas phase transition, we show that a phase transition can occur in dense asymmetric nuclear matter and it is characterized by both mechanical instability (fluctuations on the baryon density) and by chemical-diffusive instability (fluctuations on the electric charge concentration). Such thermodynamic instabilities can imply a very different electric charge fraction Z/A in the coexisting phases during the phase transition and favoring an early formation of Δ^- particles with relevant phenomenological consequences in the physics of the protoneutron stars and compact stars. Finally, we discuss the possible co-existence of very compact and very massive compact stars in terms of two separate families: compact hadronic stars and very massive quark stars.

1. Introduction

One of the very interesting aspects in nuclear astrophysics and in the heavy-ion collisions experiments is a detailed study of the thermodynamical properties of strongly interacting nuclear matter away from the nuclear ground state. In this direction, many efforts have been focused on searching for possible phase transitions in the β-stability conditions and in regime of finite temperature and density.

The new accumulating data from x-ray satellites provide important informations on the structure and formation of compact stellar objects. The recent discovery of Compact Stars (CSs) having a mass of the order of $2\,M_\odot$ [1, 2] puts rather severe constraints on the equation of state (EOS) of matter at large densities. It is clear that matter inside a compact star, i.e. β-stable and charge neutral matter, has to be stiff to allow such massive configurations. On the other hand, we know that by increasing the density new degrees of freedom come into the game, for instance hyperons and maybe deconfined quarks. These new ingredients soften the EOS close to their production threshold, but by introducing repulsive interactions the EOS can be stiff enough at large densities to support a $2\,M_\odot$ configuration [3, 4, 5].

On the other hand, the information coming from experiments with heavy ions in intermediate- and high-energy collisions is that the EOS depends on the beam energy but also sensibly on the electric charge fraction Z/A of the colliding nuclei, especially at not too high temperature

[6, 7]. Moreover, the study of nuclear matter with arbitrary electric charge fraction results to be important in radioactive beam experiments and in the physics of compact stars.

The extraction of information about the equation of state at different densities and temperatures by means of intermediate and high energy heavy ion collisions is a very difficult task and can be realized only indirectly by comparing the experimental data with different theoretical models, such as, for example, fluid-dynamical models. The EOS at density below the saturation density of nuclear matter ($\rho_0 \approx 0.16$ fm^{-3}) is relatively well known due to the large amount of experimental nuclear data available. At larger density there are many uncertainties; the strong repulsion at short distances of nuclear force makes, in fact, the compression of nuclear matter quite difficult. However, in relativistic heavy ion collisions the baryon density can reach values of a few times the saturation nuclear density and/or high temperatures. The future CBM (Compressed Baryonic Matter) experiment of FAIR (Facility of Antiproton and Ion Research) project at GSI Darmstadt, will make it possible to create compressed baryonic matter with a high net baryon density [8].

In relativistic heavy ion collisions, where finite values of density and temperature can be reached, a state of high density resonance matter may be formed and the $\Delta(1232)$-isobar degrees of freedom are expected to play a central role [9, 10, 11]. Moreover, within the non-linear Walecka model, it has been predicted that a phase transition from nucleonic matter to Δ-excited nuclear matter can take place but the occurrence of this transition sensibly depends on the Δ-meson coupling constants [12, 13, 14]. In this context, it is important to observe that also in CSs, where high baryon densities are reached, $\Delta(1232)$-resonances can be also produced in principle and it has been pointed out that the existence of Δs can be very relevant also in the core of neutron stars [15, 16, 17, 18]. The production of these particles softens the EOS and allows very compact configurations. On the other hand, this same softening forbids this hadronic family of CSs to reach very large masses [19, 20].

The purpose of this paper is twofold: first, to show that, for asymmetric warm and dense nuclear medium, the possible Δ-matter phase transition is characterized by mechanical and chemical-diffusive instabilities. Similarly to the liquid-gas phase transition [21], chemical instabilities play a crucial role in the characterization of the phase transition and can imply a very different electric charge fraction Z/A in the coexisting phases during the phase transition. Second, to investigate the relevance of Δ-isobar degrees of freedom in the bulk properties of compact stars and the possible presence of thermodynamic instability in the β-stable and charge neutral hadronic EOS. Finally, we will discuss a possible interpretation of the co-existence of compact and very massive CSs in terms of two separate families: hadronic stars, whose equation of state is soft, can be very compact, while quark stars, whose equation of state is stiff, can be very massive. In this respect an early appearance of Δ resonances is crucial to guarantee the stability of the branch of hadronic stars.

2. Nuclear Equation of State

Concerning the hadronic EOS, we use the relativistic mean field model with the inclusion of the octet of lightest baryons (nucleons and hyperons) in the framework of the GM3 non-linear Walecka type model of Glendenning-Moszkowsky [22]. In this framework the Lagrangian density $\mathcal{L}_{\rm NH}$ for nucleons and hyperons can be written as

$$\begin{aligned}\mathcal{L}_{\rm NH} &= \sum_k \overline{\psi}_k\,[i\,\gamma_\mu\,\partial^\mu - (M_k - g_{\sigma k}\,\sigma) - g_{\omega k}\,\gamma_\mu\,\omega^\mu - g_{\rho k}\,\gamma_\mu\,\vec{t}\cdot\vec{\rho}^{\,\mu}]\,\psi_k \\ &+\frac{1}{2}(\partial_\mu\sigma\partial^\mu\sigma - m_\sigma^2\sigma^2) - U(\sigma) + \frac{1}{2}\,m_\omega^2\,\omega_\mu\omega^\mu + \frac{1}{4}\,c\,(g_{\omega N}^2\,\omega_\mu\omega^\mu)^2 \\ &+\frac{1}{2}\,m_\rho^2\,\vec{\rho}_\mu\cdot\vec{\rho}^{\,\mu} - \frac{1}{4}F_{\mu\nu}F^{\mu\nu} - \frac{1}{4}\vec{G}_{\mu\nu}\vec{G}^{\mu\nu}\,, \qquad (1)\end{aligned}$$

where the sum runs over the full octet of baryons, M_k is the vacuum baryon mass of index k, the quantity $\vec{t}$ denotes the isospin operator that acts on the baryon and the field strength tensors for the vector mesons are given by the usual expressions $F_{\mu\nu} \equiv \partial_\mu\omega_\nu - \partial_\nu\omega_\mu$, $\vec{G}_{\mu\nu} \equiv \partial_\mu\vec{\rho}_\nu - \partial_\nu\vec{\rho}_\mu$. The $U(\sigma)$ is a nonlinear self-interaction potential of σ meson

$$U(\sigma) = \frac{1}{3}a\,(g_{\sigma N}\,\sigma)^3 + \frac{1}{4}b\,(g_{\sigma N}\,\sigma^4)\,. \tag{2}$$

The implementation of hyperon degrees of freedom comes from determination of the corresponding meson-hyperon coupling constants that have been fitted to hypernuclear properties. Following Refs. [23, 24], the SU(6) simple quark model can be used to obtain the relations

$$\frac{1}{3}g_{\omega N} = \frac{1}{2}g_{\omega\Lambda} = \frac{1}{2}g_{\omega\Sigma} = g_{\omega\Xi}\,, \quad g_{\rho N} = \frac{1}{2}g_{\rho\Sigma} = g_{\rho\Xi}\,, \quad g_{\rho\Lambda} = 0\,. \tag{3}$$

In addition, we can fix the scalar σ meson-hyperon ($g_{\sigma Y}$) coupling constants to the potential depth of the corresponding hyperon in normal dense matter taking into account the following recent results [25, 26]

$$U_\Lambda^N = -28\,\mathrm{MeV},\; U_\Sigma^N = +30\,\mathrm{MeV},\; U_\Xi^N = -18\,\mathrm{MeV}\,. \tag{4}$$

To incorporate Δ-isobars in the framework of effective hadron field theories, a formalism was developed to treat Δ particles analogously to the nucleons, taking only the on-shell Δs into account and the mass of the Δs are substituted by the effective one in the mean field approximation [27, 28]. The Lagrangian density of the Δ-isobars can then be expressed as [12, 13, 27]

$$\mathcal{L}_\Delta = \overline{\psi}_{\Delta\,\nu}\,[i\gamma_\mu\partial^\mu - (M_\Delta - g_{\sigma\Delta}\sigma) - g_{\omega\Delta}\gamma_\mu\omega^\mu]\psi_\Delta^\nu\,, \tag{5}$$

where ψ_Δ^ν is the Rarita-Schwinger spinor for the Δ-baryon.

Due to the uncertainty on the meson-Δ coupling constants, we limit ourselves to considering only the couplings with σ and ω meson fields, which are explored in the literature [12, 13, 29]. If the SU(6) symmetry were exact, one might adopt the universal couplings $x_{\sigma\Delta} = g_{\sigma\Delta}/g_{\sigma N} = 1$ and $x_{\omega\Delta} = g_{\omega\Delta}/g_{\omega N} = 1$. However, the SU(6) symmetry is not exactly fulfilled and one may assume the scalar coupling ratio $x_{\sigma\Delta} > 1$ with a value close to the mass ratio of the Δ and the nucleon [13]. On the other hand, QCD finite-density sum rule results show that the Lorentz vector self-energy for the Δ is significantly smaller than the nucleon vector self-energy implying therefore $x_{\omega\Delta} < 1$ [29]. In this paper we adopt different choices for the Δ-meson couplings which are consistent with the experimental flow data of heavy-ion collisions at intermediate energies [30, 10].

Because we are going to describe a dense asymmetric nuclear matter, we have to require the conservation of two charges: baryon number (B) and electric charge (C) [1]. As a consequence, the system is described by two independent chemical potentials: μ_B and μ_C, the baryon and the electric charge chemical potential, respectively. Therefore, the chemical potential of particle of index i can be written as

$$\mu_i = b_i\,\mu_B + c_i\,\mu_C\,, \tag{6}$$

[1] In heavy ion collisions strangeness can be produced only by strong-interaction via associated production and we neglect the contribution of strange hadrons because a tiny amount of strangeness can be produced in the range of temperature and density explored in this study. On the other hand, in the β-stability condition, realized in CSs, strangeness is mainly produced by weak interaction and is not conserved.

Journal of Physics: Conference Series **527** (2014) 012019 doi:10.1088/1742-6596/527/1/012019

where b_i and c_i are, respectively, the baryon and the electric charge quantum numbers of the ith hadron.

The thermodynamical quantities can be obtained from the baryon grand potential Ω_B in the standard way. More explicitly, the baryon pressure $P_B = -\Omega_B/V$ and the energy density can be written as

$$\begin{aligned} P_B &= \frac{1}{3}\sum_i \gamma_i \int \frac{\mathrm{d}^3k}{(2\pi)^3}\, \frac{k^2}{E_i^*(k)}\, [n_i(k)+\overline{n}_i(k)] \\ &- \frac{1}{2}m_\sigma^2\sigma^2 - U(\sigma) + \frac{1}{2}m_\omega^2\omega^2 + \frac{1}{4}c\,(g_{\omega N}\,\omega)^4 + \frac{1}{2}m_\rho^2\rho^2, \end{aligned} \tag{7}$$

$$\begin{aligned} \epsilon_B &= \sum_i \gamma_i \int \frac{\mathrm{d}^3k}{(2\pi)^3}\, E_i^*(k)\, [n_i(k)+\overline{n}_i(k)] \\ &+ \frac{1}{2}m_\sigma^2\sigma^2 + U(\sigma) + \frac{1}{2}m_\omega^2\omega^2 + \frac{3}{4}c\,(g_{\omega N}\,\omega)^4 + \frac{1}{2}m_\rho^2\rho^2 \,. \end{aligned} \tag{8}$$

3. Stability conditions and phase transition in dense asymmetric nuclear matter

As previously discussed, we are dealing with the study of a multi-component system with two conserved charges: baryon number and electric charge. For such a system, the Helmholtz free energy density F can be written as [21]

$$F(T,\rho_B,\rho_C) = -P(T,\mu_B,\mu_C) + \mu_B\rho_B + \mu_C\rho_C\,, \tag{9}$$

with

$$\mu_B = \left(\frac{\partial F}{\partial \rho_B}\right)_{T,\rho_C}, \qquad \mu_C = \left(\frac{\partial F}{\partial \rho_C}\right)_{T,\rho_B}. \tag{10}$$

In a system with N different particle species, the particle chemical potentials are expressed as the linear combination of the two independent chemical potentials μ_B and μ_C and, as a consequence, $\sum_{i=1}^N \mu_i\rho_i = \mu_B\rho_B + \mu_C\rho_C$.

Assuming the presence of two phases (denoted as I and II, respectively), the system is stable against the separation in two phases if the free energy of a single phase is lower than the free energy in all two phases configuration. The phase coexistence is given by the Gibbs conditions

$$\mu_B^I = \mu_B^{II}\,, \quad \mu_C^I = \mu_C^{II}\,, \quad P^I(T,\mu_B,\mu_C) = P^{II}(T,\mu_B,\mu_C)\,. \tag{11}$$

Therefore, at a given baryon density ρ_B and at a given net electric charge density $\rho_C = y\,\rho_B$ (with $y = Z/A$), the chemical potentials μ_B are μ_C are univocally determined. An important feature of this conditions is that, unlike the case of a single conserved charge, the pressure in the mixed phase is not constant and, although the total ρ_B and ρ_C are fixed, baryon and charge densities can be different in the two phases. For such a system in thermal equilibrium, the possible phase transition can be characterized by mechanical (fluctuations in the baryon density) and chemical instabilities (fluctuations in the electric charge density). As usual the condition of the mechanical stability implies

$$\rho_B\left(\frac{\partial P}{\partial \rho_B}\right)_{T,\rho_C} > 0\,. \tag{12}$$

By introducing the notation $\mu_{i,j} = (\partial\mu_i/\partial\rho_j)_{T,P}$ (with $i,j = B,C$), the chemical stability for a process at constant P and T can be expressed with the following conditions [14]

$$\rho_B\,\mu_{B,B} + \rho_C\,\mu_{C,B} = 0\,, \tag{13}$$

$$\rho_B\,\mu_{B,C} + \rho_C\,\mu_{C,C} = 0\,. \tag{14}$$

Journal of Physics: Conference Series **527** (2014) 012019 doi:10.1088/1742-6596/527/1/012019

Whenever the above stability conditions are not respected, the system becomes unstable and the phase transition take place. The coexistence line of a system with one conserved charge becomes in this case a two dimensional surface in (T, P, y) space, enclosing the region where mechanical and diffusive instabilities occur.

The chemical stability condition is satisfied if [14]

$$\left(\frac{\partial \mu_C}{\partial y}\right)_{T,P} > 0 \quad \text{or} \quad \begin{cases} \left(\frac{\partial \mu_B}{\partial y}\right)_{T,P} < 0\,, & \text{if } y > 0\,, \\ \left(\frac{\partial \mu_B}{\partial y}\right)_{T,P} > 0\,, & \text{if } y < 0\,. \end{cases} \tag{15}$$

In the above condition it has been considered possible negative values of $y = Z/A$ because of, as already observed, during a phase transition with two conserved charges, the electric charge fraction $y = \rho_C/\rho_B$ is not locally conserved in the single phase but only globally conserved. Therefore, during the compression of the system, the appearance of particles with negative electric charge (such as Δ^-) could, in principle, shift the diffusive instability region to negative values of y, even if the system is prepared with a positive y. Such a feature has no counterpart in the liquid-gas phase transition and, as we will see, it turns out to be very relevant in order to properly determine the instability region through the binodal phase diagram.

In the figure 1, we report the baryon and electric charge chemical potential isobars as a function of y, at fixed temperature $T = 50$ MeV and $x_{\sigma\Delta} = 1.3$ in the GM3 parameters set [22]. It is relevant to observe that for the value $x_{\sigma\Delta} = 1$, we do not find any mechanical or diffusive instability. Contrariwise, by increasing the $x_{\sigma\Delta}$ coupling ratio, mechanical and chemical instabilities take place. In particular, in the range $1 < x_{\sigma\Delta} \leq 1.1$, instabilities are restricted to very low values of temperature and electric charge fraction, but for $x_{\sigma\Delta} > 1.1$, such instabilities start to be much more relevant and extend to higher values of T and y.

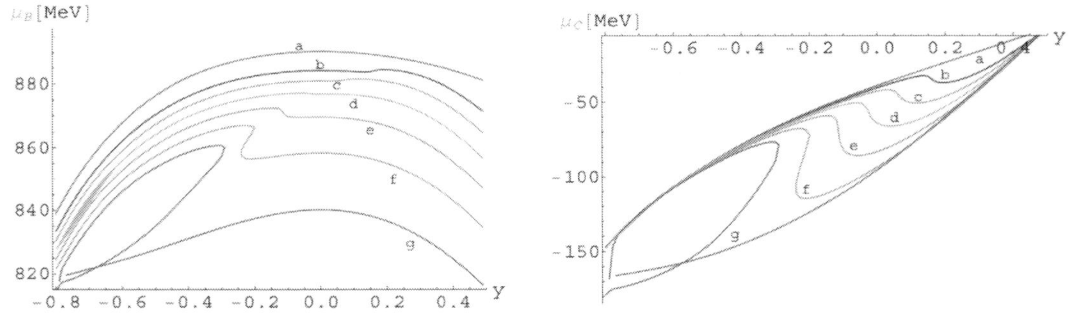

Figure 1. (Color online) Baryon (right panel) and electric charge (left panel) chemical potential isobars as a function of y at $T = 50$ MeV and $x_{\sigma\Delta} = 1.3$. The curves labeled a through g have pressure P=9,7,6,5,4,3,2 MeV/fm^3, respectively.

From the analysis of the above chemical potential isobars, we are able to construct the binodal surface relative to the nucleon-Δ matter phase transition and in figure 2, we show the relative binodal section with the two phases at different density and electric charge fraction y. The right branch (at lower density) corresponds to the initial phase (I), where the dominant component of the system is given by nucleons. The left branch (II) is related to the final phase at higher densities, where the system is composed primarily by Δ-isobar degrees of freedom (Δ-dominant phase). In presence of Δ-isobars the phase coexistence region results very different

Journal of Physics: Conference Series **527** (2014) 012019 doi:10.1088/1742-6596/527/1/012019

from what obtained in the liquid-gas case, in particular it extends up to regions of negative electric charge fraction and the mixed phase region ends in a point of maximum asymmetry with $y = -1$ (corresponding to a system with almost all Δ^{-}-particles, being antiparticles and pions contribution not relevant in this regime).

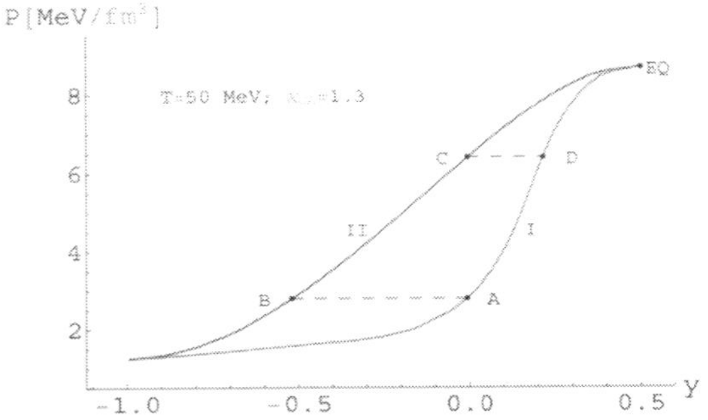

Figure 2. (Color online) Binodal section at $T = 50$ MeV and $x_{\sigma\Delta} = 1.3$.

Let us analyze the phase evolution of the system during the isothermal compression from an arbitrary initial point A, indicated in figure 2. In this point the system becomes unstable and starts to be energetically favorable the separation into two phases, therefore an infinitesimal Δ-dominant phase appears in B, at the same temperature and pressure. Let us observe that, although in B the electric charge fraction is substantially negative, the relative Δ^{-} abundance must be weighed on the low volume fraction occupied by the phase II near the point B. During the phase transition, each phase evolves towards a configuration with increasing y, in contrast to the liquid-gas case, where each phase evolves through a configuration with a decreasing value of y (with the exception of the gas phase after the maximum asymmetry point).

In figure 3, we report the phase diagram with in evidence the coexistence regions of the liquid-gas and the nucleon-Δ matter phase transition for $y = 0.3$ and 0.5 ($x_{\sigma\Delta} = 1.3$). The two coexistence regions are well separated and the features of the two phase transitions differ significantly. In fact, for the liquid-gas transition, asymmetric nuclear matter implies a reduction of the second critical density and of the critical temperature T_c. Contrariwise, for the Δ-dominant phase transition, we have a slight increase of the critical temperature and a significant reduction of the first critical density. In particular at moderate temperatures ($T \approx 30 \div 40$ MeV), the system begins the mixed phase at a baryon density of the order of ρ_0. This behavior could be phenomenologically relevant in order to identify such a phase transition in heavy ion collision experiments and in the bulk properties of protoneutron stars [31].

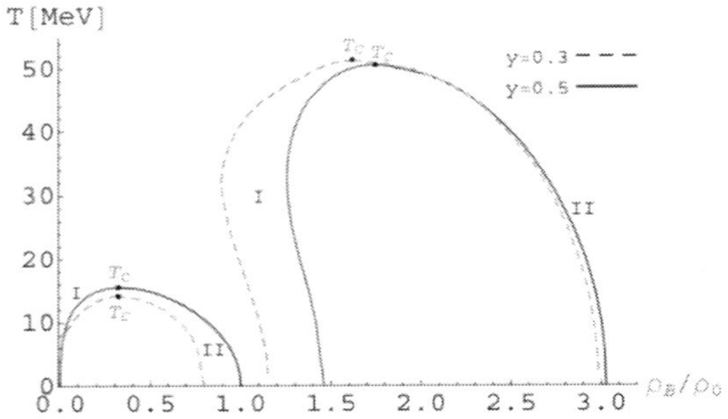

Figure 3. (Color online) Phase diagram of the liquid-gas and the nucleon-Δ matter phase transition for $y = 0.3$ (dashed curves) and $y = 0.5$ (continuous curves). The lines labeled with I and II, delimitate the first and second critical densities of the coexistence regions, respectively.

4. Mechanical and chemical instability in β-stable compact stars

In this section we are going to investigate the possible presence of thermodynamic instability in regime of β-stable hadronic matter at zero temperature. Differently by the previous section, we consider the hadronic EOS by including hyperons and electrons. Also in this case we are in presence of two conserved charge, the baryon number and the total electric charge density (fixed to $\rho_C = 0$). Such a system has the same thermodynamic stability conditions reported in the equations (12) and (15).

In figure 4 we report the pressure as a function of the baryon density for different values of the coupling $x_{\sigma\Delta}$ (and $x_{\omega\Delta} = 1$). We observe that the system does not satisfied the mechanical stability condition (12) for values of $x_{\sigma\Delta} \geq 1.25$.

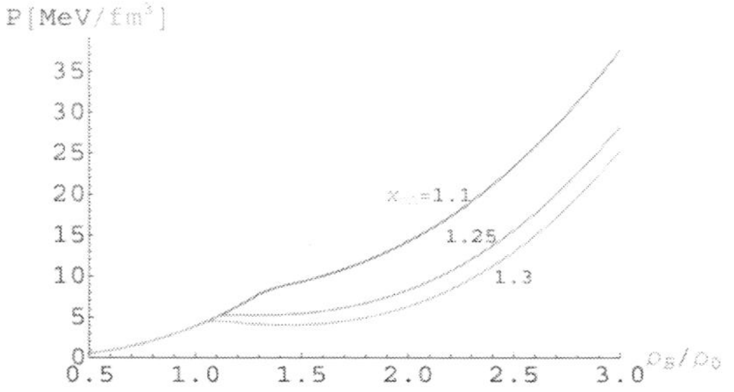

Figure 4. (Color online) Pressure as a function of baryon density for β-stable matter at zero temperature for various values of the coupling $x_{\sigma\Delta}$.

In figure 5, the electric charge chemical potential isobars μ_C are reported as a function of the net electric charge density for $x_{\sigma\Delta} = 1.25$ and at different values of pressure (the curves labeled a through e have pressure P=3, 4, 5, 8, 10 MeV/fm^3, respectively). It is possible to see that the chemical stability condition (15) is not always satisfied for fluctuations of the net ρ_C. In this case the β-stable hadronic EOS is unstable in a range of baryon density between $0.9 \div 2.5\,\rho_0$ and a mixed phase with two hadronic phases take place. It is important to outline that in this case the chemical instability condition is realized at lower pressure (and lower baryon density) with respect to the region in which the mechanical instability condition is realized. This matter of fact stresses the importance of considering both stability conditions for a binary system with two conserved charge.

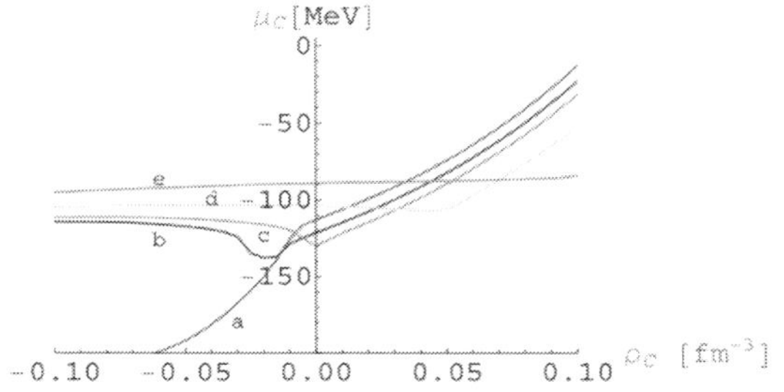

Figure 5. (Color online) Electric charge chemical potential isobars as a function of the charge density ρ_C for $x_{\sigma\Delta} = 1.25$. The curves labeled a through e have pressure P=3, 4, 5, 8, 10 MeV/fm^3, respectively.

As in the previous section, from the analysis of the two chemical potential isobars, it is possible to construct the binodal surface relative to the hadronic mixed phase at different baryon and electric charge concentrations.

The binodal section for $x_{\sigma\Delta} = 1.25$ is reported in figure 6. In this case the system has a fixed (global) net $\rho_C = 0$. Therefore, during the isothermal compression, the system in the point A (at about $\rho_B \approx 0.9\,\rho_0$) becomes unstable and starts to be energetically favorable the separation into two phases. An infinitesimal Δ-dominant phase at greater baryon density appears in B, at the same pressure. During the phase transition, each phase evolves towards a configuration with different $y = \rho_C/\rho_B$ because, although the electric charge asymmetry is globally conserved, this is not true for the single phase and it is energetically favorable to separate it into a phase at greater density with negative ρ_C, due to the presence of Δ^- particles (left branch) and a phase at lower density with $\rho_C > 0$ (right branch) rather than into two phases with equal charge fraction. Finally, the system ends the phase transition in the point C (at about $\rho_B \approx 2.5\,\rho_0$), leaving the phase with $\rho_C > 0$ in the point D.

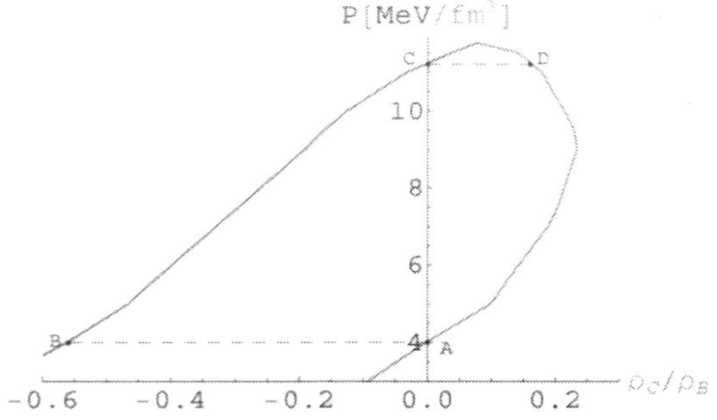

Figure 6. (Color online) Binodal section at zero temperature and $x_{\sigma\Delta} = 1.25$.

The above phase transition, strongly favors the formation of Δ^- particles at lower baryon density and implies a sensible variation in the structure of the CS. This matter of fact can be observed in figure 7, where we display the baryon density dependence of the particle's fractions. It is remarkable that the early appearance of Δ resonances, the first one being the Δ^-, considerably shifts the onset of hyperons which start to form at densities of $\sim 5\,\rho_0$.

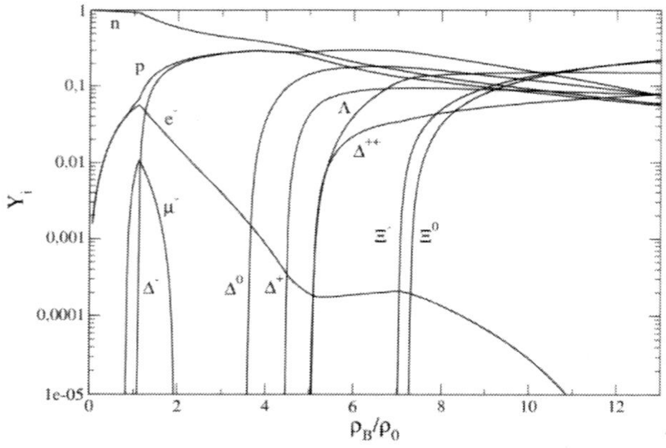

Figure 7. Particles fractions as functions of baryon density, for $x_{\sigma\Delta} = 1.25$, $x_{\omega\Delta} = 1$.

5. Can very compact and very massive neutron stars both exist?

In connection with the above results, related to the relevance of Δ-isobars in dense and β-stable nuclear matter, we are going to investigate an interpretation regarding to the possible co-existence of compact and very massive CSs in terms of two separate families of hadronic and quark stars [18].

Unfortunately the precise measurement of the radius of CSs has up to now proved to be extremely difficult, since it is in most cases based on specific assumptions concerning e.g. the atmosphere and the distance of the object under investigation. Different analysis often lead to

Journal of Physics: Conference Series **527** (2014) 012019 doi:10.1088/1742-6596/527/1/012019

opposite conclusions. There have been therefore claims of very small radii, of the order or smaller than about 10 km [32], while other analysis suggest for the same objects significantly larger radii, of the order of 12 km [33]. It is clear that a precise and model independent measurement of the radius of at least a few CSs is crucial to finally provide the necessary information which will allow the extraction of the EOS of stellar matter at large densities.

From the theoretical side the families of nucleonic, hyperonic and hybrid stars, stiff enough to reach $2\,M_\odot$, all provide radii which are not too small, typically larger than about 11.5-12 km for the canonical $1.4\,M_\odot$ star [3, 4, 34]. This seems to put a theoretical bias against the existence of stars having very small radii. No single EOS exists at the moment which is able to provide at the same time large masses for a few CSs and small radii for others. Since the situation from the observational viewpoint is still rather open, in this contribution we are going to discuss a model which satisfies those two conditions.

It is difficult to have a unique family of CSs allowing both very small radii and very massive configurations because to have small radii the EOS needs to be rather soft. Therefore, large densities are reached in the center of very compact stars, typically of the order of $5 \div 6\,\rho_0$ or larger. On the other hand, to have very massive configurations the EOS should be stiff at those same densities. No microscopic mechanism exists to allow a sudden stiffening of the EOS at those large densities. What we discuss is instead a solution based on two families of CSs, one made of hadrons and the other made of deconfined quarks stars (QSs) by assuming that the Bodmer-Witten hypothesis to hold true. While in the literature many papers exist in which two families have been discussed [35, 36], none takes into account the two constraints discussed above.

Concerning the quark matter EOS, we rely to the simple MIT bag model description in which confinement is provided by a bag constant $B_{\rm eff}$ and the perturbative QCD interactions are effectively included in the coefficient a_4 [37]. The total thermodynamical potential reads [5]:

$$\Omega = \sum_{u,d,s,e} \Omega_i + \frac{3\mu^4}{4\pi^2}(1 - a_4) + B_{\rm eff}\,, \tag{16}$$

where μ is the quark chemical potential and Ω_i are the thermodynamical potentials for non-interacting up, down, and strange quarks and electrons. The mass of the strange quark is fixed to 100 MeV while the up and down quarks are considered as massless. As shown in Refs. [5], in this scheme it is possible to obtain stellar configurations up to two solar masses or heavier. Here we will use the following parameters sets: $B_{\rm eff}^{1/4} = 142$ MeV – $a_4 = 0.9$ (set1), and $B_{\rm eff}^{1/4} = 127$ MeV – $a_4 = 0.6$ (set2) both taken from [5]. Set1 allows a maximum mass for QSs of $2\,M_\odot$, set2 has been implemented to give an example of quark EOS for which the maximum mass reaches $2.4\,M_\odot$.

The mass-radius relations for hadronic stars and for QSs are displayed in figure 8. The maximum mass of hadronic stars, containing both Δ resonances and hyperons, is close to $1.5\,M_\odot$ for the parameters' sets considered here. When excluding hyperons and Δ-resonances the maximum mass of neutron stars reaches instead a value of $\sim 2\,M_\odot$ but with a large radius. The appearance of Δ resonances is crucial to obtain very compact stellar configurations (as also shown in Ref. [15]) with radii down to 8 km: the corresponding mass-radius curves enter the area, framed by the two vertical lines, of very compact objects inferred in Ref. [32]. The appearance of hyperons in the stars provides a further softening of the EOS, reducing the maximum mass of $\sim 0.1 \div 0.2\,M_\odot$ (see solid/dashed curves related to hadronic stars). On the other hand, the mass of QSs can reach values compatible with the recent limit of $2\,M_\odot$ (solid line) or even higher values (dashed line). Notice that QSs mass-radius relations also enter the area of very compact objects but for masses $\leq 1\,M_\odot$: such light stars are difficult to produce in standard supernova simulations and moreover the lightest known neutron star has a mass of $\sim 1.2\,M_\odot$.

The interpretation we propose here is that massive stars, $M \geq 1.5 \div 1.6\, M_\odot$, are QSs with radii $R \geq 11$ km whereas stars with $R \leq 10$ km are composed mainly of nucleons and Δ resonances, with a maximum mass of $\sim 1.5 \div 1.6\, M_\odot$. The tension between measurements of large masses and small radii could be strengthened if a neutron star more massive than $2\, M_\odot$ is discovered (see, for example, Ref. [38]), favoring our interpretation of two coexisting families of CSs.

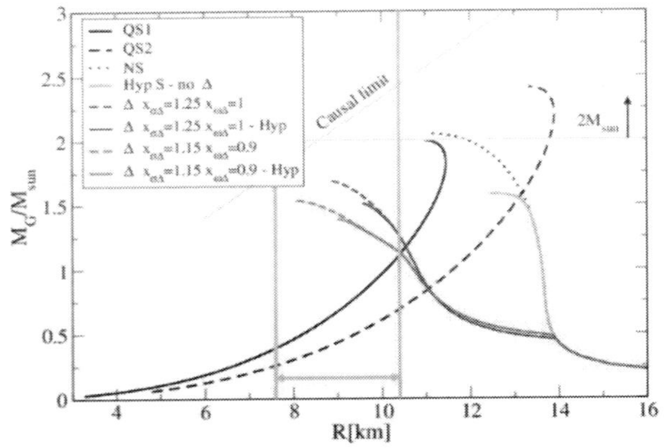

Figure 8. (Color online) Mass-radius relations of QSs (set1 and set2) and of hadronic stars. The two vertical lines correspond to the 1σ limit for the radii extracted from the analysis of quiescent low-mass X-ray binaries [32].

A crucial question concerns the astrophysical scenarios in which hadronic and QSs are formed and how QSs can generate from hadronic stars. In figure 9 we display the gravitational and baryonic masses as functions of the radius for hadronic stars and QSs. On this plot it is possible to construct a path for the formation of QSs from cold hadronic stars accreting matter from a companion. The stellar configuration labeled with B on the solid line represents the hadronic star for which hyperons start to form in the inner core (notice that at the corresponding point on the baryonic mass curves, the branch with hyperons deviates from the branch with only Δ resonances). The larger the mass of the star the larger its hyperon content. Notice that: i) only in the presence of hyperons, which carry strangeness, can droplets of strange quark matter form via nucleation; ii) the star can decay into a QS with the same baryonic mass since this process is energetically favored because the gravitational mass of the configuration D is smaller than the one of B. The energy released in the conversion of a hadronic star into a quark star has been estimated in many papers and can easily reach 10^{53} erg [35, 36, 39, 40].

All the hadronic stellar configurations between B and A can transform into QSs, the probability and velocity of conversion depending on the specific microphysics process of formation of the first droplets of quark matter and on the subsequent expansion of the newly formed phase.

There are many studies in the literature addressing these issues. In the scenario here discussed, conversion of cold hadronic stars, quantum nucleation represents a possible mechanism for the formation of the first quark matter droplet [35, 36, 39, 40]. Once a seed of quark matter is formed, the conversion of the whole hadronic star proceeds very fast, with time scales of the order of ms, due to the development of hydrodynamical instabilities [41, 42]. A detailed study of the conversion process with the new proposed EOSs is mandatory for future works.

Another scenario for the formation of quark stars is related to the supernova explosion of massive progenitors. Large densities can be reached at the moment of the collapse, soon after the bounce, due to the large fallback and hyperons can already appear at this stage, immediately triggering the formation of quark matter. There the energy released in the conversion can help Supernovae to explode [39]. In general the conversion of a hadronic star into a QS will produce

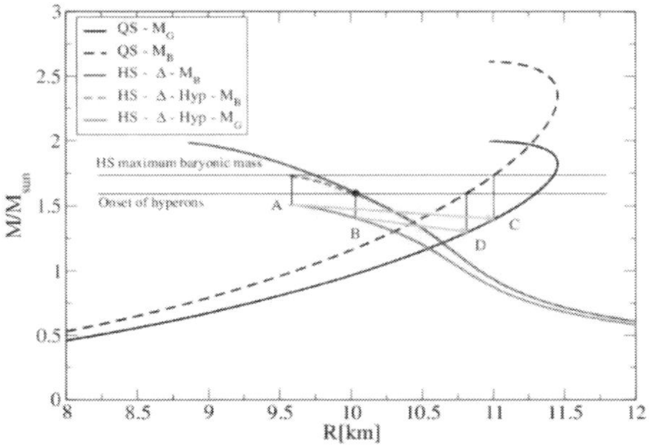

Figure 9. (Color online) Gravitational and baryonic mass-radius relations of QSs (set1) and of hadronic stars (with and without hyperons, for $x_{\sigma\Delta} = 1.25$, $x_{\omega\Delta} = 1$). A is the maximum mass of hadronic stars containing hyperons. B is the gravitational mass for which hyperons start to form. The quark stellar configurations D and C have the same baryonic masses of B and A but smaller gravitational masses.

spectacular transient events such as neutrino and gamma-ray-bursts.

There are many possible observable which could be used to test our proposal in which most of the known neutron stars (with masses close to $\sim 1.4\,M_\odot$) are hadronic stars while massive stars are more likely QSs (bare or with a crust). We predict that massive CSs also have large radii and, being composed by a different type of matter with respect to the $1.4\,M_\odot$ stars (in particular regarding strangeness), should show anomalous cooling histories and spinning frequency distributions. Moreover also quasi-periodic oscillations of very massive CSs should differ from the ones of hadronic stars [43].

6. Conclusions

The main goal of this work is to show the possible existence of chemical and mechanical instability for asymmetric and β-stable dense nuclear matter with possible phenomenological consequences in compressed baryonic matter regime, reachable in high energy heavy ion collisions, and in the physics of compact stars. At this scope we have studied the relativistic nuclear EOS with the inclusion of Δ-isobars and by requiring global conservation of baryon and electric charge densities. Similarly to the liquid-gas phase transition in nuclear matter, a nucleon-Δ matter phase transition can also occur at higher densities. We have shown that for asymmetric nuclear matter both mechanical and chemical instabilities take place and, during the phase transition, the two hadronic phases have a different baryon density and electric charge fraction. In the liquid-gas phase transition, the process of producing a larger neutron excess in the gas phase is referred to as isospin fractionation [44]. A similar effects can occur in the nucleon-Δ matter phase transition due essentially to a Δ^- excess in the Δ-matter phase with lower values of y. This feature could be phenomenologically relevant in heavy ion collision experiments and in the bulk properties of protoneutron stars. The presence of the Δ-isobar degrees of freedom plays a crucial role also in β-stable nuclear matter at zero temperature. Also in this case we are in presence of a binary system with two conserved charges (baryon and electric charge numbers) and the thermodynamic instabilities imply a phase transition favoring the formation of Δ^- particles at relative low baryon density ($\rho_B \approx \rho_0$) and shifting the formation of hyperon particles at large baryon densities ($\rho_B \approx 5 \div 6\,\rho_0$).

Finally, we have discussed the possible co-existence of very compact and very massive compact stars in terms of two separate families: compact hadronic stars in which an early appearance of Δ resonances is crucial to guarantee the stability of the branch of hadronic stars and very massive quark stars. In this context, it is proper to remember a well known argument against the co-existence of quark stars and neutron stars, based on the production of strangelets during the merging of two compact stars. If at least one of the two compact stars is a quark star it is possible that strangelets are emitted polluting the whole Galaxy and triggering the conversion of all compact stars into quark stars. However, recent numerical simulation of quark stars' mergers have shown that, in many cases, a prompt collapse to a black hole occurs and no matter is ejected [45]. Another possibility to prevent the strangelets pollution is offered by hydrodynamical simulations [42] in which the burning of a neutron star into a quark stars seems to be uncomplete and a possible thick layer of hadronic matter survives shielding the inner quark matter core.

References

[1] Demorest P, Pennucci T, Ransom S, Roberts M, and Hessels J 2010 *Nature* **467** 1081
[2] Antoniadis J *et al.* 2013 *Science* **340** 6131
[3] Bednarek I, Haensel P, Zdunik J, Bejger M, and Manka R 2012 *Astron. Astrophys.* **543** A157
[4] Weissenborn S, Chatterjee D, and Schaffner-Bielich J 2012 *Phys. Rev. C* **85** 065802
[5] Weissenborn S, Sagert I, Pagliara G, Hempel M, and Schaffner-Bielich J 2011 *Astrophys. J.* **740** L14
[6] Di Toro M *et al.* 2006 *Nucl. Phys. A* **775** 102
[7] Bonanno L, Drago A, and Lavagno A 2007 *Phys. Rev. Lett.* **99** 242301
[8] Senger P 2011 *Nucl. Phys. A* **862-863** 139
[9] Hofmann M, Mattiello R, Sorge H, Stocker H, and Greiner W 1995 *Phys. Rev. C* **51** 2095
[10] Lavagno A 2010 *Phys. Rev. C* **81** 044909
[11] Lavagno A 2013 *Eur. Phys. J. A* **49** 102
[12] Li Z, Mao G, Zhuo Y, and Greiner W 1997 *Phys. Rev. C* **56** 1570
[13] Kosov D S, Fuchs C, Martemyanov B V, and Faessler A 1998 *Phys. Lett. B* **421** 37 (1998)
[14] Lavagno A and Pigato D 2012 *Phys. Rev. C* **86** 024917
[15] Schurhoff T, Schramm S, and Dexheimer V 2010 *Astrophys. J.* **724** L74
[16] Hu Xiang and Guo Hua 2003 *Phys. Rev. C* **67** 038801
[17] Chen Y, Yuan Y, and Liu Y 2009 *Phys. Rev. C* **79** 055802
[18] Drago A, Lavagno A, and Pagliara G 2014 *Phys. Rev. D* **89** 043014
[19] Baldo M, Burgio G, and Schulze H 2000 *Phys. Rev. C* **61** 055801
[20] Vidana I, Polls A, Ramos A, Engvik L, and Hjorth-Jensen M 2000 *Phys. Rev. C* **62** 035801
[21] Müller H and Serot B D 1995 *Phys. Rev. C* **52** 2072
[22] Glendenning N K 1992 *Phys. Rev. D* **46** 1274
[23] Schaffner J, Dover C B, Gal A, Greiner C, and Stöcker H 1993 *Phys. Rev. Lett.* **71** 1328
[24] Schaffner J and Mishustin I N 1996 *Phys. Rev. C* **53** 1416
[25] Millener d J, Dover C B, and Gal A 1998 *Phys. Rev. C* **38** 2700
[26] Schaffner-Bielich J and Gal A 2000 *Phys. Rev. C* **62** 034311
[27] Boguta J 1982 *Phys. Lett. B* **109** 251
[28] de Jong F and Malfliet R 1992 *Phys. Rev. C* **46** 2567
[29] Jin X 1995 *Phys. Rev. C* **51** 2260
[30] Danielewicz P, Lacey R, and Lynch W G 2002 *Science* **298** 1592
[31] Lavagno A and Pigato D 2011 *Eur. Phys. J. A* **47** 52
[32] Guillot S, Servillat M, Webb N A, and Rutledge R E 2013 *Astrophys. J.* **772** 7
[33] Lattimer J M and Steiner A W 2013 `arXiv:1305.3242`
[34] Alford M G, Han S, and Prakash M 2013 *Phys. Rev. D* **88** 083013
[35] Berezhiani Z, Bombaci I, Drago A, Frontera F, and Lavagno A 2003 *Astrophys. J.* **586** 1250
[36] Bombaci I, Parenti I, and Vidana I 2004 *Astrophys. J.* **614** 314
[37] Alford M, Braby M, Paris M, and Reddy S 2005 *Astrophys. J.* **629** 969
[38] van Kerkwijk M, Breton R, and Kulkarni S 2011 *Astrophys. J.* **728** 95
[39] Drago A, Lavagno A, and Pagliara G 2004 *Phys. Rev. D* **69** 057505
[40] Bombaci I, Lugones G and Vidana I 2007 *Astron. Astrophys.* **462** 1017
[41] Drago A, Lavagno A, and Parenti I 2007 *Astrophys. J.* **659** 1519
[42] Pagliara G, Herzog M, and Ropke F 2013 *Phys. Rev. D* **87** 103007
[43] Watts A L and Reddy S 2007 *Mon. Not. Roy. Astron. Soc.* **379** L63
[44] Baran V, Colonna M, Di Toro M, and Greco V 2001 *Phys. Rev. Lett.* **86** 4492
[45] Bauswein A *et al.* 2009 *Phys. Rev. Lett.* **103** 011101

Journal of Physics: Conference Series **527** (2014) 012020 doi:10.1088/1742-6596/527/1/012020

The amazing properties of crystalline color superconductors

Massimo Mannarelli

INFN, Laboratori Nazionali del Gran Sasso, Assergi (AQ), Italy

E-mail: massimo@lngs.infn.it

Abstract. This paper is a brief journey into the amazing realm of crystalline color superconductors. Starting from a qualitative description of superfluids, superconductors and supersolids, we show how inhomogeneous phases may arise when the system is under stress. These basic concepts are then extended to quark matter, in which a richer variety of phases can be realized. The most interesting properties of the crystalline color superconductors are presented. This brief journey ends with a discussion of crystalline color superconductors in compact stars and related astrophysical observables. We aim at providing a pedagogical introduction for nonexpert in the field to a few interesting properties of crystalline color superconductors, without discussing the methods and the technicalities. Thus, the results are presented without a proof. However, we try to give a qualitatively clear description of the main concepts, using standard quantum field theory and analogies with condensed matter systems.

1. Introduction

The crystalline color superconducting (CCSC) phase can be qualitatively described as a system of deconfined quarks in which the strong interaction favors the formation of a quark-quark condensate that is spatially modulated as a crystal.

This rather odd phase of matter is one of the candidate phases of beta equilibrated and electrically neutral quark matter at very large baryonic density and sufficiently low temperature, see [1] for a review. The appropriate conditions may be realized within compact stellar objects (CSOs), which are stars having a radius of about 10 km and a mass of 1 - 2 $M_{\odot}$. Soon after their birth from the gravitational collapse of the remnants of a supernova explosion, the temperature of CSOs is rather hot, of the order of 10^{11}K, but then they rapidly cool down by neutrino emission to temperatures of the order of 10^7K - 10^8K that is of the order of tens of keV at most. Although very high for our daily lives, this is is a rather cold temperature as compared to the natural scale of Quantum Chromodynamics (QCD), $\Lambda_{QCD} \sim 200$ MeV. The density of the CSOs is also very large, it is so large that the average distance between nuclei is less than 1 fm $\sim 1/\Lambda_{QCD}$, which is the typical size of a nucleon. Although first principle calculations are not feasible at this scale, it is clear that for matter squeezed at such extremes, talking about nucleons is not appropriate. In these conditions the relevant degrees of freedom should be quarks and gluons.

A pictorial description of the behavior of matter with increasing baryonic density is given in figure 1. The density increases from the top to bottom of the figure and the corresponding qualitative features of matter are sketched. In the confined phase (above the horizontal dashed line), with increasing matter density the number of nucleons in a nucleus increases, reaching the so-called nuclear saturation density for heavy nuclei. Further increasing the density one expects that nuclei start to melt and neutrons are liberated, neutron drip. Up to this point QCD is a confining theory, meaning that quarks and gluons are not the correct degrees of freedom for the description of matter. For larger values of the density the strong interaction is no more able to confine quarks, and the "quark drip" should happen. Note that at this point the strong interaction is still nonperturbative. This is one of the very important aspects: there is a range of density in which quark matter if liberated is still nonperturbatively interacting. It is this range of densities that is expected to be relevant for compact stars.

Only at asymptotic densities QCD becomes perturbative and solid results based on perturbation theory can be achieved. Indeed, it is only at extreme densities that we know the state of quark matter: it becomes a color superconductor. In this phase quarks form a degenerate soup, filling the Fermi sphere up to a large Fermi energy. The color attractive interaction induces the formation of Cooper pairs, with an average correlation length larger then the average distance between two quarks (bottom line of figure 1).

The hope is that from the asymptotic density down to CSOs densities no other phase pops up. This is one of the uncontrollable hypothesis about color superconductors. In QCD, nonperturbative physics is typically studied by lattice QCD simulations. Unfortunately these simulations cannot be easily done with a large baryonic chemical potential, because of the so-called sign problem [2], see for example [3] for a recent review on progress in the field.

In the following we shall simply assume that reducing the density from the asymptotic value down to the nuclear matter phase transition, the color superconducting phase is the only energetically favored phase.

Before turning to a more detailed description of color superconductors, let us first define what are superfluids and superconductors.

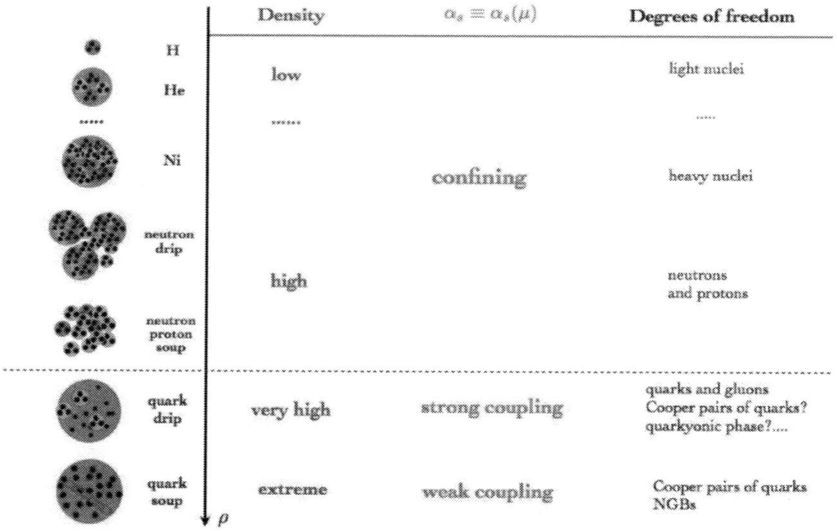

Figure 1. (Color online) Schematic representation of the various phases of baryonic matter at various densities.

2. Superfluids, superconductors and supersolids

We give two different definitions of superfluids and superconductors: one operative and one formal.

Definition 2.1. *Superfluid. Operative: Frictionless fluid with potential flow that when rotated is threaded by quantized vortices. Formal: System with the spontaneous breaking of a global symmetry below a critical temperature.*

Definition 2.2. *Superconductor. Operative: Almost perfect diamagnet, screening the magnetic field in restricted domains. Formal: System with the "spontaneous breaking" of a gauge symmetry below a critical temperature.*

In both cases, the two definitions are equivalent. In superfluids the theoretical definition is based on the Goldstone theorem implying that there is at least a massless mode, ϕ, allowing the easy transport of the quantum numbers associated with the broken global symmetries. The fluid velocity $\boldsymbol{v}$ can be written as $\boldsymbol{v} = \nabla\phi$. It follows that the vorticity, $\nabla \times \boldsymbol{v}$, vanishes almost everywhere. It can be shown that the circulation of $\boldsymbol{v}$ is quantized, meaning that each vortex carries a quantum of angular momentum. In superconductors, the magnetic field is screened over a distance $1/M$, with M the gauge fields magnetic mass acquired by the Higgs-Anderson mechanism.

In the analysis of the properties of a system one must verify that all the above requirements are met. As a counterexample, consider the ballistic propagation of a dilute system of neutral particles, say neutrons or neutrinos. One can describe it as a frictionless fluid with vanishing vorticity. However, it is certainly not a superfluid. Indeed, if one could spin it, for example by confining neutrons in a trap and then putting the trap in rotation, one would not observe quantized vortices. Moreover, no condensation occurs and no global symmetry of the system is spontaneously broken.

Consider instead the case of neutrons forming a nn condensate. This is likely to happen in the inner crust of neutron stars. In this case the global symmetry corresponding to neutron number conservation is broken (the neutron number is not conserved because neutrons can "disappear" in the condensate). This system is a genuine superfluid. That is, the spinning neutron superfluid is believed to be threaded by quantized vortices.

There is a number of subtleties that we have not addressed in the definitions of superfluids and superconductors, however, our definitions should be sufficiently general to encompass most of the known forms of superfluids and superconductors and, more important, to avoid confusion.

Finally, we say that a system has a superflow mode whenever there is a breaking of a symmetry of the system which allows the easy transport of a quantum number. We now wonder whether superfluids and superconductors are the only forms of superflows that are realizable in nature.

2.1. Supersolids

It might sound unreasonable that a solid could become superfluid or that a superfluid could solidify keeping zero viscosity. The two concepts: superfluidity and rigidness seem to be in conflict. A solid is difficult to deform (it has by definition a nonvanishing shear modulus), a fluid has no shear modulus and a superfluid has also a vanishing shear viscosity! Moreover, a superfluid is a highly delocalized system while a solid is characterized by long-range order [4].

Transforming a superfluid in a solid is in general possible. A notable example is helium. Both He^4 (bosonic) and He^3 (fermionic) become solids with increasing pressure. The way in which helium becomes solidifies is by compression. When compressed, the repulsion between helium atoms forces the system to a closely packed configuration, typically an hexagonal close-packed (HCP) crystal, a body centered cubic (BCC) crystal or a face-centered cubic (FCC) crystal. It is clear that once one of these configurations is reached, helium atoms cannot easily move around.

Journal of Physics: Conference Series **527** (2014) 012020 doi:10.1088/1742-6596/527/1/012020

Indeed the system loses its superfluid property. However, if defects are present, atoms might be able to move. For this reason it has been supposed that "nonperfect" solid helium could be a supersolid. However, experimental results with helium atoms have not unambiguously shown that supersolid helium exists, see [5, 6] for reviews. Different systems consisting of ultracold bosonic atoms with a particular long-range behavior [7] seem to be good candidates [8, 9].

Here is a definition of supersolids (see [5] for more details):

Definition 2.3. *A supersolid is a system in which the spontaneous breaking of rotational symmetries down to a discrete symmetry and the spontaneous breaking of an internal symmetry take place simultaneously and for the same type of particles.*

In a supersolid particles are delocalized throughout the whole system, and simultaneously there is long-range order [5] (meaning that the diffraction pattern of a supersolid has narrow peaks). The former property is a consequence of the breaking of an internal symmetry of the system; the latter refers to the breaking of space symmetries. What is crucial in the above definition is that the two spontaneous symmetry breaking must appear simultaneously and for the same type of particles [5].

3. Color superconductors

We now turn to quark matter, see [10, 1] for reviews. Only up, down and strange quarks are relevant, because other quark flavors have masses larger than the typical chemical potential realizable in CSOs, of the order of 400 MeV. The formation of quark Cooper pairs can be described by the condensate

$$\langle 0|\psi^\alpha_{iL}\psi^\beta_{jL}|0\rangle = -\langle 0|\psi^\alpha_{iR}\psi^\beta_{jR}|0\rangle \propto \sum_{I=1}^{3} \Delta_I \varepsilon^{\alpha\beta I}\epsilon_{ijI}\,, \tag{1}$$

where $\psi^\alpha_{iL/R}$ represents a left/right handed quark fields with color α and flavor i. The quantities Δ_1, Δ_2 and Δ_3, describe d-s, u-s and u-d Cooper pairing, respectively, and are proportional to the energy needed to break the corresponding Cooper pair; $\varepsilon^{\alpha\beta\gamma}$ and ϵ_{ijk} are the completely antisymmetric symbols in color and flavor space, respectively. The color structure is determined by the fact that the completely antisymmetric $\bar{3}$ channel is attractive, then requiring that the condensate has zero total angular momentum the flavor structure must be completely antisymmetric as well. For the sake of notation we have suppressed spinorial indices and for the sake of simplicity we have neglected pairing in the symmetric color sextet channel. In principle, pairing in the color sextet channel should be included in the color structure [11, 12], but the condensate in this channel is much smaller than in the $\bar{3}$ color channel [13, 14].

The best studied three-flavor quark phase is the so-called color-flavor locked (CFL) phase [11], which is believed to be the energetically favored form of three-flavor quark matter at asymptotic densities. It is characterized by $\Delta_1 = \Delta_2 = \Delta_3 = \Delta_{\rm CFL}$. The reason of the name "color-flavor locked" is that only simultaneous transformations in color and in flavor spaces leave the condensate invariant. The corresponding symmetry breaking pattern is the following

$$SU(3)_{color} \otimes SU(3)_L \otimes SU(3)_R \otimes U(1)_B \to SU(3)_{color+L+R} \otimes Z_2\,, \tag{2}$$

where $SU(3)_{color+L+R}$ is the diagonal global subgroup of the three $SU(3)$ groups. The presence of the Z_2 symmetry in the broken phase means that multiplication of the quark fields by -1 leaves the vacuum invariant. According to the symmetry breaking pattern, the 17 generators of chiral symmetry, color symmetry and $U(1)_B$ symmetry are spontaneously broken.

The 8 broken generators of the color gauge group correspond to the 8 longitudinal degrees of freedom of the gluons, thus these gauge bosons acquire a Meissner mass by the Higgs-Anderson

Journal of Physics: Conference Series **527** (2014) 012020 doi:10.1088/1742-6596/527/1/012020

mechanism. According to our definition 2.2 the system is a color superconductor. The attribute "color" referes to the fact that the broken gauge symmetry is $SU(3)_{color}$.

The diquark condensation induces a Majorana-like mass term in the fermionic sector which is not diagonal in color and flavor indices. Thus, the fermionic excitations consist of gapped "quasiquark" modes, that is modes nondiagonal in color and flavor indices, having mass proportional to $\Delta_{\rm CFL}$.

The low-energy spectrum consists of 9 Nambu-Goldstone bosons (NGBs) organized in an octet, associated with the breaking of the flavor group, and in a singlet, associated with the breaking of the baryonic number, the so called H-phonon.

For nonvanishing quark masses the octet of NGBs becomes massive, but the H-phonon is protected by the $U(1)_B$ symmetry and remains massless. From our definition 2.1 it follows that the CFL phase is a baryonic superfluid.

4. Crystalline color superconductors

The CFL phase is a particular realization of the standard homogenous BCS pairing, in which fermions of the whole Fermi surface contribute coherently to the formation of a condensate. In general, in a BCS phase the pairing results in a free energy gain of the order of the homogeneous paring gap, Δ_0. This ideal situation assumes that the Fermi spheres of the interacting fermions have an equal radius. However, β-equilibrated and electrically neutral quark matter might not be such an ideal system.

For definiteness, let us turn off the strong interactions, meaning that we consider an electroweakly interacting gas of u, d and s quarks. We treat the effective strange quark mass, M_s, as a parameter and consider time scales for which weak equilibrium and electrical neutrality are relevant. The weak processes $u \to d + \bar{e} + \nu_e$, $u \to s + \bar{e} + \nu_e$ and $u + d \leftrightarrow u + s$ imply that

$$\mu_u = \mu - \frac{2}{3}\mu_e \qquad \mu_d = \mu + \frac{1}{3}\mu_e \qquad \mu_s = \mu + \frac{1}{3}\mu_e \,, \tag{3}$$

where $\mu = (\mu_u+\mu_d+\mu_s)/3$ is the average chemical potential. The system is (globally) electrically neutral when the number densities obey the following equation

$$\frac{2}{3}n_u - \frac{1}{3}n_d - \frac{1}{3}n_s - n_e = 0 \,, \tag{4}$$

meaning that the total electric charge is zero. From the above equations it follows a relation between the electron chemical potential and the effective strange quark mass, see for example [15],

$$2\,\delta\mu \equiv \mu_e \simeq \frac{M_s^2}{4\mu} \,, \tag{5}$$

where we have also defined the chemical potential mismatch, $\delta\mu$. Certainly, the presence of the strong interaction can change the quark chemical potentials, but the point is that the strong interaction is diagonal in flavor, thus it can only change the intraflavor chemical potential difference between quarks of different colors. Only the locking of the flavor and color degrees of freedom by a condensate can change significantly the interflavor chemical potential difference. It is possible to show that the effect is small and does not significantly change the hierarchy of the chemical potentials [1].

If the Fermi surfaces have different radii, there is a free energy penalty proportional to $\delta\mu$ for creating Cooper pairs. For small values of $\delta\mu$, the homogenous BCS phase is still energetically favored. However, it is clear that $\delta\mu$ is producing a stress on the BCS pairing and this stress cannot be arbitrarily large. In general, we expect that when $\delta\mu > c\Delta_0$, where c is some number[1],

[1] The actual value of c depends on the detailed form of the interaction; in weak coupling $c = 1/\sqrt{2}$ [16, 17], in strong coupling larger values are allowed, see for example the analysis in [18].

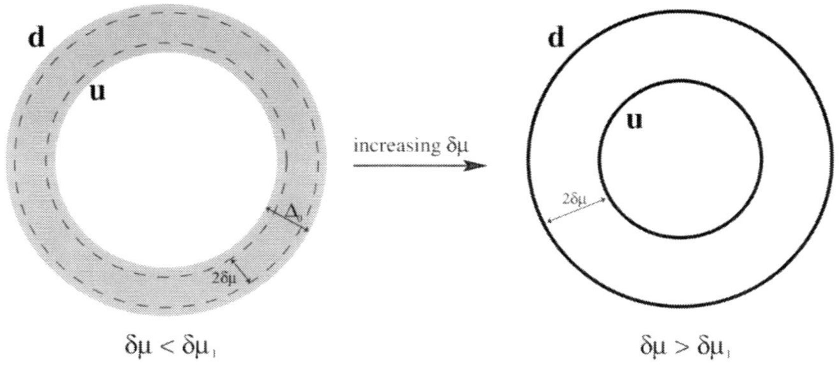

Figure 2. (Color online) Sketch of the Fermi spheres of two populations of fermions, with up (**u**) and down (**d**) spins. Left panel: the dashed black lines correspond to the Fermi spheres of two noninteracting populations for a moderate chemical potential mismatch. Turning on the attractive interaction between **u** and **d** fermions, the BCS pairing takes place producing the smearing of the Fermi spheres. Right panel: for a mismatch above the critical value $\delta\mu_1$ the Fermi spheres of the two populations are widely separated and the BCS pairing is not energetically favored.

the BCS homogenous pairing cannot take place. The naive expectation is that for large stresses induced by $\delta\mu$ the system makes a phase transition to the normal phase. This situation is pictorially described in figure 2.

The naive picture discussed so far assumes that only a homogenous condensate can be realized. As frequently happens in physical systems under stress, the transition to inhomogeneous phases might be energetically favored. The first explorations of this direction have been done by Larkin, Ovchinnikov, Fulde and Ferrell (LOFF) [19, 20]. For a simple system consisting of fermions in two spins states, labeled by $s, t = 1, 2$, the Fulde–Ferrell (FF) condensate is given by

$$\langle \psi_s(x)\psi_t(x)\rangle \propto \Delta\, \sigma_{2,st}\, e^{2i\boldsymbol{q}\cdot\boldsymbol{x}} \,, \tag{6}$$

where σ_2 is the standard Pauli matrix. The spin structure is characteristic of a spin-0 state, what is new here with respect to BCS pairing is the presence of the plane wave space modulation. In weak coupling this phase is energetically favored with respect to the normal phase in the so-called LOFF window $\delta\mu \in (\delta\mu_1, \delta\mu_2)$, where $\delta\mu_1 = \Delta_0/\sqrt{2}$ and $\delta\mu_2 \simeq 0.75\Delta_0$. Let us list the properties of the FF condensate

(a) The condensate is inhomogeneous and is modulated as a plane wave in the $\boldsymbol{q}$ direction.
(b) The $SO(3)$ space symmetry is spontaneously broken to an $O(2)$ symmetry, corresponding to rotations around the $\boldsymbol{q}$ direction.
(c) The quantity $2\boldsymbol{q}$ is the total momentum of the pair.

The property (a) is just a definition of the FF pairing. The property (b) means that the ground state is not invariant under space rotations. As a consequence there will be extra low-energy excitations (similar to standard phonons in solids), which are associated with the breaking of the $SO(3)$ space symmetry. The property (c) (probably the less obvious one[2]), states that there is a coherent current of fermionic pairs in one spontaneously chosen direction. However, it does

[2] But the reader can easily prove it by a Fourier transformation of equation (6).

Journal of Physics: Conference Series **527** (2014) 012020 doi:10.1088/1742-6596/527/1/012020

not correspond to a net current of fermions, because there exists a counterpropagating current of unpaired fermions [20].

The Fermi momentum description of the FF pairing is illustrated in the left panel of figure 3. This figure shows that only a small fraction of the Fermi sphere (corresponding to the two ribbons) participates in pairing. These *pairing regions* sit on the top of the Fermi spheres, thus there is no free energy price proportional to $\delta\mu$ to be payed. The free energy cost of the FF pairing is instead due to the creation of counterpropagating currents, limiting the extension of the LOFF window. This fact can be qualitatively understood as follows. The value of $q = |\boldsymbol{q}|$ is determined by minimizing the free energy, however it is clear from the geometry of figure 3 that $q > \delta\mu$. Thus, increasing $\delta\mu$, q must increase as well, meaning that if the pairing region remains the same the fermionic current increases. What actually happens, with increasing $\delta\mu$, is that the pairing regions shrink, reducing the generated current, until they completely disappear at $\delta\mu = \delta\mu_2$. Thus at $\delta\mu_2$ the system makes a smooth (second order) phase transition to the normal phase.

The Fermi sphere description suggests an easy way of generalizing equation (6), by the addition of a second plane wave, as in the right panel of figure 3. Here ϕ indicates the angle between the direction of the two plane waves. The angle ϕ can be used as a variational parameter, and it is probably of little surprise the fact that the favored angle corresponds to $\phi = \pi$. Keeping the discussion qualitative, in this configuration there are two coherent counterpropagating currents of paired fermions, therefore no additional current of normal fermions has to be generated. It is also clear that adding more plane waves one can further minimize the free energy, but configurations with overlapping ribbons are disfavored because the associated currents would interfere distroying coherence.

By a further generalization of the previous reasoning, configurations with several wave vectors corresponding to nonoverlapping ribbons on the Fermi surfaces should be energetically favored. These wave vectors will thus have an ordered configuration, describing the reciprocal vectors of a crystalline structure. Let us discuss this possibility in quark matter.

Including color and flavor degrees of freedom the generalized color crystalline condensate can be obtained by "joining" equation (1) with equation (6), to give

$$\langle 0|\psi^\alpha_{iL}\psi^\beta_{jL}|0\rangle = -\langle 0|\psi^\alpha_{iR}\psi^\beta_{jR}|0\rangle \propto \sum_{I=1}^{3}\Delta_I\varepsilon^{\alpha\beta I}\epsilon_{ijI}\sum_{\boldsymbol{n}_I^m\in\{\boldsymbol{n}_I\}} e^{2iq\boldsymbol{n}_I^m\cdot\boldsymbol{x}}\,, \tag{7}$$

where we have suppressed the spinorial indices.

This condensate describes the crystalline color superconducting (CCSC) phase. Note that the breaking pattern associated with this condensate is the same of the CFL phase (2) with the additional ingredient of the breaking of the rotation symmetries. In most of the crystalline phases the $SO(3)$ space symmetry is completely broken down to a discrete symmetry characterizing the crystalline pattern. Any crystalline structure is described by the set $\{\boldsymbol{n}_I\}$, which have an I index because to each of the three interaction channel may correspond a different crystalline arrangement of vectors.

The low-energy CCSC spectrum is much richer than the CFL spectrum. We now have not only the H-phonon degree of freedom, but also 3 phonon-like modes and 9 gapless quasiquark modes. Indeed, an interesting property of the crystalline phase is that for real-valued periodic condensates there exists a fermionic gapless mode iff the set $\{\boldsymbol{n}_I\}$ does not contain the null vector, $\boldsymbol{n} = \boldsymbol{0}$. Thus, for these configurations the energy of the fermionic excitations depends linearly on the residual momentum ξ, that is the momentum measured from the Fermi surface. At the lowest order in ξ/q one finds that the quasiquark modes have dispersion law

$$E(\boldsymbol{v},\xi) = c(\boldsymbol{v})\xi\,, \tag{8}$$

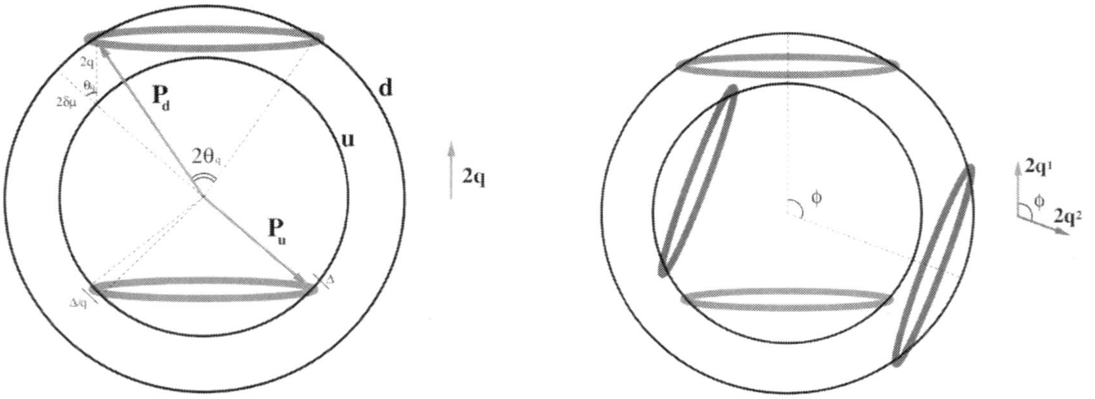

Figure 3. (Color online) Pictorial description of the CCSC pairing for two simple structures. Left panel: In the single plane wave structure pairing takes place in two ribbons on the top the Fermi spheres of up and down fermions having opening angle $\arccos(\delta\mu/q) \simeq 67°$, thickness Δ and angular width Δ/q. Right panel: Structure obtained with two plane waves with relative angle ϕ. The size and the opening angle of each ribbon is as in the single plane wave case. More complicated two-flavor CCSC structures can be obtained adding nonoverlapping ribbons on the top of the Fermi spheres. For illustrative purposes, we have greatly exaggerated the splitting between the Fermi surfaces.

where $c(\boldsymbol{v})$ is the direction dependent velocity of the excitations. In figure 4 we show $c(\boldsymbol{v})$ for two crystalline structures in two-flavor quark matter with u-d paring, that is with a condensate given in equation (7) with $\Delta_1 = \Delta_2 = 0$ and $\Delta_3 \neq 0$ and with the set of vectors $\{\boldsymbol{n}_3\}$ pointing to the vertices of a BCC structure (left panel) and a FCC structure (right panel). Note that the presence of gapless fermions does not forbid the existence of superconductivity or superfluidity [21]. Notable condensed matter examples are type II superconductors, which have gapless fermionic modes for sufficiently large magnetic fields [21, 22]. Note also that in the definitions 2.2 and 2.1 there is no mention of the spectrum of fermions.

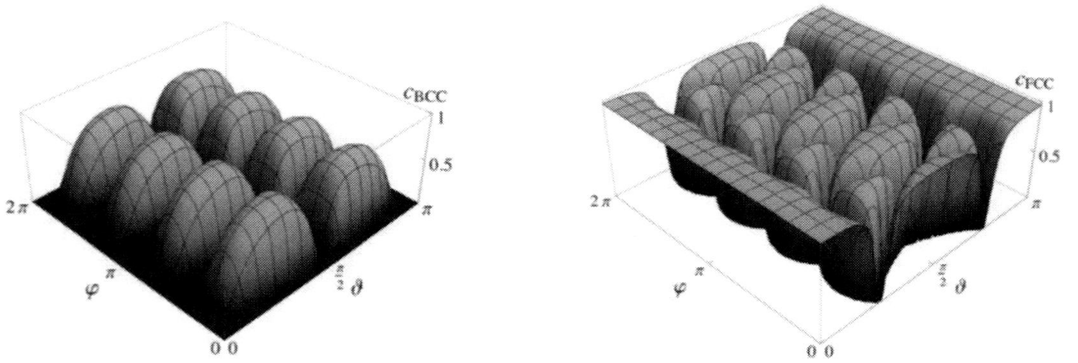

Figure 4. (Color online) Velocity of the fermions in the BCC and FCC two-flavor crystalline phases.

In three-flavor quark matter there is a richer variety of crystalline structures that can be realized. The number of possibilities is actually too large. One useful simplification is related to the fact that the chemical potential arrangement is as in figure 5. Thus, it is reasonable to expect that u-s and d-u pairing occurs with almost equal condensates $\Delta_2 \simeq \Delta_3 \equiv \Delta$, while the d-s pairing should be suppressed, meaning that $\Delta_1 \ll \Delta$, and can be neglected. In figure 5 we illustrate the Fermi spheres of a simple two plane waves condensate, obtained modulating the Δ_2 condensate by a plane wave with wave vector $\boldsymbol{q}_2$ and the Δ_3 condensate by a plane wave with wave vector $\boldsymbol{q}_3$. Using the argument of the associated currents, it should be easy to understand that the state with $\phi = 0$ is energetically favored. For a proof see [23, 1].

Many three-flavor crystalline structures have been considered in [24]; among them the CubeX and the 2Cube45z structures have the lowest free energy. The CubeX crystal consists of two sets $\{\boldsymbol{n}_2\} = \{\boldsymbol{n}_2^1, \boldsymbol{n}_2^2, \boldsymbol{n}_2^3, \boldsymbol{n}_2^4\}$ and $\{\boldsymbol{n}_3\} = \{\boldsymbol{n}_3^1, \boldsymbol{n}_3^2, \boldsymbol{n}_3^3, \boldsymbol{n}_3^4\}$ with

$$\boldsymbol{n}_2^1 = \sqrt{\tfrac{1}{3}}(1,1,1) = -\boldsymbol{n}_2^2, \qquad \boldsymbol{n}_2^3 = \sqrt{\frac{1}{3}}(-1,-1,1) = -\boldsymbol{n}_2^4,$$
$$\boldsymbol{n}_3^1 = \sqrt{\tfrac{1}{3}}(-1,1,1) = -\boldsymbol{n}_3^2, \qquad \boldsymbol{n}_3^3 = \sqrt{\frac{1}{3}}(1,-1,1) = -\boldsymbol{n}_3^4.$$

Thus, the vectors of each set point to the vertices of a rectangle; the eight vectors together point toward the vertices of a cube. In the 2Cube45z crystal, $\{\boldsymbol{n}_2\}$ and $\{\boldsymbol{n}_3\}$ each contains eight wave vectors, pointing to the corners of a cube; the two cubes are rotated by 45 degrees about one of the 4-fold axes.

So far, the evaluation of the free energy for these structures has only been done by means of a Ginzburg-Landau expansion. This method is not quantitatively reliable, however it is useful to have an estimate of the free energy of the various phases. Moreover, at least for some simple structures for which different methods are available, it underestimates the condensation energy giving a conservative value of the free energy [23]. The Ginzburg-Landau computations of [24] show that the CCSC phase is favored with respect to the homogeneous CFL and the unpaired

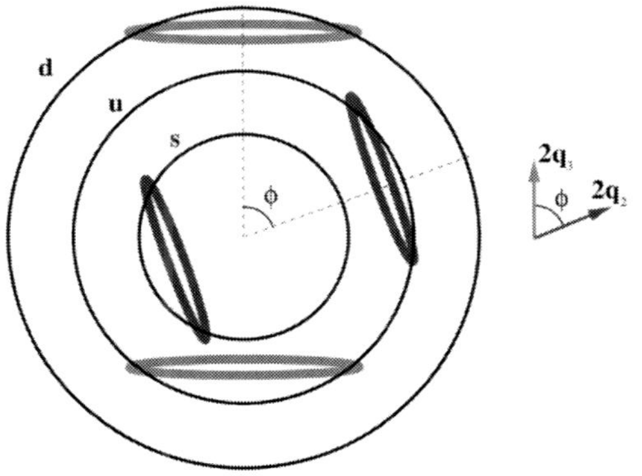

Figure 5. (Color online) Sketch showing the hierarchy of the Fermi momenta for three-flavor quark matter and a simple "crystal", a particular realization of equation (7). The aligned ribbons on the d and u Fermi surfaces indicate those quarks that contribute to the $\langle ud \rangle$ condensate with gap parameter Δ_3. The aligned ribbons on the u and s Fermi surfaces indicate those quarks that contribute to the $\langle us \rangle$ condensate with gap parameter Δ_2. Associated with Δ_2 and Δ_3 are two wave vectors with relative angle ϕ. The size and the opening angle of each ribbon is as in figure 3. More complicated three-flavor CCSC structures can be obtained adding nonoverlapping ribbons on the top of the Fermi spheres. For illustrative purposes we have greatly exaggerated the splitting between the Fermi surfaces.

Journal of Physics: Conference Series **527** (2014) 012020 doi:10.1088/1742-6596/527/1/012020

phases in the range

$$2.9\Delta_{\rm CFL} < \frac{M_s^2}{\mu} < 10.4\Delta_{\rm CFL} \,. \tag{9}$$

Using the self-consistent treatment of [25] one can transform this range in

$$442 \;\; MeV \lesssim \mu \lesssim 515 \; MeV \,. \tag{10}$$

This result is certainly model dependent, however it shows that the actual extension of the LOFF chemical potential window might not be very large.

As a consequence of the spontaneous breaking of the rotation symmetries, in the low energy spectrum there are phonon-like excitations describing the vibrations of the crystalline modulation. From the effective Lagrangian describing these modes one can extract the shear modulus of the system [26]. For the the two favored structures, 2Cube45z and CubeX, the shear modulus is a 3×3 nondiagonal matrix in coordinate space with entries proportional to

$$\nu_{\rm CQM} = 2.47\,{\rm MeV/fm^3} \left(\frac{\Delta}{10\ {\rm MeV}}\right)^2 \left(\frac{\mu}{400\ {\rm MeV}}\right)^2 . \tag{11}$$

This is a very large shear modulus, about a factor of $20-1000$ larger than in standard neutron star crust [27, 26].

Summarizing, the three-flavor CCSC phase

- Is a color superconductor
- Is a baryonic superfluid
- Has many low energy degrees of freedom: the H-phonon, the phonon-like modes (bosonic), the quark quasiparticles (fermionic)
- Is characterized by an extremely rigid space modulation of the condensate
- Is a supersolid.

The last property follows from the definition 2.3. In the CCSC phase there is indeed the simultaneous breaking of the gauge color symmetry, of the global $U(1)_B$ symmetry and of the rotational symmetries. These symmetry breakings happen simultaneously because of the condensate in equation (7).

5. Astrophysics

Given the exceptional properties of the CCSC phase, one would naively expect being easy to verify whether or not this phase is realized in nature. However, testing the existence of color superconductors is extremely nontrivial. The region of low-temperature and high-density is not easily reachable in terrestrial high-energy experiments [28]. Thus, to date the only few indirect information which we could put in relation with the color superconducting phase come from astronomical observations of compact stellar objects (CSOs).

The reason we expect color superconductors in CSOs is the following. In CSOs matter is compressed at densities about a factor 3-5 larger than in a heavy nucleus. A simple geometrical reasoning suggests that baryons are likely to lose their identity and dissolve into deconfined quarks [29]. In this case compact (hybrid) stars featuring quark cores would exist. A different possibility is that *uds* quark matter is the ground state of the hadrons [30]. In this case, the so-called strange stars [31] completely composed of deconfined matter should exist.

In CSOs, quark matter is long-lived, charge neutral and in β-equilibrium, which we have seen are good conditions for the realization of the CCSC phase. Moreover, the critical temperature (in weak coupling) is given by $T_c \simeq 0.57\Delta$ and reasonable values of the CCSC gap parameter range in between 5 MeV and 25 MeV [26]. Since for the greatest part of the CSO lifetime the temperature is much lower than this critical temperature the CCSC state should be thermodynamically favored.

Recent astronomical observations of very massive CSOs [32] seem to disfavor the possibility that a color superconducting phase is realized [33], but the results depend on the equation of state of matter at high density, and the possibility that hybrid stars of about $2M_\odot$ have a CCSC core [34, 1] cannot be excluded. Basically the available mass-radius observations do not allow us to infer in a unique way the internal structures of CSOs because hybrid stars featuring quark matter could masquerade as standard neutron stars [35].

Any other astronomical observation, related to cooling, radial and nonradial oscillations, glitches, strong magnetic fields etc. is based on the internal structure of the star and on its evolution after its birth. Therefore, it is very hard to extract useful and model independent information from these observations. Given the large uncertainty in the estimate of the parameters characterizing the CCSC phase, to date astronomical observations could only been used to restrict the parameter space of the model. Nevertheless, the investigation of these astrophysical signatures is the only means we have to connect theoretical models with astronomical observations.

One of the most promising direction is the study of the astrophysical properties related to the extraordinary rigidity of the CCSC phase. This is indeed a peculiar property of the CCSC phase; no other phase is known to have a comparable shear modulus. Suppose that the CCSC phase is realized in the core of a CSO. Then, the rigidity of the CCSC phase may allow the presence of large deformation of the core. If this deformation has a nonzero quadrupole moment with respect to the rotation axis and if this axis-asymmetric mass distribution is not compensated by the overlying nuclear envelope, a spinning CSO would efficiently emit gravitational waves [36, 37, 1]. A second possibility is that the emission of gravitational waves happens by excitation of the torsional oscillations [38]. These are particular toroidal oscillations, see for example [39], which when triggered (say by a stellar glitch), produce an emission of gravitational waves with an emitted power roughly proportional to ν^4 with frequencies in the kHz range. In principle these gravitational waves could be detected by the second-generation gravitational-wave detectors. A positive result would be, for many reasons, a great achievement and a clear indication of the existence of the CCSC phase. A negative result would be useful for further reducing the parameter space of the CCSC phase.

Acknowledgments
I acknowledge discussion with G Pupillo, who triggered my interest in supersolids and the realization of the supersolid phase in compact stars.

References
[1] Anglani R *et al.* 2013 `arXiv:1302.4264`
[2] Barbour I *et al.* 1986 *Nucl. Phys. B* **B75** 296
[3] Aarts G 2013 `arXiv:1312.0968`
[4] Penrose O and Onsager L 1956 *Phys. Rev.* **104** 576
[5] Boninsegni M and Prokof'ev N V 2012 *Rev. Mod. Phys.* **84** 759
[6] During G, Josserand C, Pomeau Y, and Rica S 2011 `arXiv:1110.1323`
[7] Henkel N, Nath R, and Pohl T 2010 *Phys. Rev. Lett.* **104** 195302
[8] Cinti F, Jain P, Boninsegni M, Micheli A, Zoller P, and Pupillo G 2010 *Phys. Rev. Lett.* **105** 135301
[9] Cinti F, Macrì T, Lechner W, Pupillo G, and Pohl T 2013 `arXiv:1302.4576`
[10] Alford M G, Schmitt A, Rajagopal K, and Schafer T 2008 *Rev. Mod. Phys.* **80** 1455
[11] Alford M G, Rajagopal K, and Wilczek F 1999 *Nucl. Phys. B* **537** 443
[12] Alford M G, Berges J, and Rajagopal K 1999 *Nucl. Phys. B* **558** 219
[13] Schafer T 2000 *Nucl. Phys. B* **B575** 269
[14] Shovkovy I and Wijewardhana L 1999 *Phys. Lett. B* **470** 189

[15] Alford M G and Rajagopal K 2002 *JHEP* **0206** 031
[16] Chandrasekhar B S 1962 *Appl. Phys. Lett.* **1** 7
[17] Clogston A M 1962 *Phys. Rev. Lett.* **9** 266
[18] Gubankova E, Mannarelli M, and Sharma R 2010 *Ann. Phys.* **325** 1987
[19] Larkin A I and Ovchinnikov Y N 1964 *Zh. Eksp. Teor. Fiz.* **47** 1136
[20] Fulde P and Ferrell R A 1964 *Phys. Rev.* **135** A550
[21] de Gennes P G 1966 *Superconductivity of Metals and Alloys* (New York: Benjamin)
[22] Saint-James D, Sarma G and Thomas E J 1969 *Type II superconductivity* (Pergamon Press Oxford, New York) ISBN 0080123929
[23] Mannarelli M, Rajagopal K, and Sharma R 2006 *Phys. Rev. D* **73** 114012
[24] Rajagopal K and Sharma R 2006 *Phys. Rev. D* **74** 094019
[25] Ippolito N D, Nardulli G, and Ruggieri M 2007 *JHEP* **0704** 036
[26] Mannarelli M, Rajagopal K, and Sharma R 2007 *Phys. Rev. D* **76** 074026
[27] Strohmayer T, van Horn H M, Ogata S, Iyetomi H, and Ichimaru S 1991 *Astrophys. J.* **375** 679
[28] Klahn T, Blaschke D, and Weber F 2012 *Phys. Part. Nucl. Lett.* **9** 484
[29] Weber F (ed) 1999 *Pulsars as astrophysical laboratories for nuclear and particle physics*
[30] Witten E 1984 *Phys. Rev. D* **30** 272
[31] Alcock C, Farhi E, and Olinto A 1986 *Astrophys.J.* **310** 261
[32] Demorest P, Pennucci T, Ransom S, Roberts M, and Hessels J 2010 *Nature* **467** 1081
[33] Logoteta D, Providencia C, Vidana I, and Bombaci I 2012 *Phys. Rev. C* **85** 055807
[34] Ippolito N, Ruggieri M, Rischke D, Sedrakian A, and Weber F 2008 *Phys. Rev. D* **77** 023004
[35] Alford M G, Braby M, Paris M W, and Reddy S 2005 *Astrophys.J.* **629** 969
[36] Lin L M 2007 *Phys. Rev. D* **76** 081502
[37] Haskell B, Andersson N, Jones D I, and Samuelsson L 2007 *Phys. Rev. Lett.* **99** 231101
[38] Lin L M 2013 `arXiv:1311.2654`
[39] McDermott P N, van Horn H M, and Hansen C J 1988 *Astrophys. J.* **325** 725

Journal of Physics: Conference Series **527** (2014) 012021 doi:10.1088/1742-6596/527/1/012021

Hybrid neutron stars with the field correlator method

Domenico Logoteta [1] **and Ignazio Bombaci** [2]

[1]Centro de Física Computacional, Department of Physics, University of Coimbra, 3004-516 Coimbra, Portugal

[2]Dipartimento di Fisica "Enrico Fermi", Università di Pisa and INFN Sezione di Pisa, Largo Bruno Pontecorvo 3, I-56127 Pisa, Italy

Abstract. We study the quark deconfinement phase transition in cold ($T = 0$) neutron star matter and we calculate various structural properties of hybrid stars. For the quark phase, we use an equation of state (EOS) based on the Field Correlator Method (FCM) extended to the case of nonzero baryon density. For the confined hadronic phase we use a relativistic mean field model considering both pure nucleonic and hyperonic matter. We constrain the values of the gluon condensate G_2, which is one of the EOS parameter within the FCM, making use of the measured mass, $M = 1.97 \pm 0.04\,M_\odot$, of the neutron star in PSR J1614-2230. Our results show that the values of G_2 extracted from the mass measurement of PSR J1614-2230 are consistent with the values of the same quantity derived, within the FCM, from recent lattice QCD calculations of the deconfinement transition temperature at zero baryon chemical potential.

1. Introduction

Neutron stars are the densest macroscopic objects in the universe. A large variety of calculations of neutron star structures [1, 2, 3, 4] predict a maximum stellar central density (the one for the maximum mass star configuration) in the range of $4-8$ times the saturation density ($\sim 2.8 \times 10^{14}$ g/cm^3) of nuclear matter. Therefore, neutron stars, can be viewed as natural laboratories to explore the low temperature T and high baryon chemical potential region of the phase diagram of quantum chromodynamics (QCD) [5, 6]. Under these conditions nonperturbative aspects of QCD are expected to play an essential role, and a transition to a phase with deconfined quarks and gluon is expected to occur and to effect a number of interesting astrophysical phenomena [7, 8, 9, 10, 11, 12, 13, 14, 15].

Current high-precision numerical calculations of QCD on a space-time lattice at zero baryon chemical potential μ_b (zero baryon density) seem to indicate that at high temperature and for physical values of the quark masses, the transition to quark gluon plasma is a crossover [16] rather than a real phase transition.

Unluckily, present lattice QCD calculations at finite baryon chemical potential are plagued with the so called "sign problem", which makes them unrealizable by all presently known lattice methods. Thus, in order to explore the QCD phase diagram at low temperature T and high μ_b, it is necessary to adopt some approximations in QCD or to apply some QCD effective model.

Along these lines the MIT bag model [17] and the Nambu Jona-Lasinio (NJL) model [18] have been widely used to calculate the quark matter EOS. However the MIT bag and the NJL models (as other QCD effective models) can not make predictions in the high T and zero baryon chemical potential region and, thus, can not be tested with present lattice QCD calculations.

Recently the deconfinement phase transition has been described using an EOS of quark gluon plasma derived within the Field Correlator Method (FCM) [19, 20] extended to finite baryon chemical potential [21, 22]. The Field Correlator Method is a nonperturbative approach to QCD which includes from first principles the dynamics of confinement in terms of color electric and color magnetic correlators. The model contains two parameters: the gluon condensate G_2 and the large distance static quark-antiquark ($Q\bar{Q}$) potential V_1. These two quantities control the EOS of the denconfined phase at fixed quark masses and temperature. A very interesting aspect of the FCM is the possibility to describe the whole QCD phase diagram, from high temperature and low baryon chemical potential, to low T and high μ_b limit.

Another interesting feature of this method is that the value of the gluon condensate can be obtained from lattice QCD calculations of the deconfinement transition temperature T_c, at $\mu_b = 0$. This establishes a useful link to directly relate lattice QCD simulations and neutron star physics. To explore this link is the main aim of this work.

2. Equation of state of neutron star matter

The quark matter equation of state we used in this work is based on the Field Correlator Method (FCM) [19, 20]. The method has been recently extended to the case of nonzero baryon density [21, 22] making possible its application to neutron star matter.

The FCM provides a natural explanation and treatment of the dynamics of confinement in terms of Color Electric $D^E(x)$, $D_1^E(x)$ and Color Magnetic $D^H(x)$, $D_1^H(x)$ Gaussian correlators. In this work we use the approach described in Refs. [23, 24].

The EOS of the quark phase is governed by two parameters namely the static long distance $Q\bar{Q}$ potential V_1 and the gluon condensate G_2. All our results have been obtained in terms of these two quantities considered as independent on density [25, 26]. Recently in Ref. [27] a density dependence on G_2 has been proposed.

Using QCD sum rules [28], the gluon condensate G_2 has been determined, with large uncertainty, to be in the range $G_2 = (0.012 \pm 0.006)$ GeV^4. Therefore we have varied G_2 within the range determined in [28]. We used the following values of the current-quark masses: $m_u = m_d = 5$ MeV and $m_s = 150$ MeV. In summary, the quark matter EOS contains two parameters: G_2 and $V_1 = V_1(T = 0)$.

For the EOS of confined hadronic matter we adopt a nonlinear relativistic mean field model [29, 30] and we make use of the GM1 parametrization given by Glendening-Moszkoski [31, 32].

The nucleon coupling constants ($g_\sigma, g_\omega, g_\rho$) are fitted to the bulk properties of nuclear matter. In particular, for the GM1 parametrization [31, 32] the incompressibility of symmetric nuclear matter and the nucleon effective mass at the empirical saturation density are respectively $K = 300$ MeV and $M^* = 0.7\ M$ (being M the bare nucleon mass). The inclusion of hyperons involves new couplings, which can be written in terms of the nucleonic ones: $g_{\sigma Y} = x_\sigma g_\sigma$, $g_{\omega Y} = x_\omega g_\omega$, and $g_{\rho Y} = x_\rho g_\rho$. In this model it is assumed that all the hyperons in the baryon octet have the same coupling. We will consider x_σ in the range 0.6–0.8. In addition, following Ref. [31], we will take $x_\rho = x_\sigma$. Finally the binding energy of the Λ particle in symmetric nuclear matter, $B_\Lambda = -28$ MeV, is used to determine x_ω in terms of x_σ. Notice that the case with $x_\sigma = 0.6$ produces stars with a larger hyperon population (for a given stellar gravitational mass) with respect to the case $x_\sigma = 0.8$ [32, 33]. In addition to these two parametrizations for hyperonic matter (hereafter called NY matter), we will consider the case of pure nucleonic matter (hereafter called N matter).

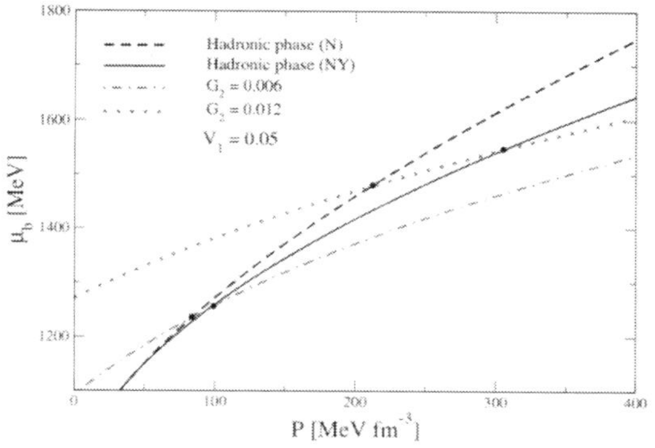

Figure 1. (Color online) Baryon chemical potential μ_b versus pressure P in β-stable matter at $T = 0$. Curves for the quark phase are relative to two different values of the gluon condensate G_2, reported in GeV^4, and $V_1 = 0.05$ GeV. Curves for the hadronic phase are relative to hyperonic matter (NY) with $x_\sigma = 0.7$ and to pure nucleonic matter (N).

3. Phase transition in beta-stable neutron star matter

The composition of neutron star matter is settled by the requirements of electric charge neutrality and equilibrium under the weak interaction processes (β-stable matter). Under such conditions and in the case of neutrino-free matter [1, 34], the chemical potential μ_i of each particle species i can be written in terms of two independent quantities, the baryonic and electric chemical potentials μ_b and μ_q respectively: $\mu_i = b_i\mu_b - q_i\mu_q$, where b_i is the baryon number of the species i and q_i denotes its charge in unit of the electron charge magnitude.

In the pure hadronic phase $\mu_b = \mu_n$, the neutron chemical potential, and $\mu_q = \mu_e$, the electron chemical potential. In the pure quark phase the quark chemical potentials μ_f (f=u, d, s) are related to μ_b and μ_q by the formulae $\mu_u = (\mu_b - 2\mu_q)/3$ and $\mu_d = \mu_s = (\mu_b + \mu_q)/3$.

We next assume a first order deconfinement phase transition [35] and require global electric charge neutrality of bulk beta-stable stellar matter [36]. An important consequence of imposing global charge neutrality is that the hadronic and the quark phases can coexist for a finite range of pressures. This treatment of the phase transition is known in the literature as the Gibbs construction for the mixed phase [36].

In the following we consider the case of $T = 0$ matter, which is appropriate to describe neutron stars interiors at times larger than about a few minutes after their formation [1, 34].

In figure 1 we plot the relation between the baryon chemical potential μ_b and the total (*i.e.* baryonic plus leptonic contributions) pressure P, in β-stable matter, for the hadron and the quark phases. For the hadronic phase we consider hyperonic matter (NY) with $x_\sigma = 0.7$ (continuous line) and pure nucleonic matter (N) (dashed line). For the quark phase we use two different values of the gluon condensate G_2 = 0.006, 0.012 GeV^4 and a common value $V_1 = 0.05$ GeV for the large distance static $Q\bar{Q}$ potential. The phase transition occurs at the intersection point between the curves describing the two different phases. This crossing point is significantly affected by the value of the gluon condensate, in particular when G_2 increases the onset of the deconfinement transition is shifted to higher pressure (higher baryon chemical potential). A similar qualitative behaviour is typical of bag-like models in terms of the bag constant B. In addition we note that the presence of hyperons in the hadronic phase moves the phase transition point to larger pressures. Similar results have been found using different values for the hyperon-nucleon couplings (x_σ = 0.6, 0.8) for the hadronic EOS.

Keeping a fixed value of the gluon condensate, $G_2 = 0.006$ GeV^4, we show in figure 2 the effects of V_1 on the phase transition point. As one can see, increasing the value of V_1 the

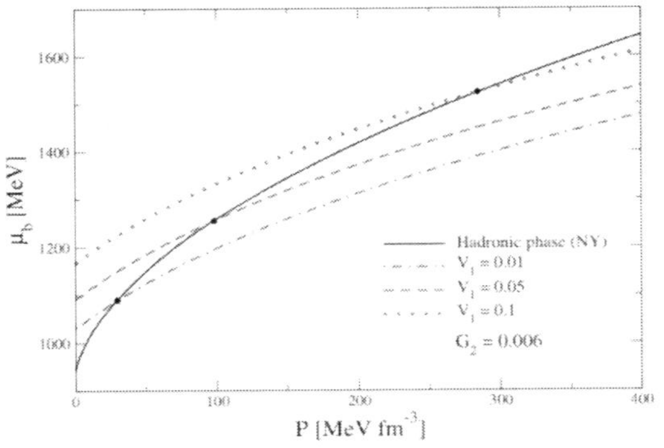

Figure 2. (Color online) Baryon chemical potential μ_b versus P in β-stable matter at $T = 0$. Curves for the quark phase are relative to three different values of V_1, reported in GeV, and for $G_2 = 0.006$ GeV^4. The curve for the hadronic phase is relative to hyperonic matter (NY) with $x_\sigma = 0.7$.

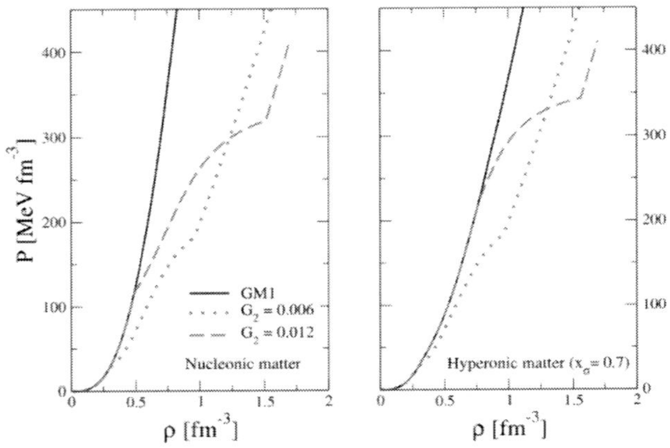

Figure 3. (Color online) Total pressure P of cold β-stable matter as a function of the baryon number density ρ, for different values of G_2 (reported in GeV^4 units) and for $V_1 = 0.05$ GeV. In the left (right) panel we show results for pure nucleonic (hyperonic) matter in the hadronic phase. For the hyperonic matter EOS we set $x_\sigma = 0.7$.

transition point is shifted to values of the pressure higher and higher.

In figure 3, we show the pressure for β-stable matter as a function of the baryon number density ρ in the case of pure nucleonic matter (left panel) and hyperonic matter (right panel) with $x_\sigma = 0.7$ for the hadronic phase. For the quark phase we consider two different values of the gluon condensate G_2. Increasing the value of G_2 we note a shift of the phase transition to larger baryon densities. This is consistent with the behaviour of figure 1.

4. Hybrid star structure

In this section we show the results of our calculations of hybrid stars structure. To this purpose we integrate the well known Tolman, Oppenheimer and Volkov equations (see *e.g.* [32, 37]) to get various stellar properties and report the results of a systematic study in which we vary the value of the gluon condensate G_2 between the constraints imposed by QCD sum rules [28]. To

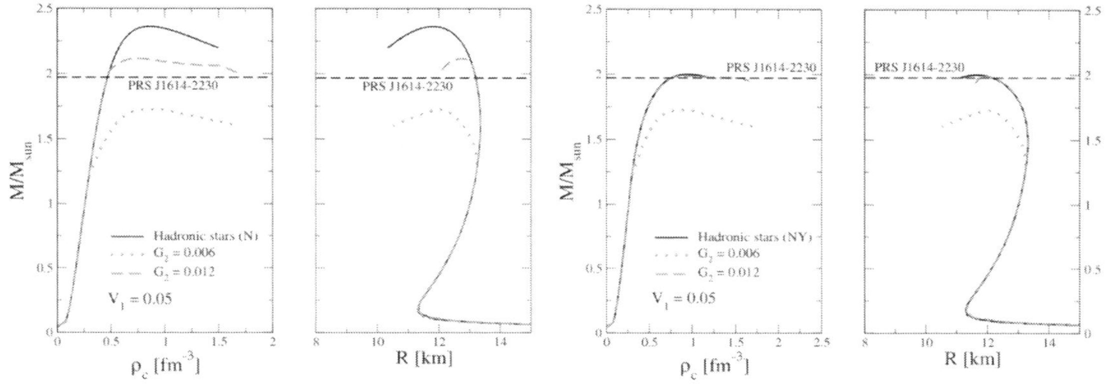

Figure 4. (Color online) Stellar mass M versus central baryon number density ρ_c (first and third panels) and versus stellar radius R (second and fourth panels) for hybrid stars for two values of G_2 (reported in GeV^4 units) and for $V_1 = 0.05$ GeV. Continuous lines in each panel denote compact stars with no quark matter content. The first two panels refer to the pure nucleonic matter case (N) while in third and fourth panels refer the hyperonic matter case (NY) with $x_\sigma = 0.7$. The dashed horizontal line, in each panel, represents the measured mass of PSR J1641-2230.

model the stellar crust we have used the EOS of Ref. [38].

In figure 4 we report the stellar gravitational mass M (in unit of the solar mass $M_\odot = 1.99 \times 10^{33}$g) versus the central baryon number density ρ_c (first and third panels) and M versus radius R (second and fourth panels) in the case of pure nucleonic stars (continuous lines) and of hybrid stars for different G_2. The dashed horizontal line, in each panel, represents the mass of the PSR J1641-2230 ($M = 1.97 \pm 0.04\, M_\odot$) [39]. We first focus on the first two panels of figure 4, where in the hadronic phase just nucleons are included. We obtain stable hybrid star configurations for all the considered values of the gluon condensate, with maximum masses ranging from $M_{max} = 1.72\, M_\odot$ (case with $G_2 = 0.006$ GeV^4) to $M_{max} = 2.11\, M_\odot$ ($G_2 = 0.012$ GeV^4). Notice that the hybrid star branch of the stellar equilibrium configurations shrinks as G_2 is increased. This is in full agreement with the results for the EOS reported in figure 3.

We now turn to the case (third and fourth panels in figure 4) in which the hadronic phase contains hyperons (NY matter) and we set $x_\sigma = 0.7$. Notice that the presence of hyperons reduces the value of the maximum mass of pure hadronic star (*i.e.* compact stars with no quark matter content) from $M_{max} = 2.33\, M_\odot$ in case of pure nucleonic stars (continuous line in the first and second panels of figure 4) to $M_{max} = 1.99\, M_\odot$ in case of hyperonic stars with $x_\sigma = 0.7$ (continuous line in the third and fourth panels in figure 4). We obtain stable hybrid star configurations also in the case of NY matter for all the considered values of the gluon condensate and, similarly to the nucleonic case, we note again that the hybrid star branch shrinks as G_2 is increased. In this case we get maximum hybrid star masses of 1.72 $M_\odot$ for $G_2 = 0.006$ GeV^4 and 1.98 $M_\odot$ for $G_2 = 0.012$ GeV^4. We note that for values of the gluon condensate $G_2 > 0.012$ GeV^4 our models are able to predict the mass of the PSR J1614-2230. More precisely, the minimum value of G_2 required for the pure nucleonic case is $G_2 = 0.010$ GeV^4 while for the hyperonic case such value is $G_2 = 0.012$ GeV^4. It should be noted that these values of G_2 depend besides the EOS of hadronic matter, on the value adopted for the large distance $Q\bar{Q}$ potential.

In table 1 we report the properties of the maximum mass configuration for hybrid stars for $V_1 = 0.01$ GeV and varying G_2 and the hyperon coupling constants. We note that in the case $x_\sigma = 0.6$, hybrid stars are possible only for $G_2 < 0.013$ GeV^4. In fact, for $G_2 > 0.013$ GeV^4 the

Table 1. Properties of the maximum mass configuration for hybrid stars for different values of the gluon condensate G_2 in GeV^4 (second column) and for $V_1 = 0.01$ GeV. The parameter x_σ (first column) fixes the hyperons coupling constants as described in section 2. The entry N in the first column refers to the case of pure nucleonic matter for the hadronic phase. The mass M_{max}, the central baryon number density ρ_c^{Hyb} and the radius R of the maximum mass configuration are reported respectively in the third, fourth and fifth column. The quantities with the label HS refer to the case of purely hadronic stars. Stellar masses are reported in unit of the solar mass $M_\odot = 1.99 \times 10^{33}$g, central densities are given in fm^{-3}, and stellar radii in km.

x_σ	G_2	M_{max}	ρ_c^{Hyb}	R	M_{max}^{HS}	ρ_c^{HS}	R^{HS}
	0.006	1.44	1.55	9.54			
N	0.012	1.89	0.77	12.55	2.33	0.87	11.70
	0.016	2.05	0.75	12.66			
	0.006	1.44	1.56	9.52			
0.8	0.012	1.89	0.77	12.53	2.15	0.94	11.50
	0.016	2.04	0.81	12.40			
	0.006	1.43	1.56	9.51			
0.6	0.010	1.73	0.89	12.00	1.80	1.00	11.49
	0.013	1.80	1.00	11.49			

baryon number density for the onset of the mixed phase is larger than the central baryon number density of the maximum mass pure hadronic star. Notice that for $x_\sigma = 0.6$ is not possible to account for the measured mass of PSR J1614-2230 for any value of G_2.

5. Lattice QCD calculations and measured neutron star masses

Within the FCM the deconfinement transition temperature T_c at $\mu_b = 0$ reads [21]

$$T_c = \frac{a_0}{2} G_2^{1/4} \left(1 + \sqrt{1 + \frac{V_1(T_c)}{2 a_0 G_2^{1/4}}} \right), \tag{1}$$

with $a_0 = (3\pi^2/768)^{1/4}$ in the case of three flavors.
In their analysis authors of Ref. [21] assume $V_1(T_c) = 0.5$ GeV, thus T_c in equation (1) is a simple function of G_2, and is represented in figure 5 by the curve labeled $V_1(T_c) = 0.5$ GeV. This result can hence be compared with lattice QCD calculations of T_c giving the possibility to extract the range of values for the gluon condensate compatible with lattice results. This comparison has been done in Ref. [21], and it is done in the present work in figure 5, where we consider recent lattice QCD calculations of T_c [40, 41]. Details to the specific lattice QCD calculations are given in the figure 5 caption. As one can see, the comparison with lattice QCD calculations of T_c restricts the gluon condensate in a rather narrow range $G_2 = 0.0025$–0.0050 GeV^4.

Next to verify if these values of G_2 are compatible with those extracted in section 4 from hybrid star calculations and measured neutron star masses, we need to relate the parameter $V_1 \equiv V_1(0)$, entering in the zero temperature EOS of the quark phase, with $V_1(T_c)$ in equation (1).

In Ref. [23] it has been shown that such relation is given by:

$$V_1(T) = V_1(0) \left\{ 1 - \frac{3}{2} \frac{\lambda T}{\hbar c} + \frac{1}{2} \left(1 + 3 \frac{\lambda T}{\hbar c} \right) e^{-\frac{\hbar c}{\lambda T}} \right\}, \tag{2}$$

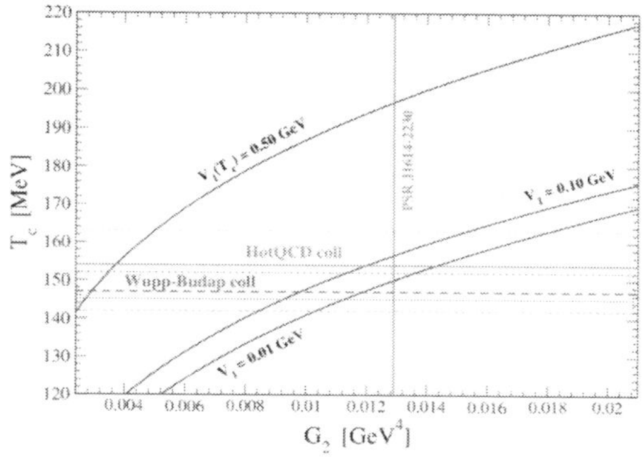

Figure 5. (Color online) Deconfinement transition temperature T_c at $\mu_b = 0$. The curve labeled with $V_1(T_c) = 0.5$ GeV reproduces the FCM results of Ref. [21] for a fixed value $V_1(T_c) = 0.5$ GeV of the large distance static $Q\bar{Q}$ potential. The curve labeled with $V_1 = 0.01$ GeV ($V_1 = 0.10$ GeV) corresponds to the transition temperature at $\mu_b = 0$ obtained solving numerically equations (1) and (2) for the case $V_1(0) = 0.01$ GeV ($V_1(0) = 0.10$ GeV). The horizontal heavy and thin lines represent respectively the central value and the error estimate of lattice QCD calculations. In particular, the (red) continuous lines refer to the calculations [41] of the HotQCD collaboration $T_c = (154 \pm 9)$ MeV; the (blue) short-dashed lines refer to the calculations [40] of the Wuppertal–Budapest collaboration $T_c = (147 \pm 5)$ MeV. Finally, the vertical green line represents the lower limit for G_2 which is compatible with the lower bound of the measured mass of PSR J1614-2230 for the case $V_1(0) = 0.01$ GeV.

where $\lambda = 0.34$ fm [42] is the vacuum correlation length. Thus $V_1(T_c) = 0.5$ GeV corresponds to $V_1(0) = 0.85$ GeV to be used in the $T = 0$ EOS of the quark phase. In this case there is no phase transition in neutron stars for all the considered values of G_2. Thus for these values of the EOS parameters PSR J1614-2230 would be a pure nucleonic star.

We can also evaluate the FCM transition temperature at $\mu_b = 0$ corresponding to the case $V_1(0) = 0.01$ GeV. To this purpose we solve numerically equations (1)–(2) and we obtain the results represented in figure 5 by the curve labeled $V_1 = 0.01$ GeV. The comparison of these results with lattice QCD calculations [40, 41] of T_c restricts the gluon condensate in the range $G_2 = 0.0103$–0.0180 GeV^4. Coming now to the astrophysical constraints on the gluon condensate, the vertical grey line in figure 5 represents the lower limit for G_2 which is compatible, in the case $V_1(0) = 0.01$ GeV, with the lower bound of the measured mass of PSR J1614-2230 (see section 4).

A similar analysis can be done for the case $V_1(0) = 0.10$ GeV. Now the comparison between the FCM transition temperature at $\mu_b = 0$ (curve labeled $V_1 = 0.10$ GeV in figure 5) and lattice QCD calculations of the same quantity gives $G_2 = 0.0085$–0.0153 GeV^4, whereas one gets $G_2 \geq 0.006$ GeV^4 from the lower bound of the measured mass of PSR J1614-2230.

We thus find that the values of the gluon condensate extracted within the FCM from lattice QCD calculations of the deconfinement transition temperature T_c at $\mu_b = 0$ are fully compatible with the value of the same quantity extracted from measured neutron star masses.

6. Summary and conclusions

In this work we have studied the quark deconfinement phase transition in neutron star matter and the properties of hybrid stars employing an EOS for the quark phase derived from the FCM extended to finite baryon chemical potential. The EOS of the FCM is parametrized in terms of the gluon condensate G_2 and of the large distance static $Q\bar{Q}$ potential V_1 at zero temperature.

We have obtained stable hybrid star configurations for all the values of the gluon condensate fulfilling the condition that the deconfinement transition can occur in pure hadronic stars.

We have established that the values of the gluon condensate extracted within the FCM from lattice QCD calculations of the deconfinement transition temperature at $\mu_b = 0$, are fully consistent with the value of the same quantity constrained by the mass measurement of PSR J1614-2230. The FCM thus provides a powerful tool to link numerical calculations of QCD on a space-time lattice with neutron stars physics.

References

[1] Prakash M, Bombaci I, Prakash M, Ellis P J, Lattimer J M, and Knorren R 1997 *Phys. Rep.* **280** 1
[2] Datta B, Thampan A V, and Bombaci I 1998 *Astron. and Astrophys.* **334** 943
[3] Lattimer J M and Prakash M 2001 *Astrophys. J.* **550** 426
[4] Peng G X, Li A, and Lombardo U 2008 *Phys. Rev. C* **77** 065807
[5] Alford M G, Schmitt A, Rajagopal K, and Schafer T 2008 *Rev. Mod. Phys.* **80** 1455
[6] Anglani R, Casalbuoni R, Ciminale M, Ippolito N, Gatto R, Mannarelli M, and Ruggieri M 2014 *Rev. Mod. Phys.* (to be published); arXiv:1302.4264
[7] Perez-Garcia M A, Silk J, and Stone J R 2010 *Phys. Rev. Lett.* **105** 141101
[8] Li X D, Ray S, Dey J, Dey M, and Bombaci I 1999 *Astrophys. J.* **527** L51
[9] Sotani H, Yasutake N, Maruyama T, and Tatsumi T 2011 *Phys. Rev. D* **83** 024014
[10] Berezhiani Z, Bombaci I, Drago A, Frontera F, and Lavagno A 2003, *Astrophys. J.* **586** 1250; Bombaci I, Parenti I, and Vidaña I 2004 *Astrophys. J.* **614** 314
[11] Lugones G and Bombaci I 2005 *Phys. Rev. D* **72** 065021; Bombaci I, Lugones G, and Vidaña I 2007 *Astron. and Astrophys.* **462** 1017
[12] Bombaci I, Logoteta D, Providência C, and Vidaña I 2011 *Astron. and Astrophys.* **528** A71
[13] Logoteta D, Providência C, Vidaña I, and Bombaci I 2012 *Phys. Rev. C* **85** 055807
[14] Weissenborn S, Sagert I, Pagliara G, Hempel M, and Schaffner-Bielich J 2011 *Astrophys. J.* **740** L14
[15] Bhattacharyya S, Thampan A V, and Bombaci I 2001 *Astron. and Astrophys.* **372** 925
[16] Aoki Y, Endrodi G, Fodor Z, Katz S D, and Szabó K K 2006 *Nature* **443** 675
[17] Chodos A *et al.* 1974 *Phys. Rev. D* **9** 3471; Farhi E and Jaffe R L 1984 *Phys. Rev. D* **30** 272
[18] Nambu Y and Jona-Lasinio G 1961 *Phys Rev.* **122** 345
[19] Dosh H G 1987 *Phys. Lett. B* **190** 177; Dosh H G and Simonov Yu A 1988 *Phys. Lett. B* **205** 339; Simonov Yu A 1988 *Nucl. Phys. B* **307** 512
[20] Di Giacomo A, Dosch H G, Shevchenko V I, and Simonov Yu A 2002 *Phys. Rep.* **372** 319
[21] Simonov Yu A and Trusov M A 2007 *Phys. Lett. B* **650** 36
[22] Komarov E V and Simonov Yu A 2008 *Ann. Phys.* **323** 1230
[23] Bombaci I and Logoteta D 2013 *MNRAS Letters* **L79** 433
[24] Logoteta D and Bombaci I 2013 *Phys. Rev. D* **88** 063001
[25] Baldo M, Burgio G F, Castorina P, Plumari S, and Zappalà D 2008 *Phys. Rev. D* **78** 063009
[26] Pereira F I M 2011 *Nucl. Phys. A* **860** 102; 2013 *Nucl. Phys. A* **897** 151
[27] Plumari S, Burgio G F, Greco V, and Zappalá D 2013 *Phys. Rev. D* **88** 083005
[28] Shifman M A, Vainshtein A I, and Zakharov V I 1979 *Nucl. Phys. B* **147** 385; *Nucl. Phys. B* **147** 448
[29] Walecka J D 1974 *Ann. Phys. (N.Y.)* **83** 491; Serot B D and Walecka J D 1986 *Adv. Nucl. Phys.* **16** 1
[30] Boguta J and Bodmer A R 1977 *Nucl. Phys. A* **292** 413
[31] Glendening N K and Moszkowski S 1991 *Phys. Rev. Lett.* **67** 2414
[32] Glendenning N K 2000 *Compact Stars, Nuclear Physics, Particle Physics, and General Relativity* (Springer, New York)
[33] Bombaci I, Panda P K, Providência C, and Vidaña I 2008 *Phys. Rev. D* **77** 083002
[34] Bombaci I, Prakash M, Prakash M, Ellis P J, Lattimer J M, and Brown G E 1995 *Nucl. Phys. A* **583** C623
[35] Fodor Z and Katz S D 2004 *J. High Energy Phys.* **04** 050
[36] Glendening N K 1992 *Phys. Rev. D* **46** 1274
[37] Haensel P, Potekhin A Y, and Yakovlev D G 2007 *Neutron Stars 1: Equation of State and Structure* (Springer)
[38] Baym G, Pethick C, and Sutherland D 1971 *Astrophys. J.* **170** 299; Negele J W and Vautherin D 1973 *Nucl. Phys. A* **207** 298
[39] Demorest P, Pennucci T, Ransom S, Roberts M, and Hessels J 2010 *Nature* **467** 1081
[40] Borsanyi S *et al.* 2010 *J. High Energy Phys.* **09** 073
[41] Bazavov A *et al.* 2012 *Phys. Rev. D* **85** 054503
[42] D'Elia M, Di Giacomo A, and Meggiolaro E 1997 *Phys. Lett. B* **408** 315

Journal of Physics: Conference Series **527** (2014) 012022 doi:10.1088/1742-6596/527/1/012022

Nuclear electromagnetic processes in ChEFT

L Girlanda[1,2], **L E Marcucci**[3,4], **S Pastore**[5], **M Piarulli**[6], **R Schiavilla**[6,7], **M Viviani**[4]

[1] Department of Mathematics and Physics, University of Salento, I-73100 Lecce, Italy
[2] INFN, Sezione di Lecce, I-73100 Lecce, Italy
[3] Department of Physics, University of Pisa, I-56127 Pisa, Italy
[4] INFN, Sezione di Pisa, I-56127 Pisa, Italy
[5] Department of Physics and Astronomy, University of South Carolina, Columbia, SC 29208, USA
[6] Department of Physics, Old Dominion University, Norfolk, VA 23529, USA
[7] Theory Center, Jefferson Laboratory, Newport News, VA 23606, USA

E-mail: `girlanda@le.infn.it`

Abstract. We review our recent work on the derivation of the nuclear electromagnetic charge and current operators in chiral perturbation theory, based on time-ordered perturbation theory. We then discuss the strategies for fixing the relevant low-energy constants, and compare the resulting predictions for the electric and magnetic form factors of the deuteron and trinucleons with experimental data, using as input accurate nuclear wave functions derived with realistic potentials.

1. Introduction

Besides providing a justification for the hierarchy of nuclear forces, establishing a clear contact with the underlying theory of strong interaction and its symmetries, and leading to a well defined expansion scheme, susceptible of systematic improvement, the chiral effective field theory (χEFT) is ideally suited for deriving consistent electroweak currents. Indeed, chiral perturbation theory is formulated as an effective theory of external currents, which are coupled to the degrees of freedom of the fundamental theory. The constraints imposed by chiral symmetry, so-called chiral Ward identities, are obtained promoting the global chiral symmetry to a local one [1], with the external currents r_μ and ℓ_μ, representing the corresponding gauge fields. The explicit breaking of chiral symmetry by quark masses and electromagnetic currents is naturally implemented: correlation functions are to be evaluated with the scalar source χ proportional to the quark mass matrix and the external currents set equal to the photon field, $r_\mu = \ell_\mu = QA_\mu$ with $Q = e\,\mathrm{diag}(\frac{2}{3}, -\frac{1}{3})$ in the meson sector and $Q = e\,\mathrm{diag}(1,0)$ in the nucleon sector. Indeed, soon after the Weinberg proposal [2] in the 90s, electroweak transition operators have been addressed by Park, Min and Rho [3], using covariant perturbation theory. Nuclear electromagnetic currents and charge operators have recently been rederived in time-ordered perturbation theory [4, 5, 6, 7] and in the so-called unitary transformation method [8, 9] within the same scheme as for the NN potential [10] based on the Okubo procedure [11]. The similarities and differences among these approaches are discussed in Refs. [4, 5, 9]. In what follows we describe the approach based on time-ordered perturbation theory (TOPT).

Journal of Physics: Conference Series **527** (2014) 012022 doi:10.1088/1742-6596/527/1/012022

The effective Lagrangian is the most general one invariant under chiral symmetry: it contains an infinite number of operators, classified according to the chiral counting, a combined expansion in powers of quark masses and small momenta, with $\chi \sim O(p^2)$. The chiral counting is thus the organizing principle: it works because Goldstone bosons have derivative interactions, as dictated by the Goldstone's theorem. From the chiral Lagrangian (see Ref. [12]), the canonical formalism allows to obtain the Hamiltonian which describes the interactions of pions, nucleons and photons. For the present discussion the following contributions are relevant

$$H_{\pi N} = \int d\mathbf{x}\, N^\dagger \Big[\frac{g_A \tau_a}{F_\pi} \boldsymbol{\sigma}\cdot\boldsymbol{\nabla}\pi_a + \frac{\boldsymbol{\tau}}{F_\pi^2}\cdot(\boldsymbol{\pi}\times\partial^0\boldsymbol{\pi}) + \dots \Big] N, \tag{1}$$

$$\begin{aligned} H_{\gamma N} &= e\int d\mathbf{x}\, N^\dagger \Big[e_N A^0 + i\frac{e_N}{2m}\left(-\overleftarrow{\boldsymbol{\nabla}}\cdot\mathbf{A} + \mathbf{A}\cdot\overrightarrow{\boldsymbol{\nabla}}\right) - \frac{\mu_N}{2m}\boldsymbol{\sigma}\cdot\boldsymbol{\nabla}\times\mathbf{A} \\ &\quad - \frac{2\mu_N - e_N}{8m^2}\left(\boldsymbol{\nabla}^2 A^0 + \boldsymbol{\sigma}\times\boldsymbol{\nabla}A^0\cdot\overrightarrow{\boldsymbol{\nabla}} - \overleftarrow{\boldsymbol{\nabla}}\cdot\boldsymbol{\sigma}\times\boldsymbol{\nabla}A^0\right) + \dots \Big] N\,, \end{aligned} \tag{2}$$

$$H_{\gamma\pi} = e\int d\mathbf{x}\,\left[A^0\left(\boldsymbol{\pi}\times\partial^0\boldsymbol{\pi}\right)_z + \epsilon_{zab}\,\pi_a\,(\boldsymbol{\nabla}\pi_b)\cdot\mathbf{A} + \dots\right]\,, \tag{3}$$

$$\begin{aligned} H_{\gamma\pi N} &= e\int d\mathbf{x}\, N^\dagger \Bigg[-\frac{g_A}{F_\pi}(\boldsymbol{\tau}\times\boldsymbol{\pi})_z\,\boldsymbol{\sigma}\cdot\mathbf{A} + \frac{g_A}{2m\,F_\pi}(\boldsymbol{\tau}\cdot\boldsymbol{\pi} + \pi_z)\,\boldsymbol{\sigma}\cdot\boldsymbol{\nabla}A^0 \\ &\quad + \left(\frac{d_8'}{F_\pi}\boldsymbol{\nabla}\pi_z + \frac{d_9'}{F_\pi}\tau_a\boldsymbol{\nabla}\pi_a + \frac{d_{21}'}{F_\pi}\epsilon_{zab}\tau_b\,\boldsymbol{\sigma}\times\boldsymbol{\nabla}\pi_a\right)\cdot\boldsymbol{\nabla}\times\mathbf{A} + \dots\Bigg] N\,. \end{aligned} \tag{4}$$

Here g_A is the nucleon axial coupling, F_π the pion decay constant, e the proton electric charge and m the nucleon mass, while the parameters d_i are (unknown) low-energy constants (LECs). $\boldsymbol{\sigma}$ and $\boldsymbol{\tau}$ are the spin and isospin Pauli matrices, and the isospin operators e_N and μ_N are defined as

$$e_N = (1+\tau_z)/2\,,\quad \kappa_N = (\kappa_S + \kappa_V\,\tau_z)/2\,,\quad \mu_N = e_N + \kappa_N\,, \tag{5}$$

with κ_S and κ_V denoting the isoscalar and isovector combinations of the anomalous magnetic moments of the proton and neutron. The non-relativistic nucleon field N is an isospin doublet, while pions are described by the isovector $\boldsymbol{\pi}$ and photons by the four-vector A^μ. The chiral dimension of the resulting vertices is obtained by counting the powers of Q, the low-momentum scale, or equivalently the number of gradients, e.g. the two terms in $H_{\pi N}$ are each of order $\sim Q$, while the first terms in $H_{\gamma\pi N}$ are of order $\sim e\,Q^0$ and $\sim e\,Q$, and the remaining ones in second line of equation (4) are of order $\sim e\,Q^2$.

The two-nucleon contact Hamiltonian can be written as

$$\begin{aligned} H_{CT} &= \int d\mathbf{x}\Bigg[\frac{1}{2}C_S\,N^\dagger N N^\dagger N + \frac{1}{2}C_T\,N^\dagger\vec{\sigma}N\cdot N^\dagger\vec{\sigma}N \\ &\quad -\frac{1}{32}(16C_1 - C_2 - 3C_4 - C_7)\,\nabla^2(N^\dagger N)N^\dagger N \\ &\quad +\frac{1}{32}(C_2 + 3C_4 + C_7)\,\nabla^2(N^\dagger\tau^a N)N^\dagger\tau^a N \\ &\quad -\frac{1}{32}(16C_3 - C_2 + C_4 + C_7)\,\nabla^2(N^\dagger\sigma^i N)N^\dagger\sigma^i N \\ &\quad +\frac{1}{32}(C_2 - C_4 - C_7)\,\nabla^2(N^\dagger\sigma^i\tau^a N)N^\dagger\sigma^i\tau^a N \\ &\quad -\frac{i}{8}C_5\left[\vec{\nabla}\cdot(N^\dagger\overleftrightarrow{\nabla}\times\vec{\sigma}N)N^\dagger N - N^\dagger\overleftrightarrow{\nabla}N\cdot\vec{\nabla}\times(N^\dagger\vec{\sigma}N)\right] \\ &\quad +\frac{1}{16}(8C_6 - C_7)\,\vec{\nabla}\cdot(N^\dagger\vec{\sigma}N)\vec{\nabla}\cdot(N^\dagger\vec{\sigma}N) - \frac{1}{16}C_7\vec{\nabla}\cdot(N^\dagger\vec{\sigma}\tau^a N)\vec{\nabla}\cdot(N^\dagger\vec{\sigma}\tau^a N)\Bigg]\,, \end{aligned} \tag{6}$$

where, inside a fermion bilinear, $\overleftrightarrow{\nabla} = \overrightarrow{\nabla} - \overleftarrow{\nabla}$. Here relativistic corrections [13] have been ignored, while the definition of the LECs C_i, typically determined by fitting low-energy two-nucleon scattering data and the deuteron binding energy, is chosen so as to conform to standard notation In addition, Fierz identities have been used such that the resulting nucleon-nucleon potential can be put in local form [14]. Minimal substitution in this Hamiltonian, $\boldsymbol{\nabla} N \to (\boldsymbol{\nabla} - i\, e\, e_N \mathbf{A})\, N$ in H_{CT} leads to a (contact) Hamiltonian $H_{CT}^{\gamma\mathrm{m}}$ which includes the coupling to the EM field and implies a (two-nucleon) contact current operator. It is also necessary to consider non-minimal couplings, entering through the electromagnetic field tensor $F_{\mu\nu}$. The only two independent operator structures are

$$H_{CT}^{\gamma\mathrm{nm}} = e \int \mathrm{d}\mathbf{x} \left[C'_{15}\, N^\dagger \boldsymbol{\sigma} N\, N^\dagger N + C'_{16} \left(N^\dagger \boldsymbol{\sigma}\, \tau_z N\, N^\dagger N - N^\dagger \boldsymbol{\sigma} N\, N^\dagger \tau_z N \right) \right] \cdot \boldsymbol{\nabla} \times \mathbf{A} \ ,$$

where the isoscalar C'_{15} and isovector C'_{16} LECs (as well as the d_i's multiplying the higher order terms in the $\gamma\pi N$ Hamiltonian) can be determined by fitting photo-nuclear data in few-nucleon systems.

2. From amplitudes to potentials

A generic transition amplitude, e.g. the two-nucleon scattering amplitude, can be calculated from the above Hamiltonians in the conventional TOPT, as

$$\langle f \mid T \mid i \rangle = \langle f \mid H_1 \sum_{n=1}^{\infty} \left(\frac{1}{E_i - H_0 + i\,\eta} H_1 \right)^{n-1} \mid i \rangle \ , \tag{7}$$

with $|i\rangle$ and $|f\rangle$ representing the initial and final NN states of energy $E_i = E_f$, H_0 the free Hamiltonian of pions and nucleons and H_1 containing the interactions. The time-ordered diagrams are obtained inserting complete sets of H_0 eigenstates between successive terms of H_1. The infinite number of contributions are ordered according to the power counting in powers of $Q/\Lambda_\chi \ll 1$, where Q is the generic pion momentum and $\Lambda_\chi \simeq 1$ GeV is the typical hadronic mass scale, i.e. the mass of the states not protected by chiral symmetry.

A generic contribution in the perturbative series, equation (7), is characterized by a certain number N of vertices, each scaling as $Q^{\alpha_i} \times Q^{-\beta_i/2}$ (i=1, ..., N), where α_i is the power counting implied by the relevant interaction Hamiltonian and β_i is the number of pions attached to the vertex, a corresponding N–1 number of energy denominators, and possibly L loops. Out of these N–1 energy denominators, N_K will involve only nucleon kinetic energies, which scale as Q^2, and the remaining $N - N_K - 1$ will involve, in addition, pion energies, which are of order Q. Loops contribute a factor Q^3 each, since they imply integrations over intermediate three momenta. Hence the power counting associated with such a contribution is

$$\left(\prod_{i=1}^{N} Q^{\alpha_i - \beta_i/2} \right) \times \left[Q^{-(N-N_K-1)}\, Q^{-2N_K} \right] \times Q^{3L} \ . \tag{8}$$

Each of the $N - N_K - 1$ "large" energy denominators can in turn be expanded as

$$\frac{1}{E_i - E_I - \omega_\pi} = -\frac{1}{\omega_\pi} \left[1 + \frac{E_i - E_I}{\omega_\pi} + \frac{(E_i - E_I)^2}{\omega_\pi^2} + \dots \right] , \tag{9}$$

where E_I denotes the kinetic energy of the intermediate two-nucleon state, ω_π the pion energies, and the ratio $(E_i - E_I)/\omega_\pi$ is of order Q. As a result the two-nucleon scattering amplitude T can be expanded as

$$T = T^{(0)} + T^{(1)} + T^{(2)} + \dots \ , \tag{10}$$

where $T^{(n)} \sim Q^n$. We can define a two-nucleon potential v such that, when iterated in the Lippmann-Schwinger (LS) equation,

$$v + v\,G_0\,v + v\,G_0\,v\,G_0\,v + \dots \,, \tag{11}$$

with G_0 the free two-nucleon propagator, $G_0 = 1/(E_i - E_I + i\,\eta)$, it leads to the on-the-energy-shell ($E_i = E_f$) T-matrix in equation (10), order by order in the power counting, assuming that

$$v = v^{(0)} + v^{(1)} + v^{(2)} + \dots \,, \tag{12}$$

admits the same expansion with $v^{(n)}$ of order Q^n. This procedure implies that

$$v^{(0)} = T^{(0)} \,, \tag{13}$$

$$v^{(1)} = T^{(1)} - \left[v^{(0)}\,G_0\,v^{(0)}\right] \,, \tag{14}$$

$$v^{(2)} = T^{(2)} - \left[v^{(0)}\,G_0\,v^{(0)}\,G_0\,v^{(0)}\right] - \left[v^{(1)}\,G_0\,v^{(0)} + v^{(0)}\,G_0\,v^{(1)}\right] \,, \tag{15}$$

$$\begin{aligned} v^{(3)} = {} & T^{(3)} - \left[v^{(0)}\,G_0\,v^{(0)}\,G_0\,v^{(0)}\,G_0\,v^{(0)}\right] - \left[v^{(1)}\,G_0\,v^{(0)}\,G_0\,v^{(0)} + \text{permutations}\right] \\ & - \left[v^{(2)}\,G_0\,v^{(0)} + v^{(0)}\,G_0\,v^{(2)}\right] - \left[v^{(1)}\,G_0\,v^{(1)}\right] \,. \end{aligned} \tag{16}$$

Since the potential is thus defined from on-shell amplitudes order by order, different off-shell extensions of the potential at a given order lead to different (on-shell) potentials at higher orders. In fact, there is an infinite class of $v^{(2)}(\nu)$ non-static corrections–labeled by the parameter ν [15, 16, 17, 6]–which, while equivalent on the energy-shell, are different off the energy-shell, and therefore lead to different potentials $v^{(3)}(\nu)$ in equation (16). However, this ambiguity is of no consequence, since it can be shown that different off-the-energy-shell extrapolations $v^{(2)}(\nu)$ and $v^{(3)}(\nu)$ are unitarily equivalent [15, 6].

The inclusion (in first order) of electromagnetic interactions in the perturbative expansion of equation (7) is straightforward: the transition operator can be expanded as

$$T_\gamma = T_\gamma^{(-3)} + T_\gamma^{(-2)} + T_\gamma^{(-1)} + \dots \,, \tag{17}$$

where $T_\gamma^{(n)}$ is of order $e\,Q^n$ (e is the electric charge). The nuclear charge, ρ, and current, $\mathbf{j}$, operators follow from $v_\gamma = A^0\,\rho - \mathbf{A}\cdot\mathbf{j}$, where $A^\mu = (A^0, \mathbf{A})$ is the electromagnetic vector field, and it is assumed that v_γ has a similar expansion as T_γ. The requirement that, in the context of the LS equation, v_γ matches T_γ order by order in the power counting implies the following relations:

$$v_\gamma^{(-3)} = T_\gamma^{(-3)} \tag{18}$$

$$v_\gamma^{(-2)} = T_\gamma^{(-2)} - \left[v_\gamma^{(-3)}\,G_0\,v^{(0)} + v^{(0)}\,G_0\,v_\gamma^{(-3)}\right] \,, \tag{19}$$

$$v_\gamma^{(-1)} = T_\gamma^{(-1)} - \left[v_\gamma^{(-3)}\,G_0\,v^{(0)}\,G_0\,v^{(0)} + \text{permutations}\right] - \left[v_\gamma^{(-2)}\,G_0\,v^{(0)} + v^{(0)}\,G_0\,v_\gamma^{(-2)}\right] \,, \tag{20}$$

$$\begin{aligned} v_\gamma^{(0)} = T_\gamma^{(0)} & - \left[v_\gamma^{(-3)}\,G_0\,v^{(0)}\,G_0\,v^{(0)}\,G_0\,v^{(0)} + \text{permutations}\right] \\ & - \left[v_\gamma^{(-2)}\,G_0\,v^{(0)}\,G_0\,v^{(0)} + \text{permutations}\right] \\ & - \left[v_\gamma^{(-1)}\,G_0\,v^{(0)} + v^{(0)}\,G_0\,v_\gamma^{(-1)}\right] - \left[v_\gamma^{(-3)}\,G_0\,v^{(2)} + v^{(2)}\,G_0\,v_\gamma^{(-3)}\right] \,, \end{aligned} \tag{21}$$

$$v_\gamma^{(1)} = T_\gamma^{(1)} - \left[v_\gamma^{(-3)}\,G_0\,v^{(0)}\,G_0\,v^{(0)}\,G_0\,v^{(0)}\,G_0\,v^{(0)} + \text{permutations}\right]$$

$$- \left[v_\gamma^{(-2)} G_0 \, v^{(0)} G_0 \, v^{(0)} G_0 \, v^{(0)} + \text{permutations} \right]$$
$$- \left[v_\gamma^{(-1)} G_0 \, v^{(0)} G_0 \, v^{(0)} + \text{permutations} \right] - \left[v_\gamma^{(0)} G_0 \, v^{(0)} + v^{(0)} G_0 \, v_\gamma^{(0)} \right]$$
$$- \left[v_\gamma^{(-3)} G_0 \, v^{(2)} G_0 \, v^{(0)} + \text{permutations} \right] - \left[v_\gamma^{(-3)} G_0 \, v^{(3)} + v^{(3)} G_0 \, v_\gamma^{(-3)} \right] , \qquad (22)$$

where $v_\gamma^{(n)} = A^0 \rho^{(n)} - \mathbf{A} \cdot \mathbf{j}^{(n)}$, $v^{(n)}$ are the NN potentials constructed in equations (13)–(16) (the ν dependence of $v^{(2)}$ and $v^{(3)}$ is understood), and use has been made of the fact that $v^{(1)}$ vanishes. In the propagator G_0, the initial energy E_i includes the photon energy ω_γ (itself of order Q^2), since $E_i = E_1 + E_2 + \omega_\gamma = E_1' + E_2'$, and the intermediate energy E_I may include, in addition to the kinetic energies of the intermediate nucleons, also the photon energy, depending on the specific time ordering being considered.

3. Electromagnetic current and charge operators up to one loop

The contributions to the electromagnetic current operator up to one loop, as resulting from the interaction Hamiltonians and the perturbative expansion outlined above, are illustrated diagrammatically in figure 1. Since there is no $n = -3$ contribution to $\mathbf{j}$ the current operator is unaffected by the non-static corrections entering the potentials $v^{(2)}$ and $v^{(3)}$. The lowest order ($n = -2$) consists of the single-nucleon convection and spin-magnetization currents:

$$\mathbf{j}^{(-2)} = \frac{e}{2m} \left(2 \, e_{N,1} \, \mathbf{K}_1 + i \, \mu_{N,1} \, \boldsymbol{\sigma}_1 \times \mathbf{q} + 1 \rightleftharpoons 2 \right) , \qquad (23)$$

where $\mathbf{q}$ is the momentum carried by the external field, $\mathbf{k}_i$ and $\mathbf{K}_i$ denote the combinations of initial and final nucleon momenta $\mathbf{k}_i = \mathbf{p}_i' - \mathbf{p}_i$, $\mathbf{K}_i = (\mathbf{p}_i' + \mathbf{p}_i)/2$, and $e_{N,i}$ and $\mu_{N,i}$ have been defined in equation (5). The counting $e\,Q^{-2}$ follows from the product of a factor $e\,Q^1$ associated with the γNN vertex, and a factor Q^{-3} due to the momentum-conserving δ-function implicit in a disconnected term of this type. In this one-body contribution we also include a q^2-dependence (which would only arise at successive chiral orders) in the form of the electric and magnetic form factors, as resulting from fits of electron scattering on the proton and deuteron. We refer

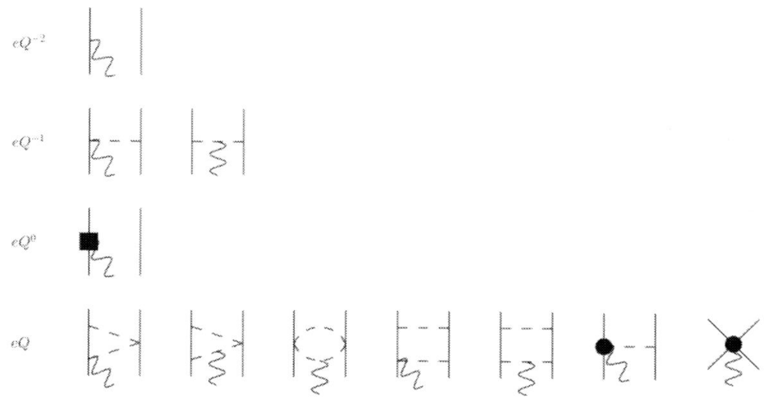

Figure 1. Diagrams illustrating one- and two-body currents entering at LO ($e\,Q^{-2}$), NLO ($e\,Q^{-1}$), N2LO ($e\,Q^0$), and N3LO ($e\,Q^1$). Nucleons, pions, and photons are denoted by solid, dashed, and wavy lines. Only the relevant topologies are indicated. Loop corrections to short-range currents turn out to vanish.

to Refs. [4, 5, 7] for the explicit expressions up to N3LO, as obtained within the formalism outlined in section 2. They depend on the known parameters g_A and F_π (NLO and N3LO), and the nucleon's magnetic moments (LO and N2LO). Loop corrections to the short-range currents turn out to cancel [7]. Unknown LECs enter the N3LO one-pion exchange (OPE) contribution involving the $\gamma\pi N$ vertex of order $e\,Q^2$ from $H_{\gamma\pi N}$,

$$\mathbf{j}^{(1)}_{\gamma\pi N} = i\,e\,\frac{g_A}{F_\pi^2}\frac{\boldsymbol{\sigma}_2\cdot\mathbf{k}_2}{\omega^2_{k_2}}\left[\Big(d_8'\tau_{2,z} + d_9'\,\boldsymbol{\tau}_1\cdot\boldsymbol{\tau}_2\Big)\mathbf{k}_2 - d_{21}'(\boldsymbol{\tau}_1\times\boldsymbol{\tau}_2)_z\,\boldsymbol{\sigma}_1\times\mathbf{k}_2\right]\times\mathbf{q} + 1\rightleftharpoons 2\ . \tag{24}$$

The contributing LECs could be fixed by relating them, in a resonance saturation picture, to the couplings in the N to Δ excitation and $\rho\pi\gamma$ transition currents, or they could be fixed by pion photo-production data on a single nucleon or photo-nuclear data at low energies. Further LECs enter the two-contact currents, from minimal and non-minimal substitution,

$$\begin{aligned}\mathbf{j}^{(1)}_{\rm CT,m} \;=\;& \frac{i}{16}\,(\boldsymbol{\tau}_1\times\boldsymbol{\tau}_2)_z\,\Big[\left[C_2 + 3\,C_4 + C_7 + (C_2 - C_4 - C_7)\,\boldsymbol{\sigma}_1\cdot\boldsymbol{\sigma}_2\right](\mathbf{k}_1 - \mathbf{k}_2)\\ & + C_7\left[\boldsymbol{\sigma}_1\cdot(\mathbf{k}_1 - \mathbf{k}_2)\,\boldsymbol{\sigma}_2 + \boldsymbol{\sigma}_2\cdot(\mathbf{k}_1 - \mathbf{k}_2)\,\boldsymbol{\sigma}_1\right]\Big]\\ & - \frac{i\,C_5}{4}\,(\boldsymbol{\sigma}_1 + \boldsymbol{\sigma}_2)\times(e_1\,\mathbf{k}_1 + e_2\,\mathbf{k}_2)\ ,\end{aligned} \tag{25}$$

where the C_i can be taken from the NN potential, and

$$\mathbf{j}^{(1)}_{\rm CT,nm} = -i\,e\left[C_{15}'\,\boldsymbol{\sigma}_1 + C_{16}'\,(\tau_{1,z} - \tau_{2,z})\,\boldsymbol{\sigma}_1\right]\times\mathbf{q} + 1\rightleftharpoons 2\ . \tag{26}$$

No three-body currents arise up to N^{3}LO included here, due to cancellations between irreducible and recoil-corrected reducible diagrams, similarly to what happens for the three-nucleon force.

Diagrams contributing to the charge operators $\rho^{(n)}$ up to order $e\,Q^1$ (N4LO) included are illustrated in figure 2. The leading contribution, of order $e\,Q^{-3}$, is a one-body operator and results from the first term of the γN interaction Hamiltonian in equation (2),

$$\rho^{(-3)} = e\,e_{N,1} + 1\rightleftharpoons 2\ , \tag{27}$$

and, as before, electric form factors can be included, even if they would arise at successive chiral orders. There are no NLO ($e\,Q^{-2}$) contributions, whereas at N2LO there is a relativistic correction of order $(Q/m)^2$ to the LO charge operator, which results from the second line of equation (2). At this order, there are in principle also a pion-in-flight term, which, however, turns out to vanish when the contributions of the six time-ordered diagrams, evaluated in the static limit, are summed up, and a OPE contribution, which vanishes due to a similar cancellation. The power counting is different from the current operator, for which the LO term is of order $e\,Q^{-2}$ (in the two-nucleon system), i.e. it is suppressed by an extra power of Q relative to $\rho^{(-3)}$, and where there are NLO ($e\,Q^{-1}$) corrections involving seagull and in-flight contributions associated with OPE, which have no counterpart in the present case. The pion-in-flight and OPE diagrams also lead to N3LO contributions due to non-static corrections resulting from the expansion of the denominators involving pion energies as in equation (9). In particular, the specific form of the N3LO charge operator depends on the (non-unique) off-the-energy shell prescription adopted for the non-static piece in the OPE potential [6]. The same applies to part of the N4LO contributions. This ambiguity in the non-static OPE and two-pion exchange (TPE) potentials and accompanying charge operators is of no consequence, however, since different form for these are related to each other by a unitary transformation [6, 7]. Thus, provided a consistent set is adopted, predictions for physical observables, such as the few-nucleon charge form factors, will

Journal of Physics: Conference Series **527** (2014) 012022 doi:10.1088/1742-6596/527/1/012022

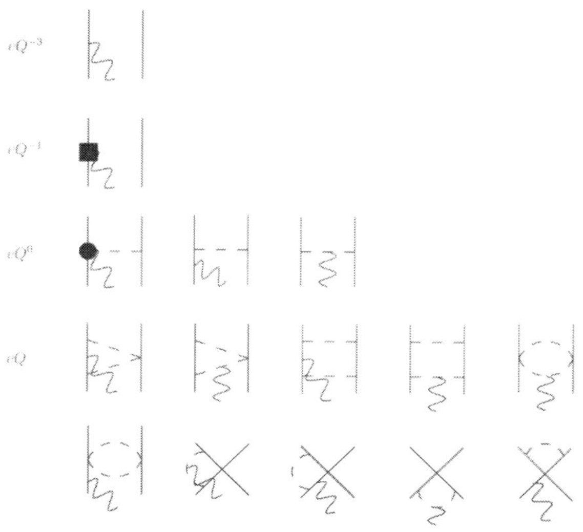

Figure 2. Diagrams illustrating one- and two-body charge operators entering at LO ($e\,Q^{-3}$), N2LO ($e\,Q^{-1}$), N3LO ($e\,Q^{0}$), and N4LO ($e\,Q^{1}$). The square represents the $(Q/m)^2$, or $(v/c)^2$, relativistic correction to the LO one-body charge operator, whereas the solid circle is associated with a $\gamma\pi N$ charge coupling of order $e\,Q$. Only the relevant topologies are indicated.

remain unaffected by the non-uniqueness associated with off-the-energy-shell effects. Finally, charge conservation,

$$\rho(\mathbf{q}=0) = \int \mathrm{d}\mathbf{x}\, \rho(\mathbf{x}) = e\,(e_1+e_2)\ , \tag{28}$$

where $\rho(\mathbf{x}$ and $\rho(\mathbf{q})$ denote, respectively, the charge density and its Fourier transform, implies that the charge operators $\rho^{(n\geq -2)}(\mathbf{q})$ vanish at $\mathbf{q}=0$. This latter requirement is satisfied by the operators illustrated in figure 2, regardless of the adopted off-the-energy shell prescription [6]. We also emphasize that, up to N4LO included, there are no unknown LECs. As a consequence, the loop integrals entering diagrams at N4LO, although individually ultra-violet divergent, sum up to give a finite result.

4. Fixing the LECs

The operators described in the previous section depend on the LECs C_i entering the two-nucleon contact Lagrangian through the minimal coupling procedure, on the non-minimal coupling LECs C'_{15} and C'_{16} and on the subleading pion-nucleon couplings d'_i entering in the one-pion exchange N3LO current. The latter could be fitted to pion photo-production data on a single nucleon, which however involve photon energies much higher than those relevant for the threshold processes under consideration here, or related to hadronic coupling constants through resonance saturation. We prefer to treat them as fitting parameters. In Refs. [18, 7] we explored different strategies to such fitting procedure. It turns out that the most effective one is to use Δ-resonance saturation for the isovector pion-nucleon couplings in equation (24),

$$d'_8 = 4d'_{21} = \frac{4\mu_{\gamma N\Delta} h_A}{9m_N(m_\Delta - m_N)} \tag{29}$$

where $\mu_{\gamma N\Delta} \sim 3\mu_N$ is the $N\Delta$ transition magnetic moment and h_A the $\pi N\Delta$ coupling constant, $h_A/f_\pi = f_{\pi N\Delta}/m_\pi$ with $f^2_{\pi N\Delta}/(4\pi) = 0.35$. We are then left with 3 parameter, one isovector and two isoscalar,

$$C'_{15} = d_1^S/\Lambda^4, \quad d'_9 = d_2^S/\Lambda^2, \quad C'_{16} = d_1^V/\Lambda^4, \tag{30}$$

The isoscalar LECs can be fixed from the magnetic moments of the deuteron and the isoscalar combination of the trinucleon. The isovector LEC can be fixed from the isovector trinucleon magnetic moment or from the np thermal radiative capture cross section. Nuclear wave functions are taken from the Hyperspherical Harmonic method [19], using either the Argonne v_{18} (AV18) [20] or the Idaho chiral N3LO [21] two-nucleon potentials for $\Lambda = 500$, 600 MeV [22]. For $A = 3$ we also included, respectively, the Urbana IX [23] and chiral N2LO [24, 25] with the intervening LECs c_D and c_E fixed from the trinucleon binding energies and the ^{3}H Gamow-Teller matrix element [26]. Electromagnetic matrix elements are calculated with MonteCarlo methods, with statistical uncertainties of less than 1%.

The cumulative contributions to the observables used to fix the LECs are shown in figures 3 and 4. It can be seen that the convergence pattern of the chiral series is reasonable. In the

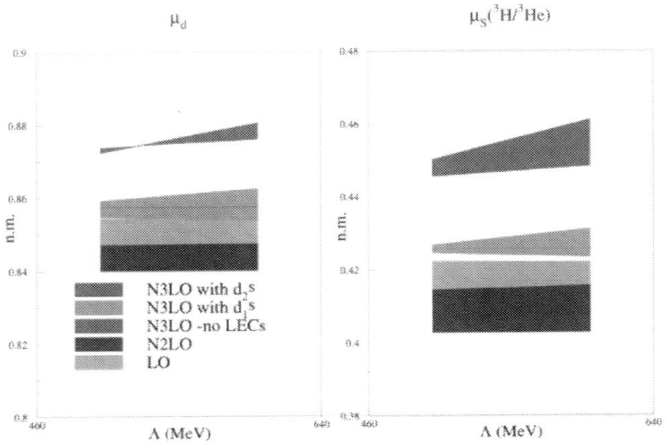

Figure 3. (Color online) Cumulative contributions to the deuteron and trinucleon (isoscalar) magnetic moments. The bands represent the variation with the two nuclear Hamiltonian model.

isovector case, depending on the strategy, either one of the two observables is in fact a prediction, which then turns out to be accurate to 2% for μ_V and 1% for σ_{np}, with a marginal dependence on the cutoff and on the Hamiltonian model. We will describe results with the latter strategy (using μ_V to constrain the LEC), which also proved effective in a recent investigation of magnetic moments and transitions in nuclei with $A \leq 9$ [27], using quantum Montecarlo techniques.

5. Results

In figure 5 we show the charge monopole and quadrupole form factors of the deuteron, confronted to the results extracted from unpolarized and tensor-polarized deuteron data.

For each interaction model also the leading order results in the expansion of the charge operator are shown. The chiral expansion stops at N3LO in this case, since the loop contribution is of isovector character. We observe very good agreement with data up to quite large momentum transfers. The cutoff dependence is larger for the chiral potentials, and does not decrease as we increase the order in the chiral expansion of the charge operator. This is a general feature that we observe. We therefore conclude that the Λ-dependence is mainly due to the interaction potentials, and that the scheme differences in the derivation of the potentials and the transition operator are important. This means that also the calculations with the chiral potentials should

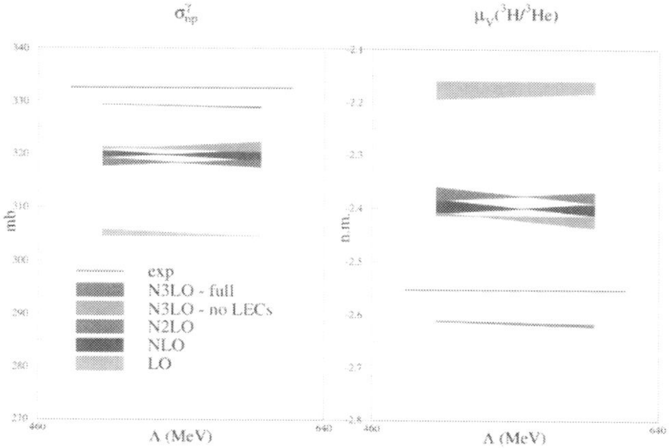

Figure 4. (Color online) Cumulative contributions to the isovector trinucleon magnetic moment and to the np radiative capture cross sections. The black line indicates the experimental value. Notations as in figure 3.

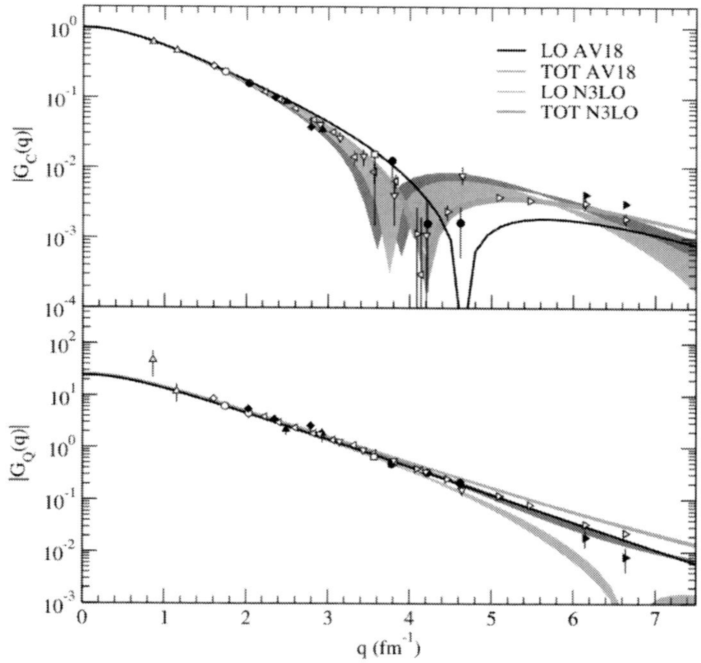

Figure 5. (Color online) Deuteron charge and quadrupole form factors at leading order (LO) and including all subleading terms up to N3LO (TOT), for the two interaction models. The bands represent the variation with the cutoff Λ between 500 and 600 MeV. The experimental data are from Refs. [28, 29, 30, 31, 32, 33, 34, 35, 36, 37, 38, 39, 40, 41, 42, 43, 44, 45, 46, 47, 48, 49].

be considered as being of hybrid character. On the contrary, we checked that the impact of the

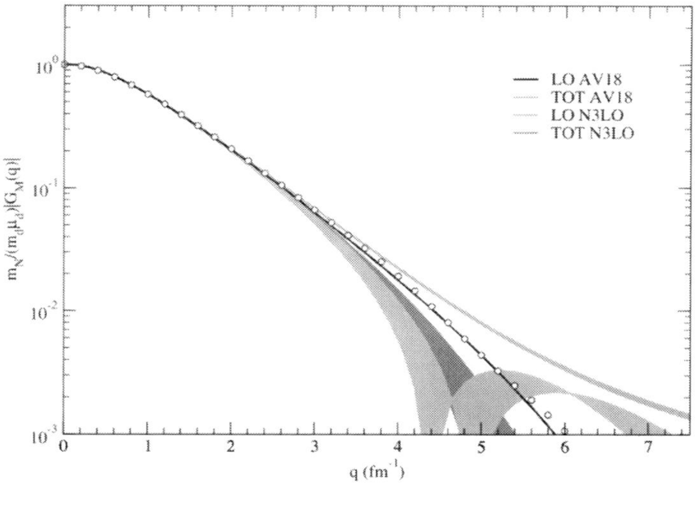

Figure 6. (Color online) Deuteron magnetic form factor, notations as if figure 5. Experimental data are from Refs. [28, 34, 50, 35, 51, 52].

off-shell ambiguity, discussed earlier is negligible.

The magnetic form factor is plotted in figure 6. The correct normalization is imposed in this case by our procedure for fixing the isoscalar LECs. The agreement between theory and experiment extends to momenta of order 2-3 fm^{-1}, depending on the interaction model. In figure 7 we show the charge form factors for ^{3}He and ^{3}H. In this case also the chiral loops contribute at N4LO. The interaction potentials also include the corresponding three-nucleon forces. We observe a large effect of two-body contributions, which bring theory closer to experiment in the diffraction region. By examining separately the individual contributions in the chiral series, however, there is no clear sign of convergence for momenta larger than 3 fm^{-1}, as the contributions at N3LO and N4LO are of comparable size, and much larger than the N2LO. Finally, in figure 8, we show the magnetic form factors of trinucleons, whose isoscalar combination is used to fix one the relevant LECs. The agreement with data is very good up to momenta $\sim$ 2 fm^{-1}. Also in this case the role of the two-body contributions is crucial in bringing theory closer to data, but the diffraction region is still problematic.

6. Conclusions

Chiral perturbation theory, in conjunction with accurate ab-initio techniques to describe light nuclei, allows to make sharp predictions for nuclear electromagnetic observables in the low-energy domain. We have described our formalism to derive the nuclear current and charge operators up to one loop order of the chiral expansion, based on the requirement that, when iterated in the Lippman-Schwinger equation, they lead to the same transition amplitude obtained within time-ordered perturbation theory, order by order in the chiral expansion.

The issue of the convergence pattern of the chiral expansion and of the naturalness of the relevant LECs, however, deserves further investigation, in particular for what concerns the role of the Δ resonance and its possible inclusion within the effective theory.

References

[1] Gasser J and Leutwyler H 1984 *Annals Phys.* **158** 142
[2] Weinberg S 1990 *Phys. Lett. B* **251** 288; 1991 *Nucl. Phys. B* **363** 3; 1992 *Phys. Lett. B* **295** 114

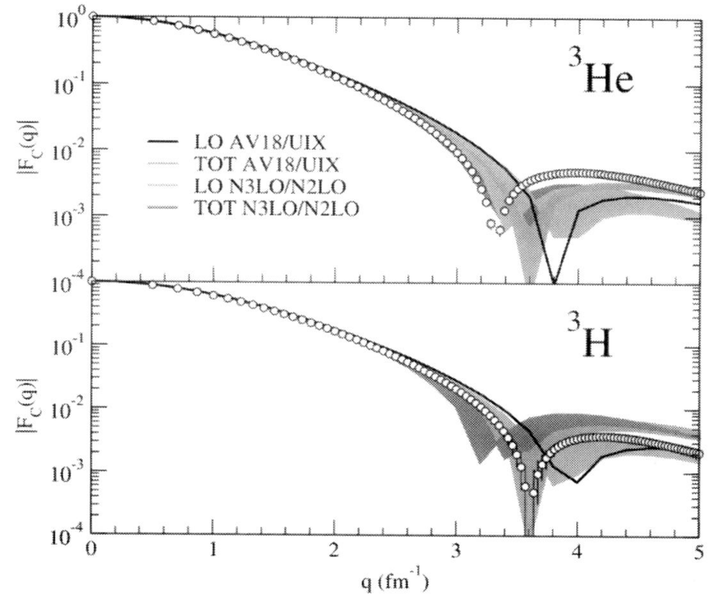

Figure 7. (Color online) Charge form factors of ^{3}He and ^{3}H. Notations as in figure 5. Experimental data are from Refs. [53].

[3] Park T -S, Min D -P, and Rho M 1996 *Nucl. Phys. A* **596** 515
[4] Pastore S, Schiavilla R, and Goity J L 2008 *Phys. Rev. C* **78** 064002
[5] Pastore S, Girlanda L, Schiavilla R, Viviani M, and Wiringa R B 2009 *Phys. Rev. C* **80** 034004
[6] Pastore S, Girlanda L, Schiavilla R, and Viviani M 2011 *Phys. Rev. C* **84** 024001
[7] Piarulli M, Girlanda L, Marcucci L E, Pastore S, Schiavilla R, and Viviani M 2013 *Phys. Rev. C* **87** 014006
[8] Kölling S, Epelbaum E , Krebs H, and Meissner U -G 2009 *Phys. Rev. C* **80** 045502
[9] Kölling S, Epelbaum E, Krebs H, and Meissner U -G 2011 *Phys. Rev. C* **84** 054008
[10] Epelbaum E, Glöckle W, and Meissner U -G 1998 *Nucl. Phys. A* **637** 107; 2000 *Nucl. Phys. A* **671** 295; 2003 *Nucl. Phys. A* **714** 535; 2005 *Nucl. Phys. A* **747** 362
[11] Okubo S 1954 *Prog. Theor. Phys.* **12** 603
[12] Bernard V 2008 *Prog. Part. Nucl. Phys.* **60** 82
[13] Girlanda L, Pastore S, Schiavilla R, and Viviani M 2010 *Phys. Rev. C* **81** 034005
[14] Gezerlis A, Tews I, Epelbaum E, Gandolfi S, Hebeler K, Nogga A, and Schwenk A 2013 *Phys. Rev. Lett.* **111** 032501
[15] Friar J L 1977 *Annals Phys.* **104** 380
[16] Friar J L 1980 *Phys. Rev. C* **22** 796
[17] Adam J, Goller H, and Arenhövel H 1993 *Phys. Rev. C* **48** 370
[18] Girlanda L, Kievsky A, Marcucci L E, Pastore S, Schiavilla R, and Viviani M 2010 *Phys. Rev. Lett.* **105** 232502
[19] Kievsky A *et al.* 2008 *J. Phys. G* **35** 063101
[20] Wiringa R B, Stoks V G J, and Schiavilla R 1995 *Phys. Rev. C* **51** 38
[21] Entem D R and Machleidt R (2003) *Phys. Rev. C* **68** 041001
[22] Machleidt R and Entem D R 2011 *Phys. Rep.* **503** 1
[23] Pudliner B S *et al.* 1997 *Phys. Rev. C* **56** 1720
[24] Epelbaum E, Nogga A, Gloeckle W, Kamada H, Meissner U -G, and Witala H 2002 *Phys. Rev. C* **66** 064001
[25] Navratil P 2007 *Few-Body Syst.* **41** 117
[26] Marcucci L E, Kievsky A, Rosati S, Schiavilla R, and Viviani M 2012 *Phys. Rev. Lett.* **108** 052502
[27] Pastore S, Pieper S C, Schiavilla R, and Wiringa R B 2013 *Phys. Rev. C* **87** 035503
[28] Buchanan C D and Yearian M R 1965 *Phys. Rev. Lett.* **15** 303
[29] Benaksas D, Drickey D, and Frèrejacque D 1966 *Phys. Rev.* **148** 1327
[30] Elias J E *et al.* 1969 *Phys. Rev.* **177** 2075

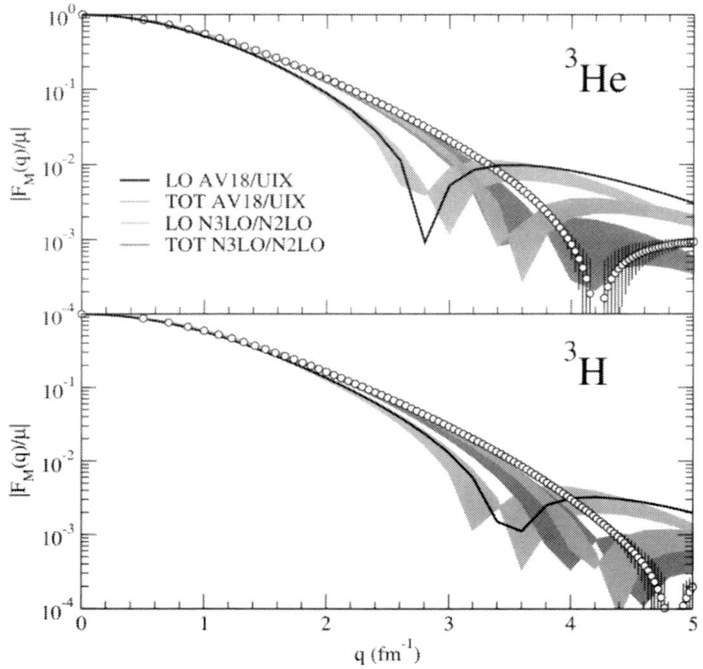

Figure 8. (Color online) Magnetic form factors of ^{3}He and ^{3}H. Notations as in figure 5. Experimental data are from Refs. [53].

[31] Galster S *et al.* 1971 *Nucl. Phys. B* **32** 221
[32] Berard R W *et al.* 1973 *Phys. Lett. B* **47** 355
[33] Arnold R G *et al.* 1975 *Phys. Rev. Lett.* **35** 776
[34] Simon G G , Schmitt C, and Walther V H 1981 *Nucl. Phys. A* **364** 285
[35] Cramer R *et al.* 1985 *Z. Phys. C* **29** 513
[36] Platchkov S *et al.* 1990 *Nucl. Phys. A* **510** 740
[37] Abbott D *et al.* 1999 *Phys. Rev. Lett.* **82** 1379
[38] Alexa L C *et al.* 1999 *Phys. Rev. Lett.* **82** 1374
[39] Schulze M E *et al.* 1984 *Phys. Rev. Lett.* **52** 597
[40] The I *et al.* 1991 *Phys. Rev. Lett.* **67** 173
[41] Zhang C *et al.* 2011 *Phys. Rev. Lett.* **107** 252501
[42] Dmitriev V F *et al.* 1985 *Phys. Lett. B* **157** 143
[43] Wojtsekhowski B B *et al.* 1986 *Pis'ma Zh. Eksp. Teor. Fiz.* **43** 567
[44] Gilman R *et al.* 1990 *Phys. Rev. Lett.* **65** 1733
[45] Nikolenko D M *et al.* 2003 *Phys. Rev. Lett.* **90** 072501
[46] Boden B *et al.* 1991 *Z. Phys. C* **49** 175
[47] Ferro-Luzzi M *et al.* 1996 *Phys. Rev. Lett.* **77** 2630
[48] Bouwhuis M *et al.* 1999 *Phys. Rev. Lett.* **82** 3755
[49] Abbott D *et al.* 2000 *Phys. Rev. Lett.* **84** 5053
[50] Auffret S *et al.* 1985 *Phys. Rev. Lett.* **54** 649
[51] Bosted P E *et al.* 1990 *Phys. Rev. C* **42** 38
[52] Sick I 2001 *Prog. Part. Nucl. Phys.* **47** 245
[53] Amroun A *et al.* 1994 *Nucl. Phys. A* **579** 596

XIV Conference on Theoretical Nuclear Physics in Italy IOP Publishing
Journal of Physics: Conference Series **527** (2014) 012023 doi:10.1088/1742-6596/527/1/012023

The proton-proton weak capture reaction within chiral effective field theory

Laura E Marcucci

Department of Physics "E. Fermi", University of Pisa, and INFN-Pisa, 56127 Pisa, Italy

E-mail: laura.marcucci@df.unipi.it

Abstract. We review the results of the most recent calculation for the astrophysical S-factor of the weak proton-proton capture reaction, over a range for the center-of-mass relative energy of 0–100 keV. The so-called chiral effective field theory approach is used, where the chiral two-nucleon potential is derived up to next-to-next-to-next-to leading order and is augmented by the full electromagnetic interaction. The low-energy constants (LEC's) entering the weak current operators are fixed so as to reproduce the $A = 3$ binding energies and magnetic moments, and the Gamow-Teller matrix element in tritium β-decay. Contributions from S and P partial waves in the incoming two-proton channel are retained. The S-factor at zero energy is found to be $\sim$ 1% larger than the value reported in the literature, mostly due to the P-waves contributions.

1. Introduction

The proton weak capture on protons, i.e., the reaction $^1\mathrm{H}(p, e^+\nu_e)^2\mathrm{H}$ (hereafter labelled pp), is the most fundamental process in stellar nucleosynthesis: it is the first reaction in the pp chain, which converts hydrogen into helium in main sequence stars like the Sun. Its reaction rate is expressed in terms of the astrophysical S-factor, $S(E)$, where E is the two-proton center-of-mass (c.m.) energy, by the relation

$$S(E) = E\exp(2\pi\,\eta)\,\sigma(E)\ , \tag{1}$$

where $\eta = \alpha/v_{\rm rel}$, α being the fine structure constant and $v_{\rm rel}$ the pp relative velocity, and $\sigma(E)$ is the pp weak capture cross section. The energy-dependence of $S(E)$ is often parametrized as [1]

$$S(E) = S(0) + S'(0)E + S''(0)E^2/2 + \cdots\ , \tag{2}$$

where $S(0)$, $S'(0)$ and $S''(0)$ are the zero-energy value of the S-factor, its first and second derivatives, both evaluated at $E = 0$. At the center of light stars like the Sun, with temperature of the order of 1.5×10^7 K, the Gamow peak is at $E \simeq 6$ keV, while in larger-mass stars, whose central temperature becomes of the order of 5×10^7 K, the Gamow peak turns out to be $E \sim 15$ keV. At these energies, the reaction cross section cannot be measured in terrestrial laboratories, and it is necessary to rely on theoretical predictions. The many studies on $S(0)$, and the few for $S'(0)$ and $S''(0)$, have been extensively reviewed in Ref. [1]. The currently recommended value for $S(0)$, $(4.01 \pm 0.01) \times 10^{-23}$ MeV fm^2 [1], is the average of values obtained within three different approaches, the "potential model" approach (PMA), "hybrid chiral effective field theory" (χEFT*) and "pionless effective field theory" ($\not\pi$EFT). The first one uses phenomenological realistic models for the nuclear potential, fitted to reproduce the large body of two-nucleon (NN) bound and scattering state data with a χ^2/datum ~ 1. The axial current operator includes both one- and two-body terms, these latest derived from meson-exchange mechanisms and the excitation of Δ-isobar resonances. The axial coupling constant for the N-to-Δ transition (g_A^*) is constrained to reproduce the experimental value of the Gamow-Teller (GT) matrix element of tritium β-decay. In the χEFT* approach, transition operators derived in χEFT are sandwiched between initial and final wave functions generated by potential models. The only unknown low-energy constant (LEC), which parametrizes the strength of a contact-type four-nucleon coupling to the axial current, is determined by fitting the experimental GT matrix element. Finally, $\not\pi$EFT is an approach applicable to low-energy processes—such as the pp reaction under consideration here—, where pions are integrated out and the NN interaction and weak currents are described by classes of point-like contact interactions, each class corresponding to given order in a systematic expansion in powers of the characteristic momentum over the pion mass.

For $S'(0)$ and $S''(0)$ the situation is much less clear than for $S(0)$. The adopted value for $S'(0)$ in Ref. [1] is $S'(0)/S(0) = (11.2 \pm 0.1)$ MeV^{-1}, as obtained in Ref. [2] and later confirmed in Ref. [3] in a PMA. No value is reported for $S''(0)$ in Ref. [1]. In Ref. [2] it was estimated by dimensional considerations that the contribution of $S''(0)$ to the pp rate would be at the level of 1% for temperatures characteristic of the solar interior. Only very recently, $S'(0)$ and $S''(0)$ have been studied in $\not\pi$EFT [4] to the third-order in the power expansion, with the results $S'(0)/S(0) = (11.3 \pm 0.1)$ MeV^{-1} and $S''(0)/S(0) = (170 \pm 2)$ MeV^{-2}.

In parallel with the $\not\pi$EFT study of Ref. [4], a systematic calculation of $S(E)$ in (pionfull) χEFT has been reported in Ref. [5]. This latest work will be reviewed in this contribution.

2. Formalism

The pp weak capture cross section $\sigma(E)$, from which the S-factor is obtained (see equation 1), is written in the c.m. frame as

$$\sigma(E) = \int 2\pi\delta(\Delta m + E - \frac{q^2}{2m_d} - E_e - E_\nu)\frac{1}{v_{\rm rel}}F(Z,E_e) \times \frac{1}{4}\sum_{s_e\, s_\nu}\sum_{s_1\, s_2\, s_d} |\langle f|H_W|i\rangle|^2 \frac{d\mathbf{p}_e}{(2\pi)^3}\frac{d\mathbf{p}_\nu}{(2\pi)^3} , \tag{3}$$

where $\Delta m = 2\,m_p - m_d$ (m_p and m_d are the proton and deuteron masses, respectively), $\mathbf{p}_e$ ($\mathbf{p}_\nu$) and E_e (E_ν) are the electron (neutrino) momentum and energy, $\mathbf{q} = \mathbf{p}_e + \mathbf{p}_\nu$ is the momentum transfer, and $F(Z, E_e)$ is the Fermi function (with $Z = 1$), which accounts for the Coulomb distortion of the final positron wave function, including radiative corrections [6]. The transition amplitude $\langle f|H_W|i\rangle$ is given by

$$\langle f|H_W|i\rangle = \frac{G_V}{\sqrt{2}} l^\sigma \langle -\mathbf{q}; d|j_\sigma^\dagger|\mathbf{p}; pp\rangle , \tag{4}$$

where G_V is the Fermi constant, $|-\mathbf{q}; d\rangle$ and $|\mathbf{p}; pp\rangle$ represent, respectively, the deuteron bound state with recoiling momentum $-\mathbf{q}$ and the pp scattering state with relative momentum $\mathbf{p}$, and l_σ and $j^\sigma(\mathbf{q})$ are the leptonic and nuclear weak currents, respectively. A standard multipole decomposition of the nuclear weak current operator leads to [5]

$$\frac{1}{4}\sum_{s_e\, s_\nu}\sum_{s_1\, s_2\, s_d} |\langle f|H_W|i\rangle|^2 = (2\pi)^2 G_V^2 L_{\sigma\tau} N^{\sigma\tau} , \tag{5}$$

where $L^{\sigma\tau}$ and $N^{\sigma\tau}$ are the lepton and nuclear tensors, respectively. The latter can be expressed in terms of the reduced matrix elements for the Coulomb, longitudinal, transverse magnetic and transverse electric multipole operators between the initial pp state with orbital angular momentum L, channel spin S ($S = 0, 1$), total angular momentum J, and the final deuteron state with total angular momentum $J_d = 1$. The integrations over $\mathbf{p}_e$ and $\mathbf{p}_\nu$ are performed by Gaussian quadratures [7], and a moderate number of Gauss points ($\sim$ 10–20 for each integration) suffices to achieve convergence to within better than 1 part in 10^3.

The two-body wave functions have been obtained using the NN potential derived in χEFT up to next-to-next-to-next-to leading order (N3LO) in the chiral expansion by Entem and Machleidt [8, 9]. The two-body Schrödinger equation has been solved variationally with the technique described in Refs. [10, 5]. To be noticed that, in the pp sector, the nuclear NN potential has been augmented by the Coulomb interaction and the higher-order electromagnetic (EM) terms, due to two-photon exchange and vacuum polarization. These higher-order terms are the same as those of the Argonne v_{18} (AV18) NN potential [11], and therefore also retain short-range corrections associated with the finite size of the proton charge distribution. The additional distortion of the pp wave function, induced primarily by vacuum polarization, has been shown to reduce $S(0)$ by $\sim$ 1% in Ref. [3]. The technique has been tested by verifying that, in the case of the AV18, the $^1S_0$ phase shifts are in agreement with those reported for the AV18 in Ref. [11]. Of course, the N3LO phase shifts listed in Ref. [9], obtained by including only the Coulomb potential, have also been reproduced. Note that the N3LO NN potential of Ref. [8] has been used in conjunction with the three-nucleon interaction (TNI) obtained up to next-to-next-to leading order (N2LO), in the version of Ref. [12], when reactions involving $A > 2$ nuclear systems are considered.

The charge-changing weak current has been first derived in χEFT by Park and collaborators in the late nineties, using the so-called heavy-baryon chiral perturbation theory (HBχPT) approach, where the baryons are treated as heavy static sources, and the perturbative expansion is performed in terms of the involved momenta over the baryon mass (see Ref. [13] and references therein). These are the same currents used to study the pp reaction in χEFT*. In particular, the polar-vector part is related, via the conserved-vector-current constraint, to the (isovector) EM current, and at N3LO includes, apart from one- (OPE) and two-pion-exchange (TPE) terms, two contact terms—one isoscalar and the other isovector—whose strengths are parametrized by the LEC's g_{4S} and g_{4V}. To be noticed that few years ago, the problem of deriving the electromagnetic current and charge operators in χEFT has been revisited by Pastore *et al.* [14] and, in parallel, by Kölling *et al.* [15]. Pastore *et al.* have used time-ordered perturbation theory (TOPT) to calculate the EM transition amplitudes, which allows for an easier treatment of the so-called reducible diagrams than the HBχPT approach. On the other hand, Kölling *et al.* have used the method of unitary transformation, the same one used to derive the chiral potentials mentioned above. The EM operators of Pastore *et al.* have been found significantly different from those of Park *et al.*, especially for the structure of the contact terms, while they are in good agreement with those of Kölling *et al.*. However, the TOPT approach has not yet been used to derive the axial component of the weak current operator, and work along this line is currently underway.

The two-body axial-vector current are diagrammatically represented in figure 1, where they are listed according to their scaling in Q, the pions' and nucleons' momenta. The leading-order (LO) contribution consists of the well known single-nucleon axial current, and is of order Q^{-3}. At order Q^{-2} it turns out that there are no contributions, and therefore the next-to-leading (NLO) contribution is of order Q^{-1}, and arises from the $(Q/m)^2$ relativistic corrections to the LO contribution (m is the nucleon mass). The next-to-next-to leading order (N2LO) currents consist of the OPE term and a contact term with one LEC, denoted with d_R. Note that N3LO contributions arise from loop and TPE terms, and they have not been calculated yet.

All the chiral potentials and currents presented above have power-law behavior in momentum space, and must be regularized before they can be used in practical calculations. This is

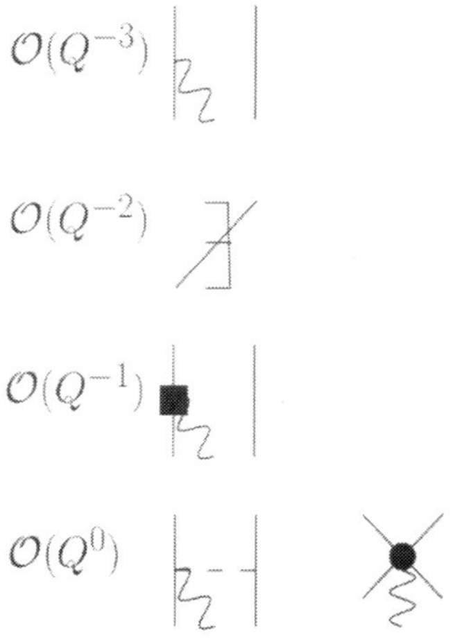

Figure 1. Diagrams illustrating one- and two-body χEFT axial currents entering at LO (Q^{-3}), NLO (Q^{-1}), and N2LO (Q^0). Nucleons, pions, and weak probes are denoted by solid, dashed, and wavy lines, respectively. The solid square represents the relativistic corrections to the one-body current, while the solid circles represent the contact terms. Only the relevant topologies are indicated. No contribution of order Q^{-2} exists.

accomplished by multiplying them by a momentum-cutoff function, whose cutoff Λ is taken to be in the range (500–600) MeV.

Some remarks on the fitting procedure of the LEC d_R are in order. As first shown in Ref. [16], d_R can be related to the LEC c_D entering one of the two contact terms present in the TNI at N2LO, via the relation

$$d_R = \frac{m}{\Lambda_\chi g_A} c_D + \frac{1}{3} m(c_3 + 2c_4) + \frac{1}{6} , \tag{6}$$

where g_A is the single-nucleon axial coupling constant, c_3 and c_4 are LEC's of the πN Lagrangian, already part of the chiral NN potential at NLO, and $\Lambda_\chi = 700$ MeV is the the chiral-symmetry-breaking scale. Therefore, it has become common practice to fit c_D (and c_E— the other LEC entering the N2LO TNI) to the triton binding energy and the Gamow-Teller matrix element in tritium β-decay. The values obtained in this way for c_D and c_E are listed in Ref. [17], where they have been used to study the muon capture on deuteron and ^{3}He, leading to predictions (with an estimated theory uncertainty of about 1%) in excellent agreement with the experimental data. Note that the first studies of $A = 3$ and 4 elastic scattering observables, as cross sections and analyzing powers, with these values for c_E and c_D have been reported in Ref. [18]. In particular, in table 1 we report the central values of c_D and c_E obtained with the N3LO NN potential and N2LO TNI mentioned above, for two different values of the cutoff Λ, 500 and 600 MeV, and the corresponding values for the ^{4}He binding energy and the $n - d$ doublet scattering length. The values of c_D and c_E obtained within this fitting procedure for two other values of the cutoff Λ, 414 and 450 MeV, have been used also to study infinite nuclear matter [19].

3. Results

In the present section we summarize the results for the pp astrophysical S-factor, as function of the c.m. pp energy E, obtained in Ref. [5]. We recall that (i) $S(E)$ is studied in the wide

Table 1. Values for the LEC's c_D and c_E obtained with the N3LO NN potential and N2LO TNI, for two different values of the cutoff Λ, 500 and 600 MeV. The corresponding values for the ^{4}He binding energy (in MeV) and the $n-d$ doublet scattering length (in fm) are also given. The row labelled "N3LO/N2LO*" indicates the results obtained with the original N2LO TNI of Ref. [12].

Model	Λ	c_D	c_E	B(^{4}He)	$^2a_{nd}$
N3LO/N2LO*	500	1.0	−0.029	28.36	0.675
N3LO/N2LO	500	−0.12	−0.196	28.49	0.666
N3LO/N2LO	600	−0.26	−0.846	28.64	0.696

energy range 2 – 100 keV. The solar Gamow peak is at $\simeq$ 6 keV, but in larger-mass stars the Gamow peak becomes 15 keV. (ii) The χEFT N3LO NN potential is augmented not only of the Coulomb interaction but also of the higher-order EM interaction contributions, like those due to two-photon exchange, Darwin-Foldy and vacuum polarization. (iii) All the $L \leq 1$ pp partial waves are considered. To our knowledge, this is the first time that the P-wave contributions are retained.

The results for the zero-energy S-factor $S(0)$ obtained retaining only the $^1S_0$ pp partial wave are given in table 2, where the one-body and the full one- plus two-body current contributions are listed. We also show the results obtained neglecting the higher-order EM contributions (i.e., with only the Coulomb interaction), and, for comparison, those obtained using the PMA. By inspection of the table we can conclude that (i) the value for $S(0)$ is in agreement with the recommended one of Ref. [1]; (ii) the higher-order EM contributions are of the order of 1% or less; (iii) the χEFT and PMA results are in excellent agreement; (iv) the cutoff dependence is extremely small ($\lesssim$ 1%). Note that the theoretical uncertainty on the results of table 2 induced by the adopted fitting procedure of the LEC d_R (axial N-to-Δ coupling constant g_A^*) in the χEFT approach (PMA) are not reported, since they are affecting the third decimal digits.

The cumulative contributions to $S(0)$ due to S- and P-waves are listed in table 3. As it can be seen by inspection of the table, the P-wave contributions sum up to a $\sim$ 1% of the total value, and the theoretical uncertainty due to the fitting procedure as well as the cutoff dependence is negligible. Again, the χEFT and PMA results are in excellent agreement.

Table 2. Zero-energy S-factor (in 10^{-23} MeV fm^2) obtained retaining only the $^1S_0$ pp partial wave. Two different values for the cutoff Λ are used, 500 and 600 MeV. The results obtained including the full EM (only the Coulomb) interaction are listed in the columns labelled "$V_{NN} + V_{EM}$" ("$V_{NN} + V_{Coul}$"). The labels IA and FULL are used when only the one-body or the full one- and two-body current contributions are retained. The PMA results are also given.

	$V_{NN} + V_{Coul}$	$V_{NN} + V_{EM}$
χEFT(500)-IA	3.96	3.94
χEFT(500)-FULL	4.03	4.01
χEFT(600)-IA	3.94	3.93
χEFT(600)-FULL	4.01	4.01
PMA-IA	3.99	3.96
PMA-FULL	4.03	4.00

Table 3. Cumulative S- and P-wave contributions to $S(0)$ in units of 10^{-23} MeV fm^2. The results labelled "χEFT(500)" and "χEFT(600)" have been obtained within the χEFT approach with two different cutoff values, 500 and 600 MeV. The results obtained within the PMA are also shown. The theoretical uncertainties are given in parentheses and are due to the fitting procedure adopted for the LEC's (or g_A^* within the PMA) in the weak current.

	$^1S_0$	$\cdots + {}^3P_0$	$\cdots + {}^3P_1$	$\cdots + {}^3P_2$
χEFT(500)	4.008(5)	4.011(5)	4.020(5)	4.030(5)
χEFT(600)	4.007(5)	4.010(5)	4.019(5)	4.029(5)
PMA	4.000(3)	4.003(3)	4.015(3)	4.033(3)

In conclusion, the χEFT results of table 3 can be summarized in the conservative range $S(0) = (4.030 \pm 0.006) \times 10^{-23}$ MeV fm^2, with a P-wave contribution of $\simeq 0.2 \times 10^{-23}$ MeV fm^2.

Finally, we show in figure 2 the energy dependence of $S(E)$ in the energy range 2 – 100 keV, as obtained within the χEFT approach. The S- and $(S+P)$-wave contributions are displayed separately, and the theoretical uncertainty is included—the curves are in fact very narrow bands. As expected, the P-wave contributions become significant at higher values of E. From these results, a least-squares polynomial fit to $S(E)$ has been performed up to order $O(E^2)$, i.e., by

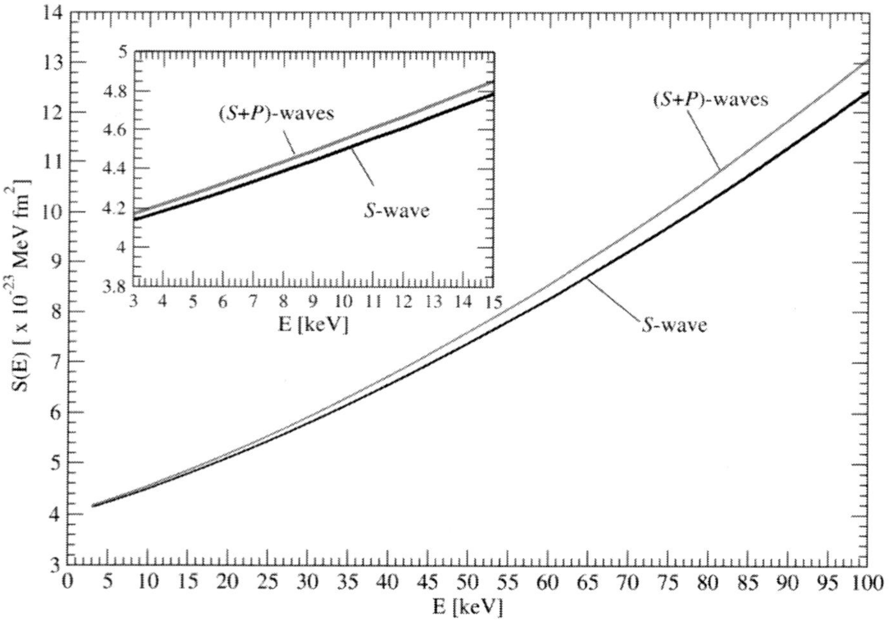

Figure 2. (Color online) Energy dependence of $S(E)$ in the range 2 – 100 keV. The S- and $(S+P)$-wave contributions are displayed separately. In the inset, $S(E)$ is shown in the range 3–15 keV.

Table 4. Values for the first $[S'(0)/S(0)]$, second $[S''(0)/S(0)]$ and third $[S'''(0)/S(0)]$ derivatives of $S(E)$ at zero energy, in units of MeV^{-1}, MeV^{-2}, MeV^{-3}, respectively, as obtained with a polynomial fit of $S(E)$ up to order $O(E^2)$ (Fit 1) and $O(E^3)$ (Fit 2), retaining all $(S+P)$-waves or only the $^1S_0$ initial partial wave. The numbers given in parentheses are the theoretical uncertainties, which account for the cutoff sensitivity and the error due to the LEC's fitting procedure. The χ^2 value is also shown, as defined in the text.

	$S'(0)/S(0)$	$S''(0)/S(0)$	$S'''(0)/S(0)$	χ^2
$S+P$ - Fit 1	12.59(1)	199.3(1)		8.8×10^{-4}
$S+P$ - Fit 2	11.94(1)	248.8(2)	−1183(8)	1.9×10^{-4}
$^1S_0$ - Fit 1	12.23(1)	178.4(3)		1.2×10^{-3}
$^1S_0$ - Fit 2	11.42(1)	239.6(5)	−1464(5)	1.9×10^{-4}

using equation (2) itself, and up to order $O(E^3)$, by adding the $S'''(0)E^3/3!$ term ($S'''(0)$ is the third derivative of $S(E)$ evaluated at $E=0$). The values for $S'(0)/S(0)$, $S''(0)/S(0)$, and $S'''(0)/S(0)$ for the two fits (up to $O(E^2)$ – Fit 1, and $O(E^3)$ – Fit 2) are listed in table 4, along with the χ^2 value, here defined as $\chi^2 = \sum_i(1 - f_i^{fit}/f_i^{calc})^2$, with f_i^{calc} (f_i^{fit}) being the calculated (fitted) $S(E)$ results. By inspection of the table we conclude that the values for the derivatives of $S(E)$ are strongly dependent on the order of the polynomial function used in the fit. However, an accurate description of the data can be obtained with a desired degree of accuracy by increasing the number of polynomial terms. Finally, we would like to remark that the values for $S'(0)/S(0)$ and $S''(0)/S(0)$ obtained in Ref. [4] are similar to those listed in the table only in the case of Fit 1, with only the $^1S_0$ contribution. The χ^2 in this case, though, assumes the highest value.

4. Conclusions

We have presented the most recent calculation of the *pp* astrophysical S-factor within the χEFT approach in a wide energy range, up to 100 keV, retaining all the S- and P-waves in the initial wave function, and including, beyond the Coulomb interaction, also higher order EM contributions, like two-photon exchange, Darwin-Foldy and vacuum polarization. The value at zero energy, $S(0)$, has been calculated also within the PMA, finding a remarkable agreement between the two different approaches. The theoretical uncertainty has been reduced below the 1% level.

As a final remark, we notice that the present χEFT model for the nuclear current operator is that of Park *et al.*, up to N2LO. Recently, the model for the EM current and charge operators of Pastore *et al.*, developed up to N3LO, has been used to study the EM structure of $A=2,3$ nuclei [20], finding a good agreement between theory and experiment. It has also been shown that the TOPT approach used by Pastore *et al.* is more suitable than the HBχPT one of Park *et al.* to treat reducible diagrams and to derive the contact terms. Therefore, it is highly desirable to derive the axial currents up to N3LO in TOPT, following the footsteps of Pastore *et al.* for the EM operators, and to repeat this study for the *pp* reaction. Finally, it should be noticed that the radiative proton-deuteron and the weak proton-^{3}He captures, also relevant in astrophysics, are at reach within the present framework. Work along these lines is currently underway.

References

[1] Adelberger E G *et al.* 2011 *Rev. Mod. Phys.* **83** 195

[2] Bahcall J N and May R M 1969 *Astrphys. J.* **155** 511
[3] Schiavilla R *et al.* 1998 *Phys. Rev. C* **58** 1263
[4] Chen J W, Liu C P, and Yu S H 2012 *Phys. Lett. B* **720** 385
[5] Marcucci L E *et al.* 2013 *Phys. Rev. Lett.* **110** 192503
[6] Kurylov A, Ramsey-Musolf M J, and Vogel P 2003 *Phys. Rev. C* **67** 035502
[7] Marcucci L E *et al.* 2000 *Phys. Rev. Lett.* **84** 5959; Marcucci L E *et al.* 2000 *Phys. Rev. C* **63** 015801
[8] Entem D R and Machleidt R 2003 *Phys. Rev. C* **68** 041001
[9] Machleidt R and Entem D R 2011 *Phys. Rep.* **503** 1
[10] Marcucci L E *et al.* 2011 *Phys. Rev. C* **83** 014002
[11] Wiringa R B, Stoks V G J, and Schiavilla R 1995 *Phys. Rev. C* **51** 38
[12] Navràtil P 2007 *Few-Body Syst.* **41** 117
[13] Park T S *et al.* 2003 *Phys. Rev. C* **67** 055206
[14] Pastore S *et al.* 2009 *Phys. Rev. C* **80** 034004; Pastore S *et al.* 2011 *Phys. Rev. C* **84** 024001
[15] Kölling S *et al.* 2009 *Phys. Rev. C* **80** 045502; Kölling S *et al.* 2011 *Phys. Rev. C* **84** 054008
[16] Gårdestig A and Phillips D R 2006 *Phys. Rev. Lett.* **96** 232301
[17] Marcucci L E *et al.* 2012 *Phys. Rev. Lett.* **108** 052502
[18] Viviani M *et al.* 2013 *Few-Body Syst.* **54** 885; Viviani M *et al.* 2013 *Phys. Rev. Lett.* **111** 172302
[19] Coraggio L *et al.* 2014 *Phys. Rev. C* **89** 044321
[20] Piarulli M *et al.* 2013 *Phys. Rev. C* **87** 014006

Journal of Physics: Conference Series **527** (2014) 012024 doi:10.1088/1742-6596/527/1/012024

Elastic and quasi-elastic electron scattering off isotopic and isotonic chains

Matteo Vorabbi, Andrea Meucci, Carlotta Giusti, Franco Davide Pacati

Dipartimento di Fisica, Università degli Studi di Pavia and
INFN, Sezione di Pavia, via A. Bassi 6, I-27100 Pavia, Italy

Paolo Finelli

Dipartimento di Fisica e Astronomia, Università degli Studi di Bologna and
INFN, Sezione di Bologna, via Irnerio 46, I-40126 Bologna, Italy

Abstract. We present theoretical predictions for electron scattering on $Z = 20$ isotopic and $N = 20$ isotonic chain. The calculations are done within the framework of the distorted-wave Born approximation for elastic scattering and the plane-wave Born approximation for the QE one. The proton and neutron density distributions are evaluated adopting a relativistic Dirac-Hartree model. We present results for the elastic and quasi-elastic cross sections and for the parity-violating asymmetry and investigate the evolution of nuclear properties as a function of the neutron and proton number. We also present a comparison with the parity-violating asymmetry parameter obtained by the PREX Collaboration on ^{208}Pb.

1. Introduction

The nuclear response to an external probe is a powerful tool to investigate the structure of hadron systems such as atomic nuclei and their constituents. In particular, electron-scattering reactions have provided the most complete and detailed information on nuclear and nucleon structure [1, 2, 3, 4, 5, 6]. A lot of experimental and theoretical work on elastic and inelastic electron scattering at different energies has provided detailed information on the charge density distribution of the nuclear ground state and on the energy, strength, and quantum numbers of the excited states produced by single particle (s.p.) or collective excitation mechanisms [2, 7, 8, 9].

The use of the electron probe can be extended to exotic nuclei. The detailed study of the properties of nuclei far from the stability line and the evolution of nuclear properties with respect to the asymmetry between the number of neutrons and protons is one of the major topics in modern nuclear physics. Of particular interest is the behavior of the s.p. properties, with a consequent modification of the shell model magic numbers.

The study of the evolution of nuclear properties along isotopic and isotonic chains requires a good knowledge of nuclear matter distributions for protons and neutrons separately. The ground state densities reflect the basic properties of effective nuclear forces and provide fundamental nuclear structure information. Elastic electron scattering allows to measure with excellent precision only charge densities and therefore proton distributions. It is much more difficult to measure neutron distributions. A model-independent probe of neutron densities is provided by parity-violating elastic electron scattering, where direct information on the neutron density can be obtained from the measurement of the parity-violating asymmetry A_{pv} parameter [10, 11, 12], that is related to the radius of the neutron distribution R_n.

In addition, quasi-elastic (QE) electron scattering can give information on the momentum distribution of nuclei. Precise measurements have been carried out from light to heavy nuclei on a wide range of kinematics. A lot of theoretical efforts have been devoted to the analysis of the data [5, 6].

In this contribution we present and discuss results of calculations of elastic and QE electron scattering and A_{pv} performed for the calcium isotopic chain and the N=20 isotonic chain. In addition, we compare our theoretical predictions for A_{pv} with the experimental results obtained by the PREX collaboration on ^{208}Pb.

2. Theoretical framework

2.1. Elastic electron scattering

In the one-photon exchange approximation and neglecting the effect of the nuclear Coulomb field on incoming and outgoing electrons, i.e. in the plane-wave Born approximation (PWBA), the differential cross section for the elastic scattering of an electron with momentum transfer q off a spherical spin-zero nucleus is given by

$$\left(\frac{\mathrm{d}\sigma}{\mathrm{d}\Omega'}\right)_{EL} = \sigma_M |F_p(q)|^2 \ , \tag{1}$$

where Ω' is the scattered electron solid angle, σ_M is the Mott cross section [4, 5] and $F_p(q)$ is the charge form factor for a spherical nuclear charge (point proton) density $\rho_p(r)$.

The PWBA is, however, not adequate for medium and heavy nuclei where the distortion produced on the electron wave functions by the nuclear Coulomb potential $V(r)$ from $\rho_p(r)$ can have significant effects. The distorted-wave Born approximation (DWBA) cross sections are obtained from the numerical solutions of the partial wave Dirac equation.

Another interesting quantity that can be measured in elastic electron scattering is the parity-violating asymmetry, which is defined as the difference between the cross sections for the elastic scattering of electrons longitudinally polarized parallel and antiparallel to their momentum. This difference arises from the interference between photon and Z^0 exchange and represents an almost direct measurement of the Fourier transform of the neutron density [10, 13]. In fact, in Born approximation, neglecting strangeness contributions and the electric neutron form factor, the parity-violating asymmetry can be expressed as [14]

$$A_{pv} = \frac{G_F\, Q^2}{4\sqrt{2}\,\pi\alpha}\left[4\sin^2\Theta_W - 1 + \frac{F_n(q)}{F_p(q)}\right] \ . \tag{2}$$

Since $4\sin^2\Theta_W - 1$ is small and $F_p(q)$ is known, we see that A_{pv} provides a practical method to measure the neutron form factor $F_n(q)$ and hence the neutron radius. For these reasons parity-violating electron scattering (PVES) has been suggested as a clean and powerful tool for measuring the spatial distribution of neutrons in nuclei.

2.2. Inclusive quasi-elastic electron scattering

The inclusive (e, e') scattering corresponds to an integral over all available nuclear states and consequently it is more directly related to the dynamics of the initial nuclear ground state.

The inclusive differential cross section for the quasi-elastic (QE) (e, e') scattering on a nucleus is obtained from the contraction between the lepton and hadron tensors as [5]

$$\left(\frac{\mathrm{d}\sigma}{\mathrm{d}\varepsilon'\,\mathrm{d}\Omega'}\right)_{QE} = \sigma_M \left[v_L R_L + v_T R_T\right] \ , \tag{3}$$

where ε' is the energy of the scattered electron and the coefficients v come from the components of the lepton tensor that, under the assumption of the plane-wave approximation for the electron wave functions, depend only on the lepton kinematics [5]. All relevant nuclear structure information is contained in the longitudinal and transverse response functions R_L and R_T. The response functions can be expressed in terms of suitable linear combinations of the components of the hadron tensor, which are given by products of the matrix elements of the nuclear current between initial and final nuclear states.

In the QE region the nuclear response is dominated by one-nucleon processes where the scattering occurs with only one nucleon, which is subsequently emitted from the nucleus by a direct knockout mechanism, and the remaining nucleons behave as spectators. Therefore, QE electron scattering can adequately be described in the relativistic impulse approximation (RIA) by the sum of incoherent processes involving only one nucleon scattering and the components of the hadron tensor are obtained from the sum, over all the s.p. shell-model states, of the squared absolute value of the transition matrix elements of the single-nucleon current [5].

A reliable description of final-state interactions (FSI) between the emitted nucleon and the residual nucleus is a crucial ingredient for the description of electron scattering data. The use of a complex optical potential (OP) in the distorted-wave impulse approximation (DWIA) [4, 5, 15, 16, 17, 18, 19, 20] provides a good description of exclusive $(e, e'p)$ data. In the case of inclusive scattering only the final lepton is detected, the final nuclear state is not determined and all elastic and inelastic channels contribute. This requires a different treatment of FSI where all final-state channels are retained and the flux, although redistributed among all possible channels, must be conserved. Different approaches have been considered to describe FSI in (e, e') scattering. In the relativistic plane wave impulse approximation (RPWIA), FSI are simply neglected. In other approaches, FSI are incorporated using purely real OP.

In a different description of FSI, the relativistic Green's function (RGF) model is adopted [21, 22, 23, 24, 25, 26, 27, 28, 29, 30, 31]; taking advantage of suitable approximation based on the IA, the explicit calculation of the single particle Green's function is avoided and the components of the hadron tensor [22, 23] are expressed in terms of matrix elements of the same type as the RDWIA ones of the exclusive $(e, e'p)$ but which involve eigenfunctions of both the OP and its Hermitian conjugate, where the imaginary part has an opposite sign and gives in one case a loss and in the other case a gain of strength. The RGF formalism makes it possible to reconstruct the flux lost into nonelastic channels in the case of the inclusive response starting from the complex OP which describes elastic nucleon-nucleus scattering data and, moreover, it provides a consistent treatment of FSI in the exclusive and in the inclusive scattering.

2.3. Relativistic model for ground-state observables

In the standard representation of relativistic mean field approaches the nucleus is described as a system of Dirac nucleons coupled to the exchange mesons and the electromagnetic field through an effective Lagrangian. The isoscalar scalar-meson (σ), the isoscalar vector-meson (ω), and the isovector vector-meson (ρ) build the minimal set of meson fields that, together with the electromagnetic one (γ), is necessary for a quantitative description of bulk and s.p. nuclear properties. The model is defined by the Lagrangian density

$$\mathcal{L} = \mathcal{L}_N + \mathcal{L}_m + \mathcal{L}_{int} \, , \tag{4}$$

where $\mathcal{L}_N$ denotes the Lagrangian of the free nucleon, $\mathcal{L}_m$ is the Lagrangian of the free meson fields and the simplest set of interaction terms is contained in $\mathcal{L}_{int}$. Bare values are used for the masses of the ω and ρ mesons: m_ω = 783 MeV and m_ρ = 763 MeV. The DD-ME2 parametrization is determined by eight independent parameters, adjusted to the properties of symmetric and asymmetric nuclear matter, binding energies, charge radii, and neutron radii of spherical nuclei [32]. The interaction has been tested in the calculation of ground state

Journal of Physics: Conference Series **527** (2014) 012024 doi:10.1088/1742-6596/527/1/012024

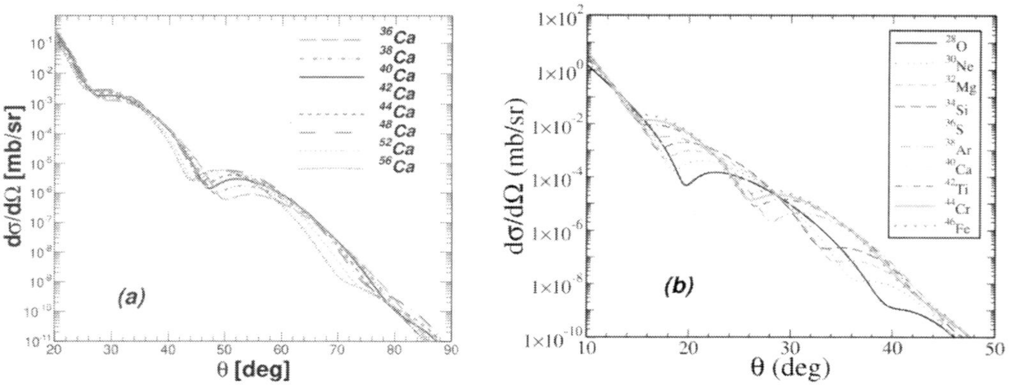

Figure 1. (Color online) Panel (a): Differential cross section for elastic electron scattering on $^{36-56}$Ca at $\varepsilon = 496.8$ MeV as a function of the scattering angle θ. Panel (b): Differential cross section for elastic electron scattering on the $N = 20$ isotonic chain at $\varepsilon = 850.0$ MeV as a function of the scattering angle θ.

properties of a large set of spherical and deformed nuclei. When used in the relativistic random-phase approximation, DD-ME2 reproduces with high accuracy data on isoscalar and isovector collective excitations. For open-shell nuclei we employed a schematic ansatz: the constant gap approximation with empirical Δ given by the 5-point formula [33]

$$\Delta^{(5)}(N_0) = -\frac{1}{8}\Big[E(N_0+2) - 4E(N_0+1) + 6E(N_0) - 4E(N_0-1) + E(N_0-2)\Big]\,. \qquad (5)$$

3. Results

In this section we present and discuss numerical predictions for elastic and QE electron scattering which can hopefully be useful for future measurements in experimental radioactive ion beam (RIB) facilities. We study the evolution of some electron scattering observables in the isotopic and isotonic chains of medium systems, which are exemplified by the cases of calcium isotopes and $N = 20$ isotons. Many of these nuclei lie in the region of the nuclear chart that is likely to be explored in future electron-scattering experiments. For each nucleus in an isotopic and isotonic chain, we compute and compare the associated elastic and QE cross sections and parity-violating asymmetry in order to obtain information on the effects of isospin asymmetry on nuclear structure.

3.1. Elastic electron scattering

The cross sections for elastic electron scattering have been calculated in the DWBA, where the distortion produced by the nuclear Coulomb field on the electron wave functions is explicitly included. We have checked that our calculations are in fair agreement with experimental data of elastic electron scattering cross sections of medium and heavy nuclei [29, 34].

The differential cross sections for elastic electron scattering on various calcium isotopes ($^{36-56}$Ca) at an electron energy $\varepsilon = 496.8.5$ MeV are shown in panel (a) of figure 1. With increasing neutron number the positions of the diffraction minima shift towards smaller scattering angles, i.e., towards smaller values of the momentum transfer. The shift of the minima towards smaller q is in general accompanied by a simultaneous increase in the height of

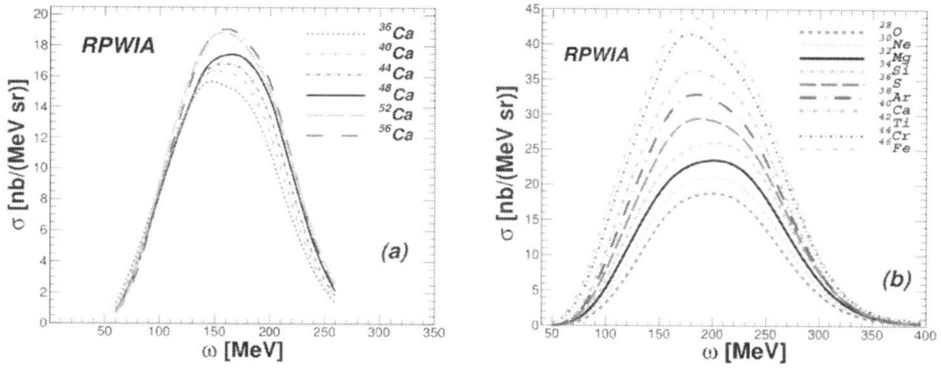

Figure 2. (Color online) Panel (a): differential RPWIA cross section for the inclusive QE (e, e') reaction on $^{36-56}$Ca at $\varepsilon = 560$ MeV and $\theta = 60^o$ as a function of ω. Panel (b): differential RPWIA cross section for the inclusive QE (e, e') reaction on $N = 20$ isotones at $\varepsilon = 1080$ MeV and $\theta = 32^o$.

the maxima. In panel (b) of figure 1 we present the same differential cross sections at $\varepsilon = 850$ MeV along $N = 20$ isotones. With increasing proton number, the positions of the diffraction minima shift towards smaller scattering angles and, correspondingly, there is an enhancement in the height of the maxima of the cross sections. Some shell effects can be seen owing to the filling of the single-particle states.

3.2. Quasi-elastic electron scattering

In the calculations of the matrix elements for QE electron scattering, the s.p. bound nucleon states are obtained from the relativistic mean-field model with density-dependent meson-nucleon vertices and the DD-ME2 parametrization as described in section 2.3.

The RPWIA cross sections for the inclusive QE (e, e') reaction on $^{36-56}$Ca isotopes at $\varepsilon = 560$ MeV and $\theta = 60^o$ are shown in panel (a) of figure 2. The results for $N = 20$ isotones at $\varepsilon = 1080$ MeV and $\theta = 32^o$ are shown in panel (b) of figure 2. Owing to the fact that FSI are neglected the differences between the results for the various isotopes and isotones are entirely due to the differences in the s.p. bound state wave functions of each nucleus. While only the charge proton density distribution contributes to the cross section of elastic electron scattering, the cross section of QE electron scattering is obtained from the sum of all the integrated exclusive one-nucleon knockout processes, due to the interaction of the probe with all the individual nucleons, protons and neutrons, of the nucleus and contains information on the dynamics of the initial nuclear ground state.

The main role is played by protons, which give most of the contribution. Increasing the neutron number along the isotopic chain there is a proportional increase of the neutron contribution, but no significant increase of the proton one. Thus the enhancement of the cross sections in panel (a) of figure 2 can be mainly ascribed to the neutrons. In panel (b) of figure 2 we show the QE (e, e') cross sections calculated in the RPWIA for isotones of the $N = 20$ chain at $\varepsilon = 1080$ MeV and $\theta = 32^o$. Increasing the proton number along the $N = 20$ chain, in panel (b) of figure 2, there is a proportional increase of the cross sections, owing to the enhancement of the proton contribution.

In general, interesting and peculiar effects are obtained in the evolution of QE inclusive cross section, but it is not easy to relate them to changes in the matter distribution, which can be

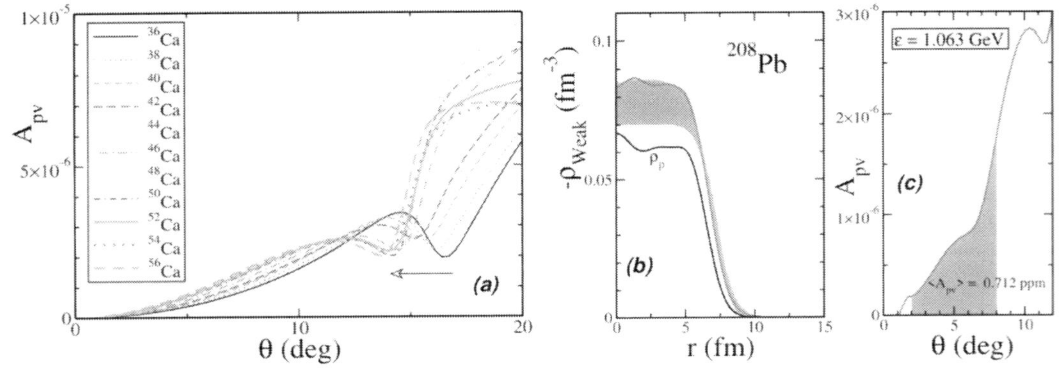

Figure 3. (Color online) Panel (a): Parity-violating asymmetry parameter A_{pv} for elastic electron scattering at $\epsilon = 850$ MeV as function of the scattering angle θ on $^{36-56}$Ca. Panel (b): Theoretical weak charge density (red line) in comparison with the experimental error band as determined in Ref. [38] for ^{208}Pb with the kinematics of the PREX experiment. The corresponding proton density profile is plotted with a black line. Panel (c): Asymmetry parameter for ^{208}Pb as a function of the angle θ (red line). The shaded area in panels (b) and (c) represents the interval covered by the acceptance function $\epsilon(\theta)$ (see Ref. [39] for more details). The asymmetry parameter averaged over the acceptance $\langle A_{pv} \rangle$ is in quite good agreement with the empirical value 0.656 ± 0.060(stat) $\pm$ 0.014(syst) ppm.

significant, particularly in the center of the nucleus.

Significant distortion effects are produced by FSI in the RGF model, but the general trend and magnitude of the cross sections, and their evolution with respect to the neutron number in the considered isotopic chain or the proton number in the considered isotonic chain are generally similar in RPWIA and RGF [29, 34].

3.3. Parity-violating asymmetry

The calculation starts with the self-consistent relativistic ground state proton and neutron densities. The charge and weak densities are calculated by folding the point proton and neutron densities. The resulting Coulomb potential $V(r)$ and weak potential $A(r)$ are used to construct $U_{\pm}(r)$ potential as it follows: $U_{\pm}(r) = V(r) \pm A(r)$. The cross sections for elastic electron scattering are obtained from the numerical solution of the Dirac equation for electron scattering in the $U_{\pm}(r)$ potential and includes Coulomb distortion effects [12, 35, 36, 37]. The cross sections for positive and negative helicity electron states are calculated and the resulting asymmetry parameter A_{pv} is plotted as a function of the scattering angle.

In panel (a) of figure 3 we plot the parity-violating asymmetry parameters A_{pv} for $^{36-56}$Ca nuclei for elastic electron scattering at $\varepsilon = 850$ MeV. At $\varepsilon = 850$ MeV the values of A_{pv} are of the order of 10^{-5}, with lower values for smaller angles and larger values for larger angles.

As suggested in Ref. [10] the asymmetry parameter A_{pv} provides a direct measurement of the Fourier transform of the neutron density. This relation has been tested and confirmed in Ref. [13] comparing asymmetries and the squares of the Fourier transforms of the neutron densities.

In addition to predictions about calcium isotopic chains we also provide calculations for recent measurements. In panels (b) and (c) of figure 3 we show our theoretical predictions for the empirical values extracted from the first run of the PREX experiment on ^{208}Pb at $\varepsilon = 1.06$ GeV. In Ref. [38] the weak charge density ($-\rho_W$) has been deduced from the weak charge form factor. The error band (shaded area) represents the incoherent sum of experimental and model errors. Our prediction, plotted by the red line in the left panel, is in rather good agreement with empirical data. In fact, if we evaluate the corresponding asymmetry parameter A_{pv} averaged over the acceptance function $\epsilon(\theta)$ [40]

$$\langle A_{pv} \rangle = \frac{\int \mathrm{d}\theta \ \sin\theta A_{pv}(\theta) \dfrac{d\sigma}{d\Omega} \ \epsilon(\theta)}{\int \mathrm{d}\theta \ \sin\theta \ \dfrac{d\sigma}{d\Omega} \ \epsilon(\theta)} \tag{6}$$

we find 0.712 ppm, in very good agreement with the empirical estimate $0.656 \pm 0.060(\mathrm{stat}) \pm 0.014(\mathrm{syst})$ ppm.

4. Conclusions

We have presented and discussed numerical predictions for the cross section and the parity-violating asymmetry in elastic and quasi-elastic electron scattering along $Z = 20$ isotopic and $N = 20$ isotonic chains with the aim to investigate their evolution with increasing neutron and proton number.

The understanding of the properties of exotic nuclei is one of the major topic of interest in modern nuclear physics. Large efforts in this directions have been done over last years and are planned for the future. Electron probes are powerful tools to achieve this goal because their interaction is well known and relatively weak with respect to the hadron force and can therefore more adequately explore the details of inner nuclear structures.

Our theoretical predictions will be useful for clarifying the different aspects of the electron scattering experiments off exotic nuclei that have been proposed in the ELISe experiment at FAIR and in the SCRIT project at RIKEN.

In this work, both elastic and inclusive quasi-elastic electron scattering have been considered. The elastic scattering can give information on the global properties of nuclei and, in particular, on the different behavior of proton and neutron density distributions. The inclusive QE scattering is affected by the dynamical properties, being the integral of the spectral density function over all the available final states, and, due to the reaction mechanism, preferably exploits the single particle aspects of the nucleus. In addition, when combined with the exclusive $(e, e'p)$ scattering, it is able to explore the evolution of the single particle model with increasing asymmetry between the number of neutrons and protons.

Our results show an evolution of the calculated cross sections without discontinuities as the asymmetry between protons and neutrons changes. The increase of the neutron number along $Z = 20$ chain enhances the nuclear and proton densities and flattens the charge densities. The increase of the proton number along each chain essentially produces an enhancement and an extension of the proton densities. The densities of the proton-rich isotones are significantly extended toward larger r with respect to those of the proton-deficient ones. An interesting way to relate the differential cross section and parity-violating asymmetry parameter for elastic scattering to the matter distribution is to consider the evolution of their first minima positions along each isotopic or isotonic chain.

The parity-violating asymmetry parameter has been also calculated to investigate the neutron skin. We have compared our calculations with the results of the first measurement of the asymmetry parameter achieved by the PREX Collaboration on ^{208}Pb and have obtained a good agreement with the empirical value.

References

[1] Hofstadter R 1956 *Rev. Mod. Phys.* **28** 214
[2] Donnelly T W and Walecka J D 1975 *Ann. Rev. Nucl. Part. Sci.* **25** 329
[3] Donnelly T W and Sick I 1984 *Rev. Mod. Phys.* **56** 461

[4] Boffi S, Giusti C, and Pacati F D 1993 *Phys. Rep.* **226** 1
[5] Boffi S, Giusti C, Pacati F D, and Radici M 1996 *Electromagnetic Response of Atomic Nuclei* (*Oxford Studies in Nuclear Physics* vol 20) (Oxford: Clarendon Press)
[6] Benhar O, Day D, and Sick I 2008 *Rev. Mod. Phys.* **80** 189
[7] De Vries H, De Jager C W, and De Vries C 1987 *Atom. Data Nucl. Data Tabl.* **36** 495
[8] Fricke G *et al.* 1995 *Atom. Data Nucl. Data Tabl.* **60** 177
[9] Heisenberg J and Blok H P 1983 *Ann. Rev. Nucl. Part. Sci.* **33** 569
[10] Donnelly T W, Dubach J, and Sick I 1989 *Nucl. Phys. A* **503** 589
[11] Donnelly T W and Peccei R D 1979 *Phys. Rep.* **50** 1
[12] Horowitz C J 1998 *Phys. Rev. C* **57** 3430
[13] Vretenar D, Finelli P, Ventura A, Lalazissis G A, and Ring P 2000 *Phys. Rev. C* **61** 064307
[14] Horowitz C J, Pollock S J, Souder P A, and Michaels R 2001 *Phys. Rev. C* **63** 025501
[15] Udías J M, Sarriguren P, Moya de Guerra E, Garrido E, and Caballero J A 1993 *Phys. Rev. C* **48** 2731
[16] Meucci A, Giusti C, and Pacati F D 2001 *Phys. Rev. C* **64** 014604
[17] Meucci A, Giusti C, and Pacati F D 2001 *Phys. Rev. C* **64** 064615
[18] Meucci A 2002 *Phys. Rev. C* **65** 044601
[19] Radici M, Meucci A, and Dickhoff W H 2003 *Eur. Phys. J. A* **17** 65
[20] Giusti C, Meucci A, Pacati F D, Co' G, and De Donno V 2011 *Phys. Rev. C* **84** 024615
[21] Capuzzi F, Giusti C, and Pacati F D 1991 *Nucl. Phys. A* **524** 681
[22] Meucci A, Giusti C, and Pacati F D 2004 *Nucl. Phys. A* **739** 277
[23] Meucci A, Capuzzi F, Giusti C, and Pacati F D 2003 *Phys. Rev. C* **67** 054601
[24] Capuzzi F, Giusti C, Pacati F D, and Kadrev D N 2005 *Ann. Phys. (N.Y.)* **317** 492
[25] Meucci A, Giusti C, and Pacati F D 2005 *Nucl. Phys. A* **756** 359
[26] Meucci A, Caballero J A, Giusti C, Pacati F D, and Udías J M 2009 *Phys. Rev. C* **80** 024605
[27] Meucci A, Caballero J A, Giusti C, and Udías J M 2011 *Phys. Rev. C* **83** 064614
[28] Meucci A, Barbaro M B, Caballero J A, Giusti C, and Udías J M 2011 *Phys. Rev. Lett.* **107** 172501
[29] Meucci A, Vorabbi M, Giusti C, Pacati F D, and Finelli P 2013 *Phys. Rev. C* **87** 054620
[30] Meucci A, Giusti C, and Vorabbi M 2013 *Phys. Rev. D* **88** 013006
[31] González-Jiménez R, Caballero J A, Meucci A, Giusti C, Barbaro M B, Ivanov M V, and Udías J M 2013 *Phys. Rev. C* **88** 025502
[32] Lalazissis G A, Nikšić T, Vretenar D, and Ring P 2005 *Phys. Rev. C* **71** 024312
[33] Moller P and Nix J R 1992 *Nucl. Phys. A* **536** 20
[34] Meucci A, Vorabbi M, Giusti C, Pacati F D, and Finelli P 2013 `arXiv:1311.4412`
[35] Rufa G 1982 *Nucl. Phys. A* **384** 273
[36] Moreno O *et al.* 2009 *Nucl. Phys. A* **828** 306
[37] Antonov A N *et al.* 2005 *Phys. Rev. C* **72** 044307
[38] Horowitz C J *et al.* 2012 *Phys. Rev. C* **85** 032501
[39] Abrahamyan S *et al.* (The PREX Collaboration) 2012 *Phys. Rev. Lett.* **108** 112502
[40] See supplemental material at `http://link.aps.org/supplemental/10.1103/PhysRevLett.108.112502`

Journal of Physics: Conference Series **527** (2014) 012025 doi:10.1088/1742-6596/527/1/012025

Nucleon tomography

Marco Radici
INFN - Sezione di Pavia, via Bassi 6, I-27100 Pavia, Italy
E-mail: marco.radici@pv.infn.it

Abstract. Starting from the interpretation of the nucleon electromagnetic form factors in terms of charge/magnetic density, we illustrate the need and the advantages of a multi-dimensional exploration of the structure of the nucleon, and in general of hadrons. We introduce new tools, like the Generalized Parton Distributions (GPDs) and the Transverse Momentum Distributions (TMDs), which enable us to explore the dynamics of partons beyond the usual collinear approximation adopted in the QCD analysis of hard processes. We emphasize the possibility of using these tools to address the (orbital) angular momentum of partons, and we highlight the latest results on the partonic decomposition of the nucleon spin. We briefly touch on the possible breaking of universality for a specific category of TMDs; as an example, we describe the Sivers effect as a fundamental test of our understanding of the color force in QCD.

1. Introduction

The nucleon is the most common element in the universe. It makes up 99.97% of visible matter. As a composite system of more elementary strongly-interacting constituents, the nucleon is difficult to be studied within QCD. In fact, its extension is of the order of femtometer; probing it requires an energy scale where α_s is large and QCD cannot be calculated with perturbative methods. We are not yet able to deduce from first principles how the nucleon structure comes about by displaying the confinement of strongly interacting quarks and gluons. At present, the most powerful attempt to solve QCD in nonperturbative regime is describing the theory on a lattice. But most of the nucleon's macroscopic properties (mass, spin, charge, magnetic moment, charge radius,..) cannot yet be explained. Let's consider for example the nucleon mass. The Higgs mechanism justifies the origin of mass of the elementary particles, like quarks, inside the Standard Model. But the mass of light quarks that determine the charge, baryon number, and isospin of the nucleon, amounts to about 1% of its total mass. What does it generate the missing 99%? QCD is a highly nonlinear theory. At the nucleon scale, its chiral symmetry is spontaneously broken; the vacuum has a nontrivial structure with nonvanishing quark and gluon condensates. Below the scale of chiral symmetry breaking, the light quarks rapidly acquire mass because of the interaction with gluon and sea-quark clouds. Therefore, almost all of the nucleon mass is given by the interaction energy involved in the QCD mechanisms that lead to confinement [1].

One of the ways to understand confinement is to study the internal structure of the nucleon, and in general of hadrons, in higher and higher detail. This requires the probe being more and more energetic. A typical tool is electron scattering in the Deep-Inelastic regime (DIS), where the 4-momentum q transferred to the target becomes larger and larger but keeping the ratio $x_B = Q^2/2P\ \ q$ fixed, with P the target 4-momentum and $Q^2 = \ \ q^2$. The kinematics is highly relativistic and it is useful to describe the system in terms of Light-Cone (LC) variables (a 4-vector a is represented as $a = \{a^+, a^-, \boldsymbol{a}_T\}$, with $a^\pm = (a^0 \pm a^3)/\sqrt{2}$). In fact, if the target is initially at rest, after absorbing the virtual photon it gets boosted along the direction of the spatial momentum transfer, say $\hat{z}$. Since $Q \to \infty$, the target momentum has a dominant component, $P^+ \sim Q$, while $P^- \sim 1/P^+$ is suppressed. The nucleon target gets squeezed to a disk along the boost direction by Lorentz transformations, and it can be conceived as an assembly of partons traveling collinearly with the parent nucleon, with LC "longitudinal" momenta $p^+ = xP^+$ and transverse momenta $|\boldsymbol{p}_T| \ll p^+$, with $x \sim x_B$.

In such a highly relativistic framework, conventionally named Infinite Momentum Frame (IMF), the usual textbook interpretation of the electromagnetic form factors of the nucleon as Fourier transforms of its charge and magnetization densities, is incorrect [2]. The main difficulty lies in the fact that the form factors are deduced from matrix elements where the initial and final nucleon states have different momenta, hence different wave functions. The Dirac and Pauli form factors parametrize the nucleon matrix elements of the electromagnetic current operator J^μ as

$$\langle N'(P', S')|\hat{J}^\mu(0)|N(P, S)\rangle = \bar{u}(N')\gamma^\mu u_N\, F_1(Q^2) + \bar{u}(N')i\frac{\sigma^{\mu\nu}q_\nu}{2M}u_N\, F_2(Q^2)\,. \tag{1}$$

It is useful to adopt the so-called Sachs form factors

$$G_E(Q^2) = F_1(Q^2) - \frac{Q^2}{4M^2}F_2(Q^2)\,, \quad G_M(Q^2) = F_1(Q^2) + F_2(Q^2)\,, \tag{2}$$

because the expression of the electron-nucleon cross section (in the one-photon approximation) depends directly on $G^2_{E/M}$ but not on $G_E G_M$. In the Breit frame where $\boldsymbol{q}/2 = \boldsymbol{P}' = -\boldsymbol{P}$ and $q^0 = 0$, G_E is simply given as the nucleon helicity-flip matrix element of $\hat{J}^0$. If the nucleon wave function could be expressed as the product of a plane wave, describing the motion of the center of mass (c.m.) with momentum $\boldsymbol{P}$, and of a function $\varphi(r)$ depending only on the internal relative coordinate r, then the Fourier transform of G_E

$$\rho(r) = \frac{2}{\pi}\int_0^\infty dQ Q^2\, j_0(Qr)\, G_E(Q^2) \tag{3}$$

would be related to $\rho(r) = |\varphi(r)|^2$, i.e. to the charge density of the nucleon. However, the above relation holds only for a nonrelativistic system. For a relativistic system like the nucleon in the IMF, the Nakanishi integral representation of the Bethe-Salpeter wave function shows that the covariant wave function cannot be simply factorized in a part describing the c.m. motion and in another one describing the internal motion: it depends explicitly and intrinsically also on the total momentum $\boldsymbol{P}$. Therefore the initial and final nucleon wave functions in the Breit frame are different, and the Fourier transform of G_E cannot be interpreted as a quantum mechanical density, namely as the modulus squared of some function [3].

We are thus confronted with the problem of identifying the proper operator whose matrix elements give the nucleon electromagnetic form factors, but that at the same time it can be related to the diagonal matrix element of a density operator.

2. The Generalized Parton Distributions

The solution to the puzzle relies on the crucial observation that the electromagnetic form factors are extracted from nucleon matrix elements of a local current operator (see equation (1)). An elegant way out is to rewrite them involving the charge density operator in IMF, $\hat{J}^+$, and re-express them in terms of a nonlocal operator along the LC x^- direction (conjugated to the dominant component P^+ in momentum space) [4, 5]:

$$\begin{aligned}\langle N'|J^+(0)|N\rangle &\equiv \langle N'|\overline{q}(0)\gamma^+ q(0)|N\rangle = \overline{u}_{N'}\gamma^+ u_N\, F_1(t) + \overline{u}_{N'}\frac{i\sigma^{+\nu}q_\nu}{2M}u_N\, F_2(t)\\ &\equiv \int dx \int \frac{dx^-}{2\pi}\, e^{ixP^+x^-}\, \langle N'|\overline{q}\Big(-\frac{x^-}{2},0,\mathbf{0}\Big)\gamma^+ q\Big(\frac{x^-}{2},0,\mathbf{0}\Big)|N\rangle\\ &= \overline{u}_{N'}\gamma^+ u_N \int dx\, H(x,\xi,t) + \overline{u}_{N'}\frac{i\sigma^{+\nu}q_\nu}{2M}u_N \int dx\, E(x,\xi,t)\,. \end{aligned} \tag{4}$$

The new objects $H(x,\xi,t)$ and $E(x,\xi,t)$ are called Generalized Parton Distributions (GPDs), because they do not only depend on the "longitudinal" LC momentum fraction x of partons, but also on the skewness $\xi = (P-P')^+/(P+P')^+$, which describes the change in the "longitudinal" LC momentum of the nucleon, as well as on $t=(P'-P)^2$. GPDs allow for a unified description of several hadronic properties [6]: from equation (4) their integral in x yields the nucleon electromagnetic form factors; while the so-called forward limit, $\xi, t \to 0$, reduces them to the conventional Parton Distribution Functions (PDFs). GPDs emerge in any DIS process with production of real photons (Deeply-Virtual Compton Scattering, DVCS) or of mesons (Deeply-Virtual Meson Production, DVMP), because corresponding factorization theorems hold at any order in α_s when $Q^2 \gg t$ [7]. Consequently, GPDs depend also on the hard scale Q^2 of the process at which they are extracted. For simplicity, this dependence will be omitted in the following, unless explicitly mentioned.

2.1. GPDs and impact parameter densities

The connection between GPDs and the spatial structure of the nucleon can be made evident by using nucleon states that are transversally localized in space. In fact, in IMF the boosts along transverse directions do not depend on interactions and form a kinematic subgroup of the Poincaré group that obeys the same commutation relations as in ordinary nonrelativistic quantum mechanics. The nucleon state with transverse c.m. $\boldsymbol{R}=0$ is built by [8]

$$|N(P^+,\boldsymbol{R}=0,S)\rangle = \mathcal{N}\int \frac{d\boldsymbol{P}_T}{(2\pi)^2}\,|N(P^+,\boldsymbol{P}_T,S)\rangle\,, \tag{5}$$

where $\mathcal{N}$ is a suitable normalization and the integral is limited to $|\boldsymbol{P}_T| \ll P^+$ in order to preserve the partonic interpretation. Having defined states at $\boldsymbol{R}=0$, it is possible to define also transverse distances $\boldsymbol{b}$ with respect to $\boldsymbol{R}$, then to make a translation by $\boldsymbol{b}$ in the operator appearing in equation (4), and finally to define the impact-parameter dependent distribution [4]

$$\begin{aligned} q(x,\boldsymbol{b}) &= \int \frac{dx^-}{2\pi}\, e^{ixP^+x^-}\, \langle N(P^+,\boldsymbol{R}=0,S)|\overline{q}\Big(-\frac{x^-}{2},0,\boldsymbol{b}\Big)\gamma^+ q\Big(\frac{x^-}{2},0,\boldsymbol{b}\Big)|N(P^+,\boldsymbol{R}=0,S)\rangle\\ &= |\mathcal{N}|^2 \int \frac{d\boldsymbol{P}'_T}{(2\pi)^2}\frac{d\boldsymbol{P}_T}{(2\pi)^2}\, e^{i(\boldsymbol{P}'_T-\boldsymbol{P}_T)\cdot\boldsymbol{b}} \int \frac{dx^-}{2\pi}\, e^{ixP^+x^-}\\ &\qquad \times\langle N(P^+,\boldsymbol{P}'_T,S)|\overline{q}\Big(-\frac{x^-}{2},0,\mathbf{0}\Big)\gamma^+ q\Big(\frac{x^-}{2},0,\mathbf{0}\Big)|N(P^+,\boldsymbol{P}_T,S)\rangle\\ &= \int \frac{d\boldsymbol{q}}{(2\pi)^2}\, e^{i\boldsymbol{q}\cdot\boldsymbol{b}}\, H(x,\xi=0,t=-\boldsymbol{q}^2)\,. \end{aligned} \tag{6}$$

That $q(x,\boldsymbol{b})$ is a partonic density, it can be recognized by taking the integral over x:

$$\begin{aligned} q(\boldsymbol{b}) &= \int dx\, q(x,\boldsymbol{b}) = \langle N(P^+,\boldsymbol{R}=0,S)|q_+^\dagger(0,0,\boldsymbol{b})\, q_+(0,0,\boldsymbol{b})\,|N(P^+,\boldsymbol{R}=0,S)\rangle\\ &= \int \frac{d\boldsymbol{q}}{(2\pi)^2}\, e^{i\boldsymbol{q}\cdot\boldsymbol{b}}\, F_1(Q^2=\boldsymbol{q}^2)\,, \end{aligned} \tag{7}$$

Journal of Physics: Conference Series **527** (2014) 012025 doi:10.1088/1742-6596/527/1/012025

and by recalling that $\bar{q}\gamma^+ q = \sqrt{2} q_+^\dagger q_+$, where $q_+ = P_+ q$ is the LC "good" component of the partonic field q, with $P_+ = 1/2\, \gamma^- \gamma^+$.

Therefore, by taking the specific limit $\xi = 0, t = -(\boldsymbol{P}'_T - \boldsymbol{P}_T)^2$, namely by letting the nucleon change only its transverse momentum while keeping the same P^+, the Fourier transform of the nonspin-flip GPD H gives the impact-parameter dependent distribution $q(x, \boldsymbol{b})$, which describes the probability density for a parton with flavor q and LC momentum x to be located in the transverse plane at distance $\boldsymbol{b}$ from the transverse c.m. $\boldsymbol{R} = 0$. Hence, for each x we can take a snapshot of the nucleon in the transverse plane: we can make a nucleon tomography.

Equation (7) gives a solution to the puzzle depicted at the end of section 1: the Fourier transform of the Dirac form factor F_1 gives access to the nucleon charge distribution in the transverse plane when the nucleon is moving relativistically with large LC "longitudinal" momentum in the IMF.

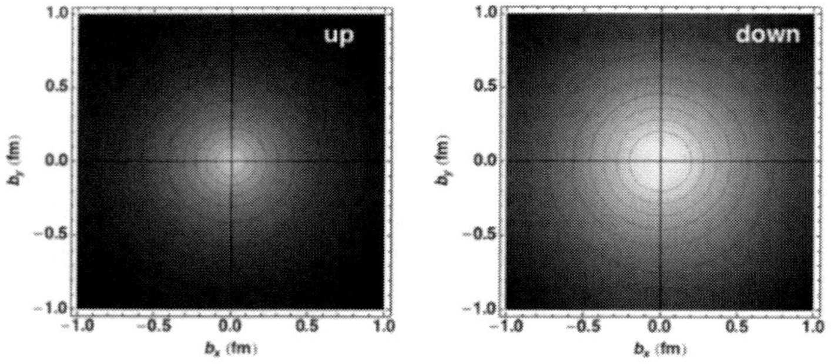

(a) (b)

Figure 1. (Color online) The charge distribution of up quark (a) and down quark (b) in the transverse plane, as obtained by Fourier transforming the Dirac form factor parametrized in Ref. [9]. Figure adapted from Ref. [10].

In figure 1, the plot for the up (a) and down (b) transverse charge distributions $q(\boldsymbol{b})$ are shown, as obtained by taking the parametrization of F_1 from Ref. [9]. We can remark that the down quark appears wider than the up by almost 30%. Interestingly, if the same exercise is repeated for the neutron, the resulting profile of $q(|\boldsymbol{b}|)$ shows that the central core is negatively charged while the peripheral cloud (supposedly made of pions) is positively charged [8], contrary to naïve expectations based on the idea that the neutron is somewhat a proton surrounded by a negatively charged pion cloud.

2.2. Polarized nucleons

In equation (6), the matrix elements do not involve the flip of the nucleon helicity S. For a transversely polarized N, e.g., along $\hat{y}$, the polarization state is $|N(P, S_y)\rangle = (|N(P,S)\rangle + i|N(P,-S)\rangle)/\sqrt{2}$. Hence, the more general version of equation (6) for a transversely polarized N involves both nonspin-flip and spin-flip matrix elements, therefore it involves both the GPDs H and E. Following the same pattern leading to equation (7), it is easy to realize that the Fourier transform of the GPD E in the limit $\xi = 0, t = -\boldsymbol{q}^2$, introduces a distortion in the transverse charge density related to the Pauli form factor. In formulae, the final expression of the transverse charge density reads [11]

$$\rho(\boldsymbol{b}) = q(\boldsymbol{b}) + \cos\phi_b \int \frac{dQ}{2\pi} \frac{Q^2}{2M} J_1(Q|\boldsymbol{b}|)\, F_2(Q^2 = \boldsymbol{q}^2) \,. \tag{8}$$

We deduce that the transverse polarization of the nucleon along $\hat{y}$ produces a distortion of its transverse charge density along $\hat{x}$ (since $\boldsymbol{b} = |\boldsymbol{b}|\{\cos\phi_b, \sin\phi_b\}$), and the size of the effect is proportional to its anomalous magnetic moment $\kappa = F_2(0)$. In figure 2, we show this effect as obtained by taking again the parametrization of F_1 and F_2 from Ref. [9]. A natural explanation

Journal of Physics: Conference Series **527** (2014) 012025 doi:10.1088/1742-6596/527/1/012025

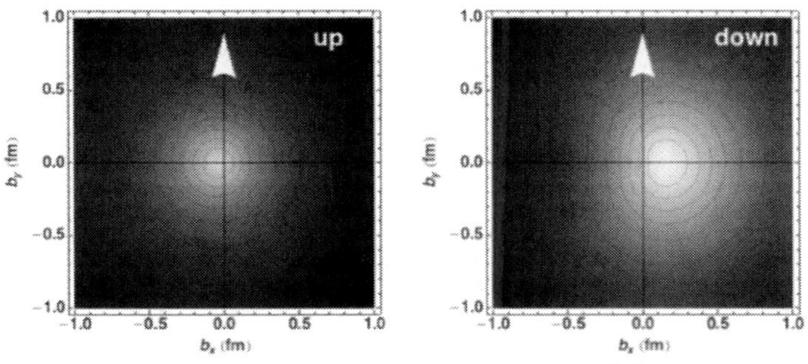

(a) (b)

Figure 2. (Color online) The distorted charge distribution of up quark (a) and down quark (b) in the transverse plane for a proton transversely polarized along $\hat{y}$, as obtained by Fourier transforming the Dirac and Pauli form factors parametrized in Ref. [9]. Figure adapted from Ref. [10].

is that a spin-orbit effect is active at the partonic level. So one may wonder if GPDs can open the doorway to explore the still unsolved puzzle about the partonic decomposition of the N spin.

3. GPDs and the N spin

Equation (4) can be generalized to calculating the nucleon matrix elements of the "m-spin" operator $\hat{O}^{\mu}_{\mu_1\ldots\mu_m} = \overline{q}(0)\gamma^\mu i\overleftrightarrow{D}_{\mu_1}\ldots i\overleftrightarrow{D}_{\mu_m} q(0)$ [6],

$$\langle N'|\hat{n}^{\mu_1}\ldots\hat{n}^{\mu_m}\hat{O}^{\mu}_{\mu_1\ldots\mu_m}|N\rangle = \overline{u}_{N'}\gamma^\mu u_N \sum_{i\,\text{even}}^{m} A_{m+1,i}(t)\,(2\xi)^i + \overline{u}_{N'}\frac{i\sigma^{\mu\nu}q_\nu}{2M}u_N \sum_{i\,\text{even}}^{m} B_{m+1,i}(t)\,(2\xi)^i$$
$$= \overline{u}_{N'}\gamma^\mu u_N \int dx\, x^m\, H(x,\xi,t) + \overline{u}_{N'}\frac{i\sigma^{\mu\nu}q_\nu}{2M}u_N \int dx\, x^m\, E(x,\xi,t)\,, \qquad (9)$$

where the light-like 4-vectors $\hat{n}^{\mu_1}, \ldots, \hat{n}^{\mu_m}$, are chosen to satisfy $\hat{n}\cdot(P+P')/2 = 1$ and $\hat{n}\cdot q = -2\xi$. Equation (9) illustrates the so-called polinomiality of GPDs, namely the fact that their m-th Mellin moment in x can be represented as an even polynomial in ξ. The polynomial coefficients $A_{m+1,i}(t)$, $B_{m+1,i}(t)$, are named Generalized Form Factors (GFFs); we have $A_{1,0}(t) \equiv F_1(t)$ and $B_{1,0}(t) \equiv F_2(t)$.

Starting from the energy-momentum density tensor of QCD, we can define the angular momentum operator $\hat{J}^f_i$ for a parton of flavor f, and expand its nucleon matrix elements by specializing equation (9) to $m = 1$ such that

$$\frac{1}{2} = \sum_{f=q,\bar{q},g} J^f_z \equiv \sum_{f=q,\bar{q},g} \langle N(P,S=1/2)|\hat{J}^f_z|N(P,S=1/2)\rangle$$
$$= \frac{1}{2}\sum_{f=q,\bar{q},g}\left[A^f_{2,0}(0) + B^f_{2,0}(0)\right] = \frac{1}{2}\int_0^1 dx\, x\left[H^f(x,0,0) + E^f(x,0,0)\right]. \qquad (10)$$

The last line of equation (10) represents the so-called Ji's sum rule [6] and illustrates the connection between GPDs and the problem of decomposing the N spin in terms of its elementary constituents. On the other hand, the Ji's sum rule sets a challenge to practical calculations of J^f_z. The forward limit $\xi, t \to 0$, of the nonspin-flip GPD H^f is simply the well known unpolarized PDF $f^f_1(x)$ (or the $q(x)$ recovered by integrating the impact parameter $\boldsymbol{b}$ in equation (6)). But the forward limit of the spin-flip GPD E^f is not accessible, because in this limit there is no handle to describe the N spin flip. So, any calculation of J^f_z based on the extraction of GPDs from, e.g. DVCS/DVMP data, needs a model-dependent extrapolation to reach the forward limit of E. Moreover, in equation (10) it is understood that any partonic object depends also on Q^2: the GPDs do explicitly depend on the hard scale of the process at which they are extracted (see section 2); hence, also the contribution to the total angular momentum by each specific flavor does. The balance between the different flavors is ruled by evolution equations of DGLAP type [12], because the Mellin moments in x do not alter the short-distance behaviour of QCD. In the asymptotic limit $Q^2 \to \infty$ of the IMF, the N spin is shared approximately in equal parts by (anti)quarks and gluons (for $n_f = 5$ flavors).

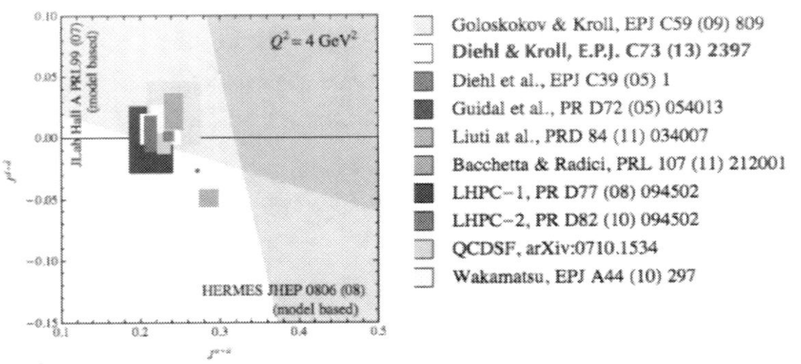

Figure 3. (Color online) The contribution to the N spin carried by $d + \bar{d}$ against $u + \bar{u}$ at $Q^2 = 4$ GeV^2. From top down, the first five boxes refer to the extraction of GPDs from exclusive DVCS data. The 6th box refers to the connection between GPD E and TMD $f_{1T}^{\perp}$. The last four boxes refer to calculations of GFFs on lattice. Figure adapted from Ref. [13].

In figure 3, the contribution to the N spin carried by the total angular momentum of the $d + \bar{d}$ flavor is plotted against the $u + \bar{u}$ one at the scale $Q^2 = 4$ GeV^2 (figure adapted from Ref. [13]). The notation is the following. From top down, the first five boxes refer to the calculation of the Ji's sum rule (10) based on the extraction of GPDs H and E from exclusive DVCS data. The result corresponding to the 6th box will be described in section 4. The last four boxes describe some of the results obtained on lattice, where the GFFs in equation (10) can be directly computed. From figure 3, we deduce that almost all calculations are compatible within errors. There is a general consensus about the fact that quarks and antiquarks carry together slightly more than half of the N spin (at the indicated scale), and that 85% of this portion is carried by the flavor u. The sea quark contribution is smaller by one order of magnitude, but the uncertainty in the calculation is still very large. Nothing is known about the gluon contribution, except that experimental measurements with polarized nucleons indicate an almost vanishing gluon helicity [14] (but with large errors).

3.1. The Orbital Angular Momentum of partons

At the end of section 2.2, we stated that the observed distortion of the transverse charge distribution of partons in transversely polarized nucleons must be attributed to a sort of spin-orbit effect generated by the orbital motion of partons. But the Ji's sum rule (10) gives access only to the total angular momentum carried by partons. So, it comes natural to wonder if GPDs are useful tools to deduce information also on the Orbital Angular Momentum (OAM) of partons.

By inspecting the analytic structure of the partonic total angular momentum that can be deduced from the QCD energy-momentum tensor, it is natural to define the OAM L_z^f for a parton with flavor f as the difference between its total angular momentum J_z^f and its helicity S_z^f. Using equation (10), we can identify it as the Mellin moment of the OAM distribution function $L_z^f(x)$ defined as [12]

$$L_z^f = J_z^f - S_z^f = \int_0^1 dx\, L_z^f(x) = \frac{1}{2}\int_0^1 dx \left\{ x\left[f_1^f(x) + E^f(x,0,0)\right] - g_1^f(x) \right\}, \tag{11}$$

where $g_1^f(x)$ is the well known helicity distribution for a flavor f (at the hard scale Q^2).

The definition (11) is gauge invariant, but above all it defines a measurable quantity: f_1 and g_1 are extracted from inclusive DIS of electrons on (polarized) nucleons, while E can be extracted from exclusive DVCS or DVMP. Unfortunately, the analytic structure of the QCD energy-momentum tensor does not allow to decompose the gluon total angular momentum in a similar manner. Moreover, the operator defined through equation (11) does not satisfy the canonical commutation relations of a general angular momentum operator. In the literature, another definition of $\hat{J}^f$ [15] appeared before the one in equation (10). It allows for a clear decomposition of both quark and gluon parts into the respective OAM and helicity contributions, each term defining an operator that satisfies the canonical commutation relations. However, this so-called Jaffe-Manohar decomposition is not gauge invariant. Several attempts to extend it to a gauge invariant quantity have been released. The discussion is still open about which choice is more convenient (for a recent review, see Ref. [16]).

An alternative definition of OAM can be built in terms of the so-called Wigner distributions $W_{S,S_q}(x, \boldsymbol{p}_T, \boldsymbol{b})$, which contain information on the correlation between the transverse momentum $\boldsymbol{p}_T$ and position $\boldsymbol{b}$ of a parton with helicity S_q and LC "longitudinal" fraction x of the momentum of the parent nucleon with helicity S [17]. The 5-dimensional W contain the maximum information about the dynamics of partons inside the nucleon moving in the IMF. By construction, they are not positive-definite, i.e. they are not densities. But they are not constrained by the Heisenberg principle because $\boldsymbol{p}_T$ is not the conjugate variable of $\boldsymbol{b}$. The impact-parameter distribution $q(x, \boldsymbol{b})$ of equation (6) can be recovered from the Wigner distribution $W(x, \boldsymbol{p}_T, \boldsymbol{b})$ by simply integrating its $\boldsymbol{p}_T$ dependence. For an unpolarized quark in a longitudinally polarized nucleon, the quark OAM can be defined as [18]

$$l_z^q = \int d\boldsymbol{b} \left[\boldsymbol{b} \times \int dx\, d\boldsymbol{p}_T\, \boldsymbol{p}_T\, W_{LU}^q(x, \boldsymbol{p}_T, \boldsymbol{b}) \right]_z = \int d\boldsymbol{b}\, \left[\boldsymbol{b} \times \langle \boldsymbol{p}_T^q \rangle(\boldsymbol{b}) \right]_z . \tag{12}$$

In the above definition, the W's are used as semi-classical phase space distributions to define the average transverse momentum $\langle \boldsymbol{p}_T^q \rangle(\boldsymbol{b})$ carried by a quark q in a certain transverse position $\boldsymbol{b}$. The OAM is then built by generalizing the classical definition. The l_z^q operator satisfies the canonical commutation relations but it is built with objects, the Wigner distributions, that cannot be extracted from any known process. Anyway, the definition (12) can be used to test models of the Wigner distribution, i.e. of the N wave function.

The definition (11) is undoubtedly more convenient also because L_z^f can be calculated on the lattice. Most lattice calculations, some of which are reported in figure 3, converge to the result that $L_z^u \approx 0.3 \sim -L_z^d$. The helicities S_z^f of quarks $f = u, d, s$, are constrained to add up to the nucleon singlet axial charge g_A^0. The lattice result for g_A^0 is rather close to its experimental value and implies that quark helicities contribute at most to 25% of the N spin, i.e. they amount to ≈ 0.125. Since the extracted gluon helicity is approximately almost vanishing and nothing is basically known about the gluon OAM, the strong cancellation of quark OAM leaves the puzzle about the N spin decomposition still open.

4. The Transverse-Momentum Distributions

In the previous section, we introduced the Wigner distributions $W_{S,S_q}(x, \boldsymbol{p}_T, \boldsymbol{b})$, that contain the most complete information on the parton dynamics inside a nucleon moving in the IMF. Even if

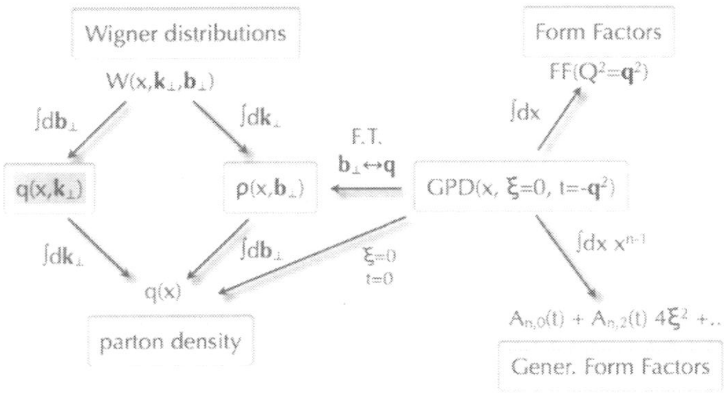

Figure 4. The roadmap to a multi-dimensional exploration of the N inner structure. The marked box shows the role of TMDs in completing the manifold picture starting from the Wigner distributions W's down to the usual parton densities PDFs (adapted from Ref. [19]).

the definition is somewhat academic because no process is currently known where the W's could be extracted from, it has been pointed out that by simply integrating the $\boldsymbol{p}_T$ dependence for an unpolarized parton we can recover the impact-parameter distribution $q(x, \boldsymbol{b})$ of equation (6), which is actually measurable through the bidimensional Fourier transform of the Dirac and Pauli form factors. Analogously, it is natural to explore the consequence of integrating the $\boldsymbol{b}$ dependence of the W's. In this way, we can build new partonic densities in momentum space, the Transverse-Momentum Distributions $q(x, \boldsymbol{p}_T)$ (TMDs), that describe the probability density of finding a parton with flavor q and polarization S_q carrying a LC "longitudinal" momentum x and a transverse momentum $\boldsymbol{p}_T$. Since the impact parameter $\boldsymbol{b}$ is the conjugate variable of $\boldsymbol{q} = \boldsymbol{P}_T' - \boldsymbol{P}_T$ and not of $\boldsymbol{p}_T$, the TMDs are not simply the Fourier transforms of $q(x, \boldsymbol{b})$. Lastly, by integrating the $\boldsymbol{p}_T$ dependence of TMDs we recover the standard PDFs as functions of x. Hence, the TMDs contain a more detailed information on the dynamics of partons in momentum space and represent a new milestone in the roadmap to the exploration of the nucleon internal structure (see figure 4).

quark pol.

nucleon pol.	U	L	T
U	f_1		$h_1^\perp$
L		g_{1L}	$h_{1L}^\perp$
T	$f_{1T}^\perp$	g_{1T}	h_1, $h_{1T}^\perp$

Twist-2 TMDs

Figure 5. The eight leading-twist TMDs organized according to the unpolarized (U), longitudinal (L), or transverse (T) polarization state of the parton (along table rows), or of the parent nucleon (along table columns).

At leading twist, there are eight TMDs. They are grouped in figure 5 according to the unpolarized (U), longitudinal (L), or transverse (T) polarization state of the parton (along the table rows) or of the parent nucleon (along the table columns). The TMDs can be typically extracted in Semi-Inclusive processes in the DIS regime (SIDIS) where one measures the azimuthal angle of the final-state products around a reference axis defined in the lab frame. If the transverse momentum of the outgoing particles is small with respect to the hard scale Q of the process, a specific factorization proof has been put forward [20]. In this framework, the cross section can be parametrized in terms of various structure functions (depending on the polarization of the beam/target/final products), each one involving a specific TMD in the table of figure 5 (and a corresponding fragmentation function, if the detected final products are hadrons) [21]. Each structure function has a peculiar dependence upon the azimuthal angle of the final products, as well as the azimuthal angle of the various polarization vectors involved. Consequently, it can be isolated through azimuthal asymmetries, sometimes involving a spin flip of a polarized particle (in which case, we deal with Single-Spin Asymmetries - SSAs). Several experimental measurements of azimuthal asymmetries were performed in the last 15 years, addressing each one of the eight TMDs in figure 5. However, the collected amount of data is not yet sufficient to grant a knowledge of TMDs comparable to what we have learnt so far about PDFs. Only recently first attempts to parametrize the unpolarized $f_1^q(x, \boldsymbol{p}_T)$ have appeared in the literature [22, 23], in order to explore if and how the $\boldsymbol{p}_T$ dependence changes with flavor q and with x.

On the other hand, the literature about theoretical investigations on TMDs is vast and the discussion is still open about setting up the proper framework for describing their evolution with the hard scale (see Refs. [21, 24] and references therein, for a comprehensive review). Here, we just mention two elements in figure 5 that historically triggered the focus on this field.

The first one is the function h_1 appearing in the lower right corner of the table. It describes the unbalance between number densities of partons with transverse polarization parallel or antiparalllel to the transverse polarization of the parent nucleon. As such, it is connected to elementary mechanisms flipping the parton helicity; in jargon, it is named transversity and referred to as a chiral-odd distribution. If we integrate the $\boldsymbol{p}_T$ dependence of TMDs, all the entries in the table of figure 5 disappear except for the three elements in the diagonal: $f_1(x)$, $g_{1L}(x) \equiv g_1(x)$, and just $h_1(x)$. All three PDFs are necessary to reach a complete partonic picture of the N (spin) structure at leading twist. But the momentum and helicity distributions are usually extracted from inclusive DIS processes, while the transversity is suppressed because it is a chiral-odd function. The h_1 can be extracted only in semi-inclusive processes involving another chiral-odd partner. The typical example, for which there are experimental data, is the SIDIS on transversely polarized protons. If a single pion is detected and its azimuthal distribution is measured, the h_1 gets convoluted with a chiral-odd fragmentation function that relates the transverse polarization of the (fragmenting) quark to the transverse momentum of the final pion, the so-called Collins function [25]. In this way, the $h_1^u(x)$ and $h_1^d(x)$ were extracted for the first time (for more details on the most updated result, see Ref. [26]). If two pions are detected with a small transverse relative momentum $|\boldsymbol{R}_T| \ll Q$ and in a collinear kinematics (i.e., with a vanishing transverse component of their total momentum), the $h_1(x)$ is simply multiplied by the di-hadron fragmentation function $H_1^{\sphericalangle}$ that relates the transverse polarization of the fragmenting quark to $\boldsymbol{R}_T$ [27, 28, 29, 30]. The advantage of this method is that the collinear framework remarkably simplifies the theoretical description. Recent experimental data for the detection of $(\pi^+\pi^-)$ pairs allowed for a direct extraction of the valence contributions to the proton $h_1(x)$ [31, 32]. Within the experimental statistical errors, this extraction is compatible with the one based on the Collins effect. This is a remarkable achievement because it opens the doorway to a more precise determination of the theoretically interesting, but yet poorly known, tensor charge, i.e. the first Mellin moment of $h_1(x)$.

The second element in figure 5, that historically drew the interest in TMDs, appears in the lower left corner of the table. It describes the difference in the momentum distribution of unpolarized partons when in IMF they move together inside a parent nucleon with transverse polarization parallel or antiparallel to a specific transverse direction [33]. For many years, this function has been thought to vanish because of time-reversal symmetry; in jargon, it is classified as a naïve T-odd function [1]. Nevertheless, it addresses the same physics of the impact-parameter distribution $\rho(\boldsymbol{b})$ in equation (8), but in momentum space. Therefore, it must have a close connection to the problem of the partonic decomposition of the N spin.

[1] There are two naïve T-odd distributions in the table of figure 5 of leading-twist TMDs: the Sivers function $f_{1T}^{\perp}$ and its "mirror" $h_1^{\perp}$, the Boer-Mulders function, that in an unpolarized hadron describes the unbalance between partons transversely polarized along or against a specified transverse direction.

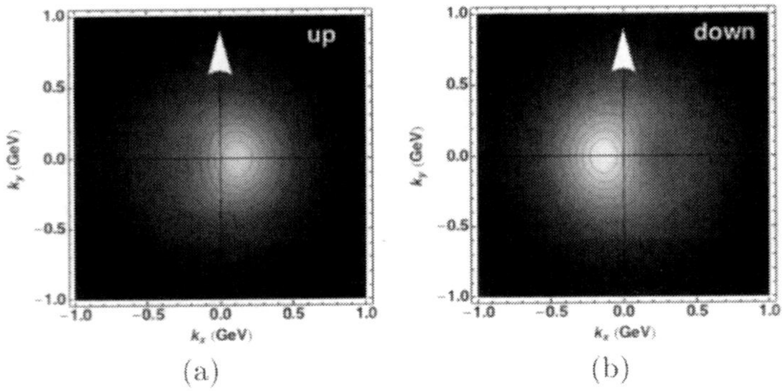

Figure 6. (Color online) The distorted momentum distribution of up quark (a) and down quark (b) in the transverse plane for a proton transversely polarized along $\hat{y}$, as obtained by parametrizing the Sivers effect [10].

4.1. The Sivers function and the N spin

In figure 6, the left panel shows the distortion of the momentum distribution of the up quark in the transverse plane when the parent proton is transversely polarized along $\hat{y}$ (indicated by the white arrow) and moves towards the reader. The right panel shows an opposite distortion for the down quark. The distortion is clearly along $\hat{x}$: the effect is similar to the one depicted in figure 2 but in momentum space, and its size is given by the Sivers function $f_{1T}^{\perp}$ [10]. Therefore, we can argue that the quarks drift in opposite directions, according to their charge, because also in this case they are subjected to a sort of spin-orbit effect. The Sivers function appears in the cross section for the same process as the Collins effect, namely the SIDIS off transversely polarized protons leading to the detection of a single pion. However, it is contained in a structure function with a different dependence on the azimuthal angles of the final pion and of the proton target polarization. Hence, it can be isolated through a different SSA, the so-called Sivers effect. Sensitive to the distortion induced by spin-orbit, the unpolarized struck parton eventually fragments into the detected pion with an azimuthally asymmetric distribution. The Sivers effect has been measured by three different collaborations: HERMES at DESY [34], COMPASS at CERN [35], and Hall A at Jefferson Lab [36]. Several parametrizations have been figured out to extract the x dependence of the Sivers function from these data. Most of them agree that the valence up and down contributions have similar size but opposite sign, the $f_{1T}^{\perp u}$ being negative [37, 38]. The distortions displayed in figure 6 are a direct consequence of these results. The contribution of the sea quarks is much more questionable, because the statistical errors are large and the size presumably very small [38]. Many theoretical papers have recently appeared also on the development of evolution equations for the Sivers function in the TMD framework. The topic is matter of intense debate and entering the details is beyond the scope of this presentation.

Since the Sivers effect originates from a spin-orbit distortion of the motion of partons, it is natural to suspect that it can play a role in the partonic decomposition of the N spin. The starting point is the Ji's sum rule (10) described in section 3. We already stressed that the forward limit of the spin-flip GPD $E(x,0,0)$ cannot be accessed, forcing to model-dependent extrapolations. However, we already noted that both the GPD E and the TMD $f_{1T}^{\perp}$ describe from two different perspectives the distortion induced by the same fact, namely that the nucleon is transversely polarized. It is natural to try to establish a link between the two pictures. But one effect is not the Fourier transform of the other: the impact parameter $\boldsymbol{b}$ is conjugate to the nucleon transverse momentum change $\boldsymbol{P}_T' - \boldsymbol{P}_T = \boldsymbol{q}$, not to the parton transverse momentum $\boldsymbol{p}_T$. A rigorous derivation of this link is very difficult, because it would imply a description of the nonperturbative color forces exerted on the active parton by the surrounding spectators. A semi-classical interpretation is possible, it is referred to as the "color lensing" effect [39] and it is described in figure 7.

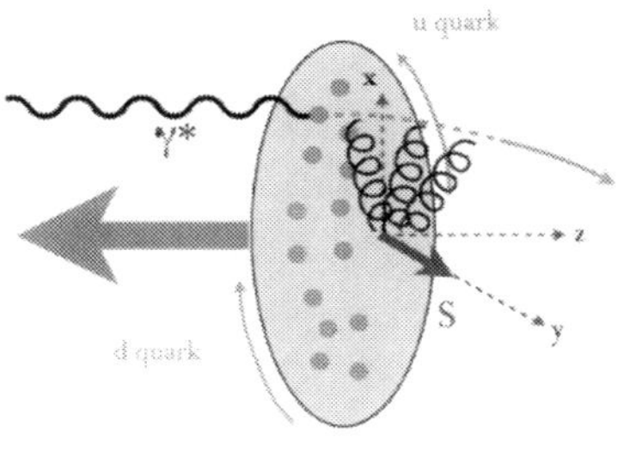

Figure 7. The color lensing effect [39].

In the Sivers effect, in the c.m. frame the nucleon is moving along the longitudinal direction against the virtual photon probe γ^*. The nucleon is transversely polarized along $\hat{y}$ with polarization S. Consequently, the charge distribution of partons is distorted along $\hat{x}$ because of a spin-orbit effect induced by the GPD E: the up quarks accumulate in the $x > 0$ hemisphere rotating anti-clockwise, and viceversa for the down quarks. When the virtual photon probe hits an up quark, it will preferably do it in the $x > 0$ hemisphere. In the DIS regime, the momentum transfer absorbed is very large and the struck quark tries to escape the nucleon. The attractive color forces deflect the trajectory of the up quark towards the $x < 0$ hemisphere. Confinement implies that the up quark eventually fragments into a π^+, preferably detected in the $x < 0$ hemisphere. Analogously, a π^- emerging from the fragmentation of a down quark would be preferably detected in the $x > 0$ hemisphere. This asymmetric distribution of pions along $\hat{x}$ is reminiscent of the distortion of quark distributions in momentum space described by the Sivers function $f_{1T}^{\perp}$. According to this semi-classical picture, the GPD E and the TMD $f_{1T}^{\perp}$ are connected by a complicated analytical formula that involves a function describing the above mentioned residual color forces [39]. Hence, an actual calculation of the color lensing effect requires model assumptions. Inspired by the spectator diquark model [40], at some scale Q_L we used the relation [38]:

$$f_{1T}^{\perp\,(0)f}(x; Q_L^2) = -L(x)\, E^f(x, 0, 0; Q_L^2)\ , \tag{13}$$

in order to compute the parton total angular momentum through the Ji's sum rule by using the parametrization of the x dependence of the Sivers function obtained from SIDIS data on the Sivers effect, and by constraining its integral to reproduce the N anomalous magnetic moment. In equation (13), the left hand side represents the first moment in $\boldsymbol{p}_T$ of the Sivers function $f_{1T}^{\perp f}(x, \boldsymbol{p}_T; Q_L^2)$. The total angular momenta obtained with this procedure produce the results identified by the 6th box (from top down) in the column of references displayed on the right of figure 3, in good agreement with the results obtained by very different procedures (see section 3).

The color lensing effect of equation (13) suggests that the content of the Sivers function $f_{1T}^{\perp}$ crucially depends on the structure of the residual color forces felt by the struck parton. In the SIDIS process of figure 7, these color forces represent a sort of "final-state interaction" effect. But $f_{1T}^{\perp}$ can appear also in other hard processes involving transversely polarized nucleons like, e.g., the hadronic collision leading to the semi-inclusive production of Drell-Yan lepton pairs. If one of the two colliding hadrons is a transversely polarized nucleon, the Sivers effect is now realized by producing a virtual photon with a transverse momentum with respect to the average collision direction and, eventually, by producing an azimuthally asymmetric distribution of the final lepton pair. However, in the hadronic collision the annihilating quark coming from the transversely polarized nucleon feels the color field generated inside the other colliding hadron as if, intuitively, it were a sort of "initial-state interaction". By a careful inspection of the color structure of the involved nonlocal operator, the net effect is that the Sivers function should appear with an opposite sign with respect to the SIDIS cross section, thus breaking in principle its universality feature [41]. Since experimental data for semi-polarized Drell-Yan processes are not yet available, this result is a true prediction based on the general nonperturbative color structure dictated by QCD gauge invariance. Hence, an experimental test of this prediction is of outmost importance [42, 19].

References

[1] Mueller B 2005 *Nucl. Phys. A* **750** 84
[2] Miller G A 2009 *Phys. Rev. C* **80** 045210
[3] Miller G A 2010 *Ann. Rev. Nucl. Part. Sci.* **60** 1
[4] Burkardt M 2003 *Int. J. Mod. Phys. A* **18** 173
[5] Diehl M 2002 *Eur. Phys. J. C* **25** 223
[6] Ji X 1997 *Phys. Rev. Lett.* **78** 610
[7] Collins J C and Freund A 1999 *Phys. Rev. D* **59** 074009
[8] Miller G A 2007 *Phys. Rev. Lett.* **99** 112001
[9] Kelly J 2004 *Phys. Rev. C* **70** 068202
[10] Bacchetta A and Contalbrigo M 2012 *Il Nuovo Saggiatore* **28** 16
[11] Carlson C E and Vanderhaeghen M 2008 *Phys. Rev. Lett.* **100** 032004
[12] Hoodbhoy P, Ji X D, and Lu W 1999 *Phys. Rev. D* **59** 014013
[13] Bacchetta A and Radici M 2012 *PoS* **QNP2012** 041
[14] Aschenauer E *et al.* 2013 `arXiv:1304.0079`
[15] Jaffe R L and Manohar A 1990 *Nucl. Phys. B* **337** 509
[16] Leader E and Lorce C 2013 `arXiV:1309.4235`
[17] Lorce C, Pasquini B, and Vanderhaeghen M 2011 *JHEP* **1105** 041
[18] Lorce C, Pasquini B, Xiong X, and Yuan F 2012 *Phys. Rev. D* **85** 114006
[19] Accardi A *et al.* 2012 `arXiv:1212.1701`
[20] Ji X, Ma J P, and Yuan F 2004 *Phys. Lett. B* **597** 299
[21] Bacchetta A, Diehl M, Goeke K, Metz A, Mulders P J, and Schlegel M 2007 *JHEP* **02** 093
[22] Signori A, Bacchetta A, Radici M, and Schnell G 2013 *JHEP* **1311** 194
[23] Anselmino M, Boglione M, J O Gonzales H, Melis S, and Prokudin A 2014 *JHEP* **1404** 005
[24] Collins J 2011 *Foundations of Perturbative QCD* (Cambridge University Press)
[25] Collins J C 1993 *Nucl. Phys. B* **396** 161
[26] Anselmino M *et al.* 2013 *Phys. Rev. D* **87** 094019
[27] Collins J C and Ladinsky G A 1994 `arXiv:hep-ph/9411444`
[28] Jaffe R L, Jin X, and Tang J 1998 *Phys. Rev. Lett.* **80** 1166
[29] Radici M, Jakob R, and Bianconi A 2002 *Phys. Rev. D* **65** 074031
[30] Bacchetta A and Radici M 2003 *Phys. Rev. D* **67** 094002
[31] Bacchetta A, Courtoy A, and Radici M 2011 *Phys. Rev. Lett.* **107** 012001
[32] Bacchetta A, Courtoy A, and Radici M 2013 *JHEP* **1303** 119
[33] Sivers D W 1990 *Phys. Rev. D* **41** 83
[34] Airapetian A *et al.* (The HERMES Collaboration) 2009 *Phys. Rev. Lett.* **103** 152002
[35] Alekseev M *et al.* (The COMPASS Collaboration) 2009 *Phys. Lett. B* **673** 127
[36] Qian X *et al.* (The Jefferson Lab Hall A Collaboration) 2011 *Phys. Rev. Lett.* **107** 072003
[37] Anselmino M, Boglione M, and Melis S 2012 *Phys. Rev. D* **86** 014028
[38] Bacchetta A and Radici M 2011 *Phys. Rev. Lett.* **107** 212001
[39] Burkardt M 2002 *Phys. Rev. D* **66** 114005
[40] Bacchetta A, Conti F, and Radici M 2008 *Phys. Rev. D* **78** 074010
[41] Collins J C 2002 *Phys. Lett. B* **536** 43
[42] Denisov O (The COMPASS Collaboration) 2012 *Nuovo Cim. C* **035N2** 239

Journal of Physics: Conference Series **527** (2014) 012026 doi:10.1088/1742-6596/527/1/012026

Investigating the homogeneous Bethe-Salpeter equation in Minkowski space

Giovanni Salmè
Istituto Nazionale di Fisica Nucleare, Sezione di Roma, P.le A. Moro 2, I-00185, Rome, Italy
E-mail: salmeg@roma1.infn.it

Abstract. A short review of a novel approach, based on the Nakanishi perturbation-theory integral representation and devised for solving the Bethe-Salpeter equation directly in Minkowski space, is presented. The two ingredients for constructing such a new tool are: i) the Nakanishi perturbation-theory integral representations of the N-leg transition amplitudes and ii) the Light-front framework, that also allows one to gain insights into the Fock content of the states of an interacting system. Numerical results, corresponding to the Bethe-Salpeter equation in ladder approximation, for a simple system, composed by two massive scalars interacting through the exchange of a massive scalar, are provided for illustrating the effectiveness of the approach.

1. Introduction

To achieve a fully covariant description of a bound system in Minkowski space, still represents a challenge, though the field theoretical framework for dealing with was settled up in the early fifties [1, 2]. In particular, one should recall that a bound system appears as a pole in the analytic structure of the T-matrix, that also has cuts, related to the scattering states. It is very important to notice that such a pole cannot be reconstructed within a perturbative framework, and therefore, in order to reproduce a bound state one has to resort to non perturbative tools, like e.g. path integrals or integral equations. The homogeneous Bethe-Salpeter equation (BSE) [1] is an integral equation, obtained within a field-theory framework, that allows one to describe bound states in Minkowski space (see [3] for an extended review and [4] for a textbook presentation). It should be pointed out that the scattering states fulfill an inhomogeneous integral equation, in strict analogy with the non relativistic approach (see Ref. [5] for a recent introduction to the topic). Summarizing, the BSE yields the possibility to take properly into account the dynamics, within a field-theoretical framework, directly in Minkowski space, differently, e.g., from the path-integral approach carried out on the lattice, that exploits a Euclidean space. Indeed, the vast literature on BSE illustrates the non trivial efforts for getting solutions. In particular, roughly speaking, the largest amount of efforts was put on devising numerical methods in Euclidean space, e.g. by applying a Wick rotation to BSE (see, e.g., Ref. [6] for a general review and also Ref. [7] for a recent application to the spinor-spinor BSE) or on reducing from 4D to 3D the formal expression of BSE, introducing suitable constraints (see, e.g., Ref. [8] for an example of a covariant reduction). Only in the last years, solutions of BSE in Minkowski space have been presented, as shown in Refs. [9, 10], and more recently in Ref. [11]. Such an improvement was made feasible by applying the so-called perturbation-theory integral representation (PTIR) introduced by N Nakanishi [12, 13] with the aim to give a novel formal description of a generic

Journal of Physics: Conference Series **527** (2014) 012026 doi:10.1088/1742-6596/527/1/012026

N-leg transition amplitude, but within the Feynman-diagram framework. If one adopts the PTIR of the 3-leg amplitude, i.e. the so-called vertex function, as a suitable Ansatz for the BS amplitude, after properly taking into account the formal relation between the BS amplitude itself and the vertex function, one can proceed toward an elaboration that leads to obtain solutions of the BSE directly in Minkowski space.

In what follows, it will be briefly illustrated the formalism for solving the homogeneous BSE developed in Refs. [5, 11] (see also Ref. [14]), where both the Nakanishi PTIR and the non explicitly covariant version of the Light-front (LF) approach (see Ref. [15] for a recent review of this specific LF framework) were adopted. Moreover, for the sake of concreteness, some relevant numerical results are also reviewed. It should be pointed out that results for the inhomogeneous case, obtained through a direct integration of the singularities present in the inhomogeneous BSE in ladder approximation, can be found in Ref. [16], while in Ref. [17] the scattering length and the phase-shifts, obtained within the Nakanishi PTIR approach, will be shown.

2. The BSE in a nutshell

In this section, the derivation of the homogeneous BSE is shortly reviewed (for a more detailed presentation see, e.g., Refs. [3] and [4]) in order to introduce some of the relevant issues of the field and the general formalism. For the sake of simplicity, let us consider a simple system composed by two massive scalars that interacts through the exchange of a massive scalar, i.e. a model with interacting Lagrangian like $\mathcal{L}_{int} = g\ \phi^2\chi$ (see Ref. [18] for the caveats about this model), but without self-energy and vertex corrections.

The path for obtaining the homogeneous BSE, that, we repeat, is the proper equation determining a bound state within a field-theory framework, starts with the 4-point Green's Function, given by

$$G(x_1, x_2; y_1, y_2) = < 0\,|\,T\{\phi_1(x_1)\phi_2(x_2)\phi_1^+(y_1)\phi_2^+(y_2)\}\,|\,0 > \quad . \tag{1}$$

The four-point Green's function fulfills an integral equation, pictorially described in figure 1 and schematically written as follows

$$G \ = \ G_0 \ + \ G_0\,\mathrm{I}\,G \ , \tag{2}$$

where G_0 is the free Green's function of the two interacting scalars, viz

$$G_0(k,p) = G_0^{(1)}G_0^{(2)} = \frac{i}{(\frac{p}{2}+k)^2 - m^2 + i\epsilon}\ \ \frac{i}{(\frac{p}{2}-k)^2 - m^2 + i\epsilon} \tag{3}$$

where $p = p_1 + p_2$ is the total momentum of the interacting system with total mass $p^2 = M^2$ and $k = (p_1 - p_2)/2$. In equation (2), the kernel I is given by the infinite sum of irreducible Feynman graphs, namely the ones that cannot be split into two more simple graphs by drawing a line that does not cut any exchanged-scalar lines. As an example, in figure 2, the first diagrams contributing to the irreducible kernel are depicted. It has to point out that all the diagrams contributing to the 4-point Green's function, i.e. the reducible and the irreducible ones, can be generated by infinitely iterating each diagram present in the irreducible kernel [1]. Inserting a complete Fock basis in G, one can single out the bound state contribution (assuming that only one bound state be present, for the sake of simplicity). In the Fourier space, the contribution appears as a pole, i.e.

$$G_B(k,q;p_B) \simeq \frac{i}{(2\pi)^{-4}}\ \frac{\phi(k;p_B)\ \bar{\phi}(k;p_B)}{2\omega_B(p_0 - \omega_B + i\epsilon)} \tag{4}$$

Journal of Physics: Conference Series **527** (2014) 012026 doi:10.1088/1742-6596/527/1/012026

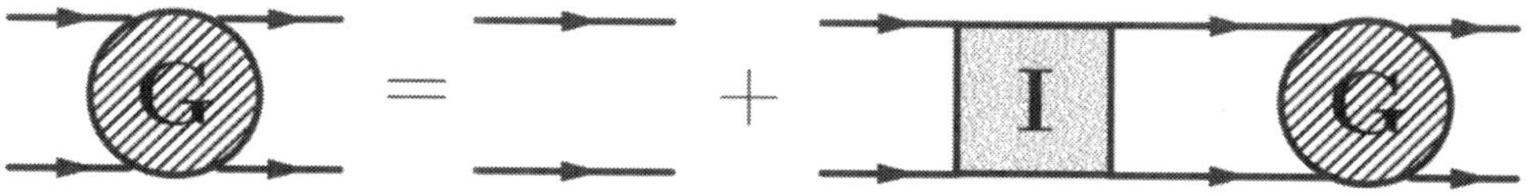

Figure 1. Pictorial representation of the integral equation fulfilled by the 4-point Green's function $G(x_1, x_2; y_1, y_2)$, see equation (2).

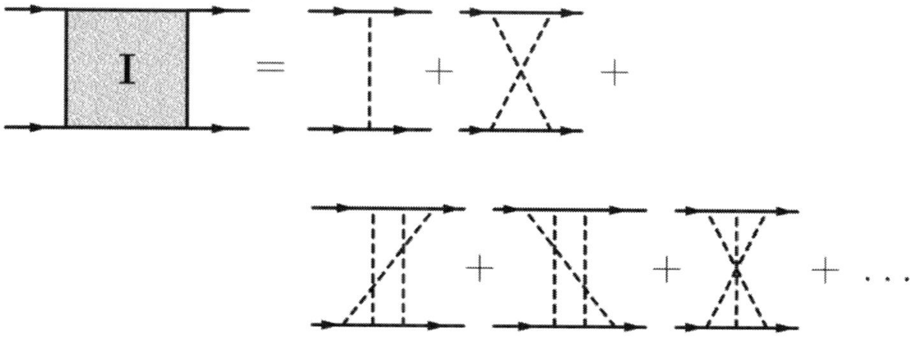

Figure 2. Feynman diagrams, up to the third order in g^2, contributing to the infinite series yielding the irreducible kernel I, see equation (2).

where $\omega_B = \sqrt{M_B^2 + |\mathbf{p}|^2}$ and $\phi(k; p_B)$ is the Bethe-Salpeter amplitude for a bound state, after factorizing the "trivial" global motion of the two-particle system (i.e. a 4D delta function). In configuration space, the full BS amplitude reads as follows

$$\Phi(X, x; p_B) = \langle 0|T\{\phi_1(x_1)\phi_2(x_2)\}|p_B\, \beta\rangle \tag{5}$$

where $X = (x_1 + x_2)/2$ and $x = x_1 - x_2$.

In conclusion, for $p_0 \to \omega_B$, i.e. in the neighborhood of the bound-state pole, the 4-point Green's function can be approximated by $G \simeq G_B +$ *regular terms*. Inserting in the integral equation (2), such an approximate expression (that becomes exact in the limit $p_0 \to \omega_B$), one can deduces the integral equation that determines the BS amplitude for a bound state, i.e. the homogeneous BSE. The equation reads

$$\phi(k; p_B, \beta) = G_0(k; p_B, \beta) \int d^4q' \; \mathrm{I}(k, q'; p_B) \; \phi(q'; p_B, \beta) \tag{6}$$

with nor self-energy neither vertex corrections, at the present stage. It is worth noting that $\mathrm{I}(k, q'; p_B)$, the irreducible kernel in the BSE, is exactly the same appearing in the integral equation for the 4-point Green's function, equation (2).

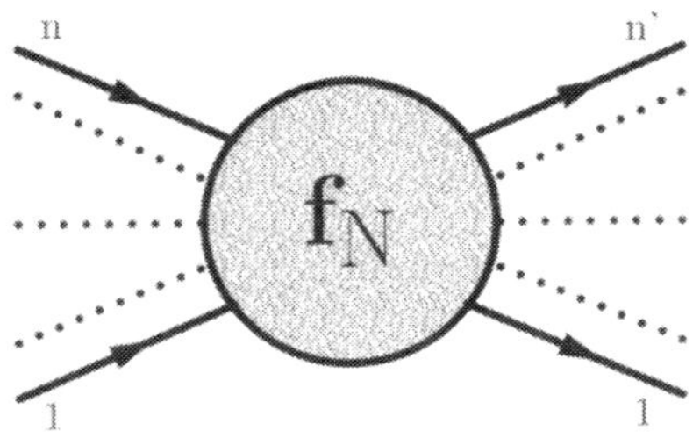

Figure 3. Pictorial representation of the N-leg transition amplitude, with $n + n' = N$.

3. The Nakanishi perturbation-theory integral representation

The Nakanishi PTIR of the N-leg transition amplitude (depicted in figure 3) is the first ingredient for constructing a very effective tool for solving the homogeneous BSE (for the treatment of the inhomogeneous BSE, see Ref. [5]). In the sixties, Nakanishi [12, 13] proposed the PTIR, after a suitable elaboration of the well-known parametric formula for any Feynman diagram. In particular, for N external legs, a generic contribution to the transition amplitude can be written as follows

$$f_{\mathcal{G}}(p_1, p_2, ..., p_N) \propto \prod_{r=1}^{k} \int d^4q_r \frac{1}{(\ell_1^2 - m_1^2)(\ell_2^2 - m_2^2) \ \ldots \ (\ell_n^2 - m_n^2)} \tag{7}$$

where n propagators and k loops are present. The label $\mathcal{G}$ in (7) is a shorthand notation for indicating a graph with a given pair (n, k). Following the standard elaboration (see, e.g., [4]), one can write

$$f_{\mathcal{G}}(s) \propto \prod_{i=1}^{n} \int_0^1 d\alpha_i \frac{\delta(1 - \sum_{j=1}^{n} \alpha_j)}{U^2(\alpha) \ [F(n, N, \alpha, s) + i\epsilon]^{n-2k}} \tag{8}$$

where $U^2(\alpha)$ is a well defined function, not relevant for the following discussion, and

$$F(n, N, \alpha, s) = -\sum_{j=1}^{n} \alpha_j \ m_j^2 + \sum_h \ \eta_h \ s_h$$

with the dependence upon the external momenta, $p_1, p_2 \ldots p_N$, traded off in favor of all the independent scalar products $s \equiv \{s_1, s_2, \ldots s_h, \ldots\}$ that one can construct.

In order to get rid of the dependence upon (n, k) in the denominator of equation (8), Nakanishi proposed a simple, but effective, change of variable, obtaining a compact and elegant expression of the full N-leg amplitude $f_N(s) = \sum_{\mathcal{G}} \ f_{\mathcal{G}}(s)$. It is relevant to emphasize that the sum on $\mathcal{G}$ is infinite. In more detail, by introducing the following identity

$$1 \doteq \prod_h \int_0^1 dz_h \delta\left(z_h - \frac{\eta_h}{\beta}\right) \int_0^\infty d\gamma \ \delta\left(\gamma - \sum_l \frac{\alpha_l m_l^2}{\beta}\right) \tag{9}$$

with $\beta = \sum \eta_i$ and after integrating by parts $n - 2k - 1$ times one gets

$$f_{\mathcal{G}}(s) \propto \prod_h \int_0^1 dz_h \int_0^\infty d\gamma \frac{\delta(1 - \sum_h z_h) \ \tilde{\phi}_{\mathcal{G}}(z, \gamma)}{(\gamma - \sum_h z_h s_h)} \tag{10}$$

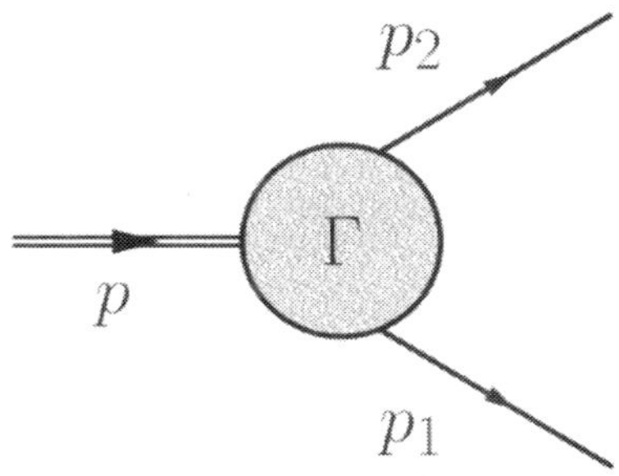

Figure 4. Pictorial representation of the 3-leg transition amplitude, namely the so-called vertex function.

where $\tilde{\phi}_{\mathcal{G}}(z,\gamma)$ is a proper weight function. The dependence upon the details of the diagram, i.e. (n,k), has moved from the denominator to the numerator. This simple achievement can be obtained for any diagram $\mathcal{G}$ that contributes to $f_N(s)$, with the final result that one meets the *same* denominator in all the expressions. Therefore the full N-leg transition amplitude can be formally written as

$$f_N(s) = \sum_{\mathcal{G}} f_{\mathcal{G}}(s) \propto \prod_h \int_0^1 dz_h \int_0^\infty d\gamma \frac{\delta(1-\sum_h z_h)\ \phi_N(z_h,\gamma)}{(\gamma - \sum_h z_h s_h)} \tag{11}$$

where the weight function $\phi_N(z_h,\gamma)$ is

$$\phi_N(z_h,\gamma) = \sum_{\mathcal{G}} \tilde{\phi}_{\mathcal{G}}(z_h,\gamma)$$

It is very important to emphasize that the Nakanishi weight function is a real function that depends upon real variables, $z_h \in [-1,1]$ and $\gamma \in [0,\infty]$ (γ is positive by definition, see equation (9)).

Within the BS framework, the Nakanishi integral representation (11) can be usefully exploited for obtaining an expression of the 3-leg transition amplitude, that in turn is related to the BS amplitude through the proper multiplication by the inverse of G_0, (see Refs. [9, 10] for an early application of the Nakanishi representation to the bound state and Ref. [5] for the extension to the scattering states). In particular, the benefit of applying the Nakanishi representation, that has been devised within the Feynman-diagram framework in other words within a perturbative scheme, is given by its explicit analytic structure. This greatly helps the formal manipulations of the BSE that one has to carry out for making feasible the calculations of the BS amplitude in Minkowski space.

In particular, the PTIR vertex function (see figure 4) is given by

$$f_3(s) = \int_0^1 dz \int_0^\infty d\gamma \frac{\phi_3(z,\gamma)}{\gamma - \frac{p^2}{4} - k^2 - zk\cdot p - i\epsilon} \quad . \tag{12}$$

In what follows, the following questions will be addressed: (i) how can the Nakanishi weight function, ϕ_3, be determined for an actual, dynamical model? (ii) Can the Nakanishi representation of the vertex function, elaborated in perturbation theory, be used in a non perturbative realm, as the BS framework does?

Journal of Physics: Conference Series **527** (2014) 012026 doi:10.1088/1742-6596/527/1/012026

4. The projection of the BSE onto the null plane and the Nakanishi PTIR

The second ingredient for obtaining a viable equation that allows one to calculate the BS amplitude in Minkowski space is represented by the so-called projection onto the null-plane (see Ref. [15] for a recent review).

Let us consider the PTIR vertex function as an Ansatz for the BS amplitude, by taking into account that a factor G_0 has to be included. This increases the power of the denominator in equation (12) by a factor of two, namely the BS amplitude is written [9, 10, 5, 11] as

$$\Phi_b(k,p) = i \int_{-1}^{1} dz' \int_0^{\infty} d\gamma' \; \frac{g_b(\gamma', z'; \kappa^2)}{[\gamma' + \kappa^2 - k^2 - p \cdot k z' - i\epsilon]^3} \tag{13}$$

where $g_b(\gamma', z'; \kappa^2)$ is the Nakanishi weight function and $\kappa^2 = m^2 - M^2/4$.

Since the analytic structure of the PTIR Ansatz for the BS amplitude is explicit, one can integrate equation (13) on the LF variable $k^- = k^0 + k_z$ (see, e.g.,[15]), obtaining the valence component of the state of the interacting system, once the state itself has been expanded onto the Fock basis (see [5, 11] for more details). The valence component reads in terms of the Nakanishi weight function g_b as follows

$$\begin{aligned} \psi_{n=2}(\xi, k_\perp) &= \frac{p^+}{\sqrt{2}} \, \xi \, (1-\xi) \int \frac{dk^-}{2\pi} \Phi_b(k,p) = \\ &= \frac{1}{\sqrt{2}} \xi \, (1-\xi) \int_0^{\infty} d\gamma' \; \frac{g_b(\gamma', 1-2\xi; \kappa^2)}{[\gamma' + k_\perp^2 + \kappa^2 + (2\xi - 1)^2 \, \frac{M^2}{4} - i\epsilon]^2} \end{aligned} \tag{14}$$

When we consider a 4D plane wave and integrate on k^-, it is easily seen that the conjugate variable $x^+ = t + z$ has to be vanishing, and therefore the null plane $x^+ = 0$ becomes the hypersurface where the initial dynamics takes place (x^+ plays the role of LF time, see Ref. [20] for an introduction to the Hamiltonian dynamics onto the null-plane). This observation explains why in the LF jargon the k^--integration is dubbed LF projection onto the null plane (see [15] for a review of relevant formal results obtained by exploiting the k^--integration within the so-called non explicitly covariant LF approach).

The LF projection onto the null plane can be applied to the homogeneous BSE, equation (6), obtaining [5, 11] (see Ref. [10] for the explicitly covariant LF treatment)

$$\begin{aligned} &\int_0^{\infty} d\gamma' \frac{g_b(\gamma', z; \kappa^2)}{[\gamma' + \gamma + z^2 m^2 + (1 - z^2)\kappa^2 - i\epsilon]^2} = \\ &= \int_0^{\infty} d\gamma' \int_{-1}^{1} dz' \; V_b^{LF}(\gamma, z; \gamma', z') g_b(\gamma', z'; \kappa^2). \end{aligned} \tag{15}$$

with $V_b^{LF}(\gamma, z; \gamma', z')$ determined by the irreducible kernel $\mathrm{I}(k, k', p)$ that appears both in equations (2) and (6).

It should be pointed out that equation (15) is an integral equation, and therefore it allows one to take into account non-perturbative effects in the BS amplitude, once the Ansatz (13) is assumed. On the other hand, within a perturbative framework, at any order, the Ansatz is exact. Therefore the problem is: does the equation in (15), formally obtained from the homogeneous BSE, have solutions (that, indeed, are non perturbative solutions)? In Refs. [9, 10, 11], a positive answer has been given through a careful numerical investigation of the so-called ladder approximation of the irreducible kernel $\mathrm{I}(k, k', p)$, represented by the first diagram in figure 2. For the calculations within a a cross-ladder approximation (i.e. the first and second diagrams in figure 2) and an explicitly covariant LF approach see Ref. [19].

Nakanishi enriched his theoretical investigation of the N-leg transition amplitudes by demonstrating a theorem on the uniqueness of the weight function $\phi_N(z_h, \gamma)$ [13]. If such a theorem holds also in the non perturbative context of the BSE, then a simpler integral equation for the weight function can be deduced. In particular, within the LF framework one gets [5, 11]

$$g_b(\gamma, z; \kappa^2) = \int_0^\infty d\gamma' \int_{-1}^1 dz' \; \mathcal{V}_b(\gamma, z; \gamma', z'; \kappa^2) g_b(\gamma', z'; \kappa^2) \tag{16}$$

where $\mathcal{V}_b(\gamma, z; \gamma', z'; \kappa^2)$ is a new kernel, properly related to $V_b^{LF}(\gamma, z; \gamma', z')$, viz

$$V_b^{LF}(\gamma, z; \gamma'', z') = \int_0^\infty d\gamma' \frac{\mathcal{V}_b(\gamma', z; \gamma'', z'; \kappa^2)}{[\gamma' + \gamma + z^2 m^2 + (1 - z^2)\kappa^2 - i\epsilon]^2} \tag{17}$$

An equation analogous to (16) was adopted in Ref. [9], but using standard canonical variables.

5. Quantitative studies

We have carried out a comprehensive investigation [11], in ladder approximation, of the simple scalar model, $\mathcal{L} = g\phi^2\chi$, (i) varying both binding energies $0 < B/m \le 2$ (with $B = 2m - M$) and the mass of the exchanged scalar, μ, (ii) using the two eigen-equations, equations (15) and (16), both obtained within the non-explicitly covariant LF approach. In particular, numerical results for eigenvalues and weight functions g_b, in ladder approximation, have been also obtained in Refs. [9, 10], but using different numerical techniques, and different frameworks, in Ref. [9] the standard variables and the uniqueness theorem and in Ref. [10] the explicitly covariant LF and equation (15). In our work [11] (see also Ref. [14]), in order to solve both equations (15) and (16), we have expanded the weight function g_b onto a basis given by the product of Gegenbauer (for the compact variable z) and Laguerre polynomials. It was very gratifying to obtain substantially the same results from both equations. Before illustrating the numerical results, it is necessary to emphasize what the results themselves indicate. First of all, the different numerical techniques adopted in Refs. [9, 10, 11] lead to the same eigenvalues, within the numerical accuracy adopted. Furthermore, as shown in Ref. [11], the uniqueness theorem by Nakanishi holds, as numerically checked at the level of a few percent.

In table 1, there is an example of the accurate agreement one can achieve (for the final results see Ref. [11]). As it can be immediately realized, the ladder approximation of the irreducible kernel depends linearly upon $\alpha = g^2/(16\pi m^2)$ and it is customary, to assign a value to the binding energy of the system and solve an eigenvalue problem, where the eigenvalues are the numerical values of the coupling α and the eigenvectors are the corresponding weight functions g_b. It should be pointed out that equation (15) is a generalized eigenequation, while equation (16) is a standard one. In figure 5 (see Ref. [11]), the comparison between the eigenvectors corresponding to the solutions of equations (15) and (16), for a given value of the binding energy and the mass of the exchanged scalar, is shown.

5.1. Valence Probabilities and LF Distributions

Once the Nakanishi weight function is evaluated, one can straightforwardly obtain the BS amplitude and normalize it. Then, the probability of the valence wave function, $\psi_{n=2}(\xi, k_\perp)$, results properly determined [11]. Moreover, one can calculate the LF distributions of the valence state, that represent relevant observables, e.g. in Hadron Physics.

The probability of the valence component is given by [11]

$$P_{val} = \frac{1}{(2\pi)^3} \int_0^1 \frac{d\xi}{2\,\xi(1-\xi)} \int d\mathbf{k}_\perp \; \psi^2_{n=2/p}(\xi, k_\perp) =$$
$$= \frac{1}{(16\pi)^2} \int_{-1}^1 dz\; (1 - z^2) \int_0^\infty d\gamma \left[\int_0^\infty d\gamma' \; \frac{g_b(\gamma', z; \kappa^2)}{[\gamma' + \gamma + z^2 m^2 + (1 - z^2)\kappa^2]^2}\right]^2 \quad . \tag{18}$$

Table 1. Preliminary values of $\alpha = g^2/(16\pi m^2)$ (see Ref. [11] for the final ones) by solving the generalized eigenequation (15) (indicated by LF-V (FSV) in the third column) and the standard eigenequation (16) (indicated by LF-U (FSV) in the fourth column). In the second column, for the sake of comparison, there are the results obtained in Ref. [10]. The chosen value for the mass of the exchanged scalar is $\mu/m = 0.50$.

B/m	α LF-V (CK)	α LF-V (FSV)	α LF-U (FSV)
0.01	1.440	1.44	1.44
0.10	2.498	2.50	2.50
0.20	3.251	3.25	3.25
0.50	4.901	4.90	4.90
1.00	6.712	6.71	6.71

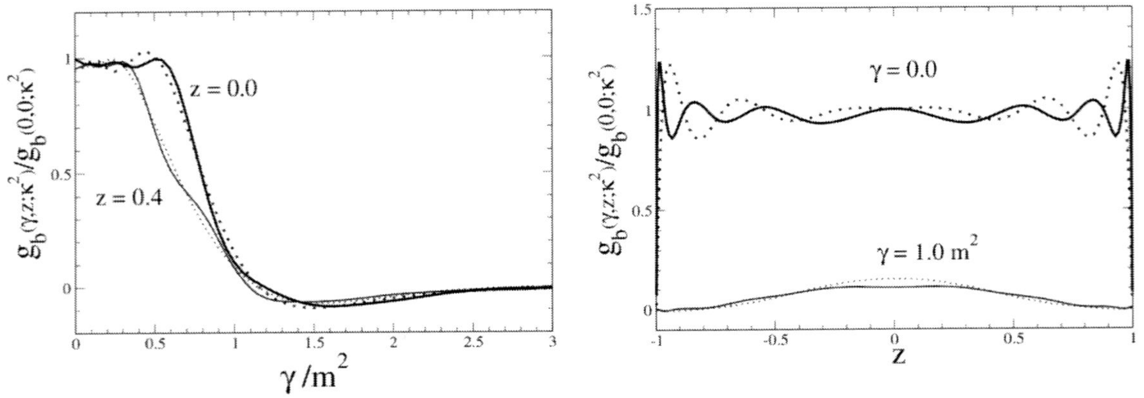

Figure 5. Left panel: the Nakanishi weight function $g_b(\gamma, z; \kappa^2)$, with $\kappa^2 = m^2 - M^2/4$, for $B/m = 1$ and $\mu/m = 0.5$ vs the adimensional variable γ/m^2 ($\gamma = k_\perp^2$) and two different values of the variable z. Right panel: the same as in the left panel, but vs the variable z and two values of γ (see Ref. [11]).

Table 2 shows preliminary results of the valence probability (see Ref. [11] for the final results and more details) for the case with $\mu/m = 0.5$. It is worth noting that such a calculation is a fully dynamical evaluation and that, as expected when $B \to 0$ (i.e. the continuum is approached), the two-particle state becomes dominated by the valence component, i.e. $P_{val} \to 1$.

The longitudinal and transverse-momentum LF distributions are defined in terms of the valence wave function as follows [11]

$$\phi(\xi) = 2\,\frac{(1-z^2)}{(16\pi)^2}\int_0^\infty d\gamma\left[\int_0^\infty d\gamma'\,\frac{g_b(\gamma', z; \kappa^2)}{[\gamma' + \gamma + z^2 m^2 + (1-z^2)\kappa^2]^2}\right]^2$$

$$\mathcal{P}(\gamma) = \frac{1}{(16\pi)^2}\int_{-1}^{1} dz\,(1-z^2)\left[\int_0^\infty d\gamma'\,\frac{g_b(\gamma', z; \kappa^2)}{[\gamma' + \gamma + z^2 m^2 + (1-z^2)\kappa^2]^2}\right]^2 \qquad (19)$$

In figure 6 an example of the previous distributions is shown. It is very interesting that the calculations performed by using the Nakanishi weight function obtained from equation (15) and

Table 2. Valence probability for $\mu/m = 0.50$.

B/m	α	P_{val}
0.001	1.167	0.98
0.01	1.440	0.96
0.10	2.498	0.87
0.20	3.251	0.83
0.50	4.900	0.77
1.00	6.711	0.74
2.00	8.061	0.72

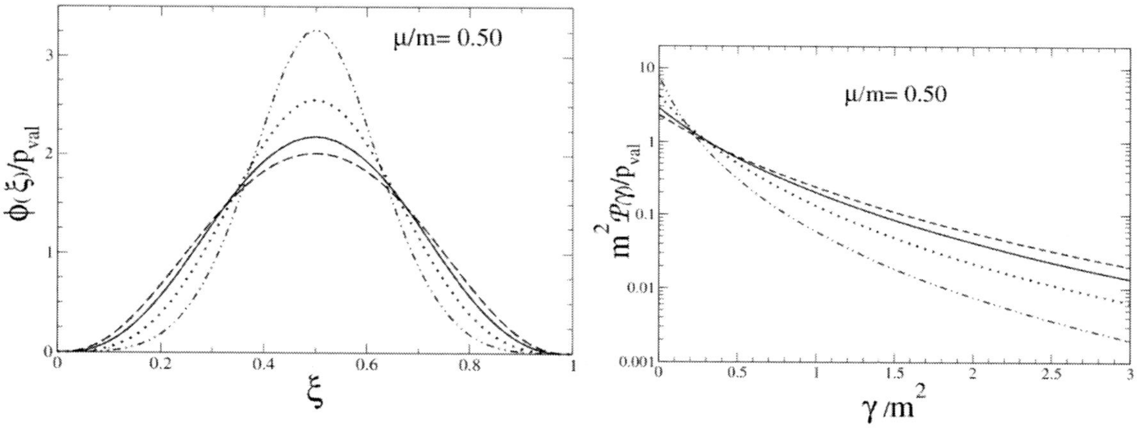

Figure 6. Left panel: the longitudinal LF-distribution, $\phi(\xi) = \int dk_\perp^2 \; |\psi_{n=2}(\xi, k_\perp)|^2$, vs the longitudinal-momentum fraction $\xi = k^+/M$. Dash-double-dotted line: $B/m = 0.20$. Dotted line: $B/m = 0.50$. Solid line: $B/m = 1.0$. Dashed line: $B/m = 2.0$. Note that $\int_0^1 d\xi \; \phi(\xi) = P_{val}$. Right panel: the same as the left panel, but for the transverse LF-distribution $\mathcal{P}(\gamma) = \int d\xi \; |\psi_{n=2}(\xi, k_\perp)|^2$ vs the adimensional variable γ/m^2 ($\gamma = k_\perp^2$). Note that $\int_0^\infty d\gamma \; \mathcal{P}(\gamma) = P_{val}$ (see Ref. [11]).

the one from equation (16) are indistinguishable in the figure. The right panel of figure (6) puts in evidence the similar behavior of the transverse distributions, corresponding to different values of the binding energy (notice that the same holds for different values of μ, as well). In the log plot, the tails can be easily made overlapping by a suitable shift. showing a universal behavior. This feature suggests that it is possible to extract information on the dynamics (clearly dictated by the adopted kernel) by analyzing the tail of the transverse distribution at high momentum.

It is worth noting that the strong agreement between the results from equations (15) and (16) holds also for the valence probability, that, after all, represents the normalization of the LF distributions.

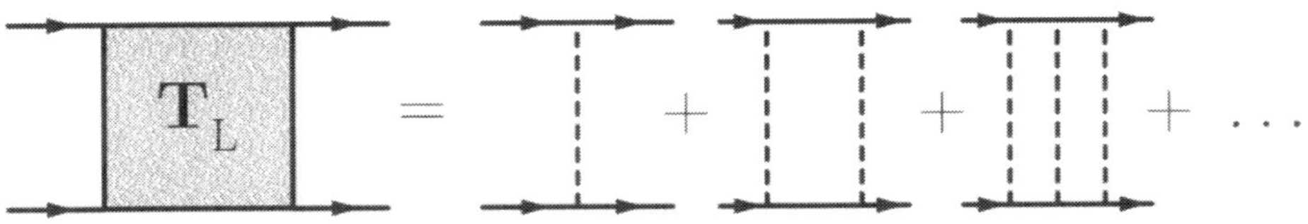

Figure 7. T-matrix in ladder approximation.

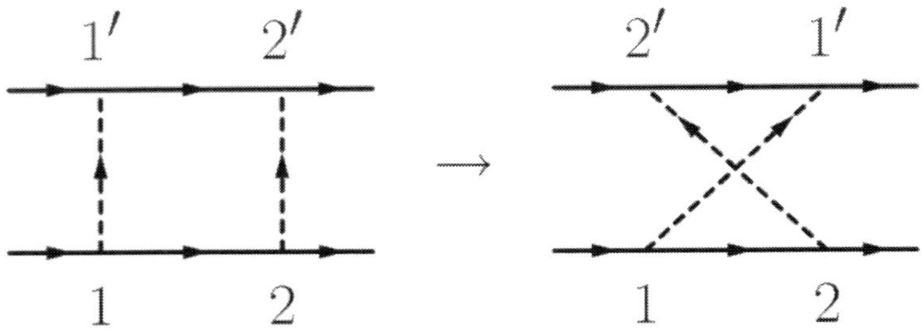

Figure 8. Symmetry transformation to be applied to the box-diagram for obtaining the lowest order crossed-box diagram.

6. Conclusions and perspectives

The cross-fertilization between the Light-Front framework and the Nakanishi perturbation-theory integral representation paves the path toward a new class of non perturbative calculations, carried out within a rigorous field-theoretical approach, namely the solutions of the Bethe-Salpeter equation in Minkowski space, both for bound and scattering states.

The LF framework has well-known advantages, with respect to the canonical approach, very helpful in performing the analytical integrations. This is a useful feature that can be exploited for making more easy to find the solutions of the BSE.

Our numerical investigations, performed in ladder approximation at the present stage (the cross-ladder contributions will be investigated in [21]), confirm both the robustness of the Nakanishi Ansatz for the BS amplitude and the uniqueness theorem for the weight function g_b. Moreover, we extended the numerical analysis of an actual dynamical model by studying the valence probability and the LF distributions, that represent appealing observables for a deep insight into the dynamics of an interacting system.

Calculations are in progress for both (i) scattering length and (ii) phase-shifts, still in ladder approximation [17]. This task will make complete the first stage of the investigation of this new tool. In the meantime, in order to face with the crossed-box contribution, a simple symmetry transformation has been explored, that in turn should allow one to resum an infinite set of cross-ladder diagrams. A sketch of the main steps of this new investigation are given in figures 8, 7 and 9. Staring from the T-matrix in ladder approximation, as shown in figure 7, one could apply the transformation illustrated in figure 8, and obtain the cross-ladder series, as depicted in figure 9, that contribute to the irreducible kernel I in equations (2) and (6). Embedding

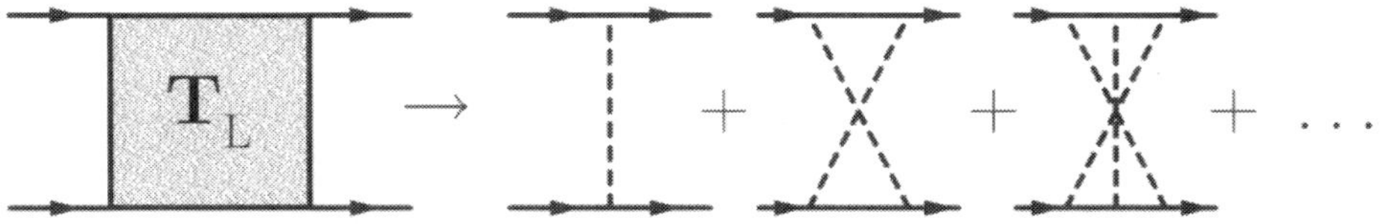

Figure 9. The infinite cross-ladder terms that contribute to the irreducible kernel I in equations (2) and (6)

the previous schematic pattern within the PTIR framework will be the aim of a forthcoming work [21].

References

[1] Salpeter E E and Bethe H A 1951 *Phys. Rev.* **84** 1232
[2] Gell-Mann M and Low F 1951 *Phys. Rev.* **84** 350
[3] Nakanishi N 1969 *Prog. Theor. Phys. Suppl.* **43** 1
[4] Itzykson C and Zuber J B 2006 *Quantum Field Theory* (Dover Publications)
[5] Frederico T, Salmè G, and Viviani M 2012 *Phys. Rev. D* **85** 036009
[6] Nakanishi N 1972 *Prog. Theor. Phys. Suppl.* **48** 2066
[7] Dorkin S M, Kaptari L P, Ciofi degli Atti C, and Kämpfer B 2011 *Few-Body Syst.* **49** 233
[8] Gross F 2013 *Few-Body Sist.* **54** 39
[9] Kusaka K, Simpson K, and Williams A G 1997 *Phys. Rev. D* **56** 5071
[10] Karmanov V A and Carbonell J 2006 *Eur. Phys. J. A* **27** 1
[11] Frederico T, Salmè G, and Viviani M 2014 *Phys. Rev. D* **89** 016010
[12] Nakanishi N 1963 *Phys. Rev.* **130** 1230
[13] Nakanishi N 1971 *Graph Theory and Feynman Integrals* (Gordon and Breach, New York)
[14] Frederico T, Salmè G, and Viviani M 2014 *Few-Body Syst.* in press and ArXiv:1402.1656
[15] Frederico T and Salmè G 2011 *Few-Body Syst.* **49** 163
[16] Carbonell J and Karmanov V A 2013 *Phys. Lett. B* **727** 319
[17] Frederico T, Salmè G, and Viviani M 2014 in preparation
[18] Baym G 1960 *Phys. Rev.* **117** 886
[19] Carbonell J and Karmanov V A 2006 *Eur. Phys. J. A* **27** 11
[20] Leutwyler H and Stern J 1978 *Ann. Phys.* **112**, 94
[21] Frederico T, Iannone A, Salmè G, and Viviani M in preparation

Journal of Physics: Conference Series **527** (2014) 012027 doi:10.1088/1742-6596/527/1/012027

Nuclear effects and neutron structure in deeply virtual Compton scattering off ^{3}He

Matteo Rinaldi

Dipartimento di Fisica e Geologia, Università degli studi di Perugia and INFN sezione di Perugia, Via A. Pascoli 06100 Perugia, Italy

E-mail: `matteo.rinaldi@pg.infn.it`

Abstract. The study of nuclear generalized parton distributions (GPDs) could be a crucial achievement of hadronic physics since they open new ways to obtain new information on the structure of bound nucleons, in particular, to access the neutron GPDs. Here, the results of calculations of ^{3}He GPDs in Impulse Approximation (IA) are presented. The calculation of the sum of GPDs $H + E$, and $\tilde{H}$, with the correct limits, will be shown. These quantities, at low momentum transfer, are largely dominated by the neutron contribution so that ^{3}He is an ideal target for these kind of studies. Nevertheless the extraction of neutron information from future ^{3}He data could be non trivial. A procedure, which takes into account nuclear effects encoded in IA, is presented. The calculation of H, E and $\tilde{H}$ allows also to evaluate the cross section asymmetries for deeply virtual compton scattering (DVCS) at Jefferson Lab kinematics. Thanks to these observations, DVCS off ^{3}He could be an ideal process to access the neutron information in the next future.

1. Introduction

In the last few years, generalized parton distributions (GPDs) draw the attention of the hadronic physics community since they allow to obtain new information on the parton structure of hadrons [1, 2, 3]. GPDs encode the non perturbative hadron structure in some peculiar hard exclusive processes. In particular, these quantities could give fundamental information on hadrons, such as the 3-dimensional structure of these systems at parton level [4]. In this work we focus on the future possibility of shedding some light on the so called "Spin crisis" problem. In principle this could be realized thanks to the relation between the GPDs and the total angular momentum of partons inside the hadrons: by subtracting from it the quark helicity contribution to the hadron spin, it would be possible to access, for the first time, their orbital angular momentum (OAM) [2, 3].

The golden process to study these quantities is the deeply virtual compton scattering (DVCS), which is described, at leading twist, mainly by H, E and $\tilde{H}$ GPDs. This reaction could be sketched in the following way: $eH \longmapsto e'H'\gamma$ when $Q^2 \gg M^2$ ($Q^2 = -q \cdot q$ is the momentum transfer between the leptons beam e and e', Δ^2 the one between hadrons H and H' with momenta P and P', and M is the nucleon mass). Another important kinematical variable is the so called skewedness, $\xi = -\Delta^+/(P^+ + P'^+)$ [1]. Even though DVCS is the cleanest process to access the

[1] In this paper, $a^\pm = (a^0 \pm a^3)/\sqrt{2}$.

Journal of Physics: Conference Series **527** (2014) 012027 doi:10.1088/1742-6596/527/1/012027

GPDs, the measurement of these quantities is still a theoretical and experimental challenge. Despite of these difficulties, data for proton and nuclear targets are being analyzed [5, 6].

The study of hard exclusive processes off nuclear targets, and the relative measurement of their GPDs, can give new information on possible medium modifications of the structure of bound nucleons (see Ref. [7]). In other words, these studies are very important to have new information on the origin of the so called EMC effect, an impossible task with the usual deep inelastic scattering (DIS) studies. To this aim few-body systems play a special role since, for these targets, realistic treatments are possible and exotic effects could be, in principle, distinguished from the conventional ones. The use of nuclear targets are also necessary, clearly, to study and to obtain information on the neutron GPDs and, in particular, this investigation is the main purpose of this work.

2. ^{3}He GPDs in impulse approximation

In order to have a complete flavor decomposition of the GPDs, the proton data are clearly not sufficient so that the study of the neutron GPDs has become a priority of the hadron physicists and, as mentioned in the last section, this could be realized by using nuclear targets. To this aim, among the light nuclei, ^{3}He is an ideal ones due its spin structure, in fact almost the 90% of the ^{3}He spin comes from the neutron one [8, 9]. In particular we expect that this nuclear system could be considered a unique target to extract the neutron E_q^n GPD. In fact this quantity is, at low momentum transfer, related to the anomalous magnetic moments which, for proton and neutron assumes the following values $k_p \sim 1.79\ \mu_N$ and $k_n \sim -1.91\ \mu_N$ (where μ_N is the nuclear magneton) respectively so they are similar in size but with opposite sign. This properties make isoscalar nuclei, such as ^{2}H and ^{4}He, not useful to extract the E_q^n GPD of the neutron since the proton contribution basically cancels the neutron one (see Ref. [7, 10] for relevant work on isoscalar light nuclei). In the ^{3}He case, the situation is completely different and to understand this feature it is sufficient to consider the dipole magnetic moment of ^{3}He and of the neutron: $\mu_3 \simeq -2.13\ \mu_N$ and $\mu_n \simeq -1.91\ \mu_N$, which are very similar and would be equal if an independent particle model were valid for ^{3}He. Thanks to these properties, ^{3}He simulates an effective polarized free neutron target so that it could be used to extract neutron information by properly taking into account nuclear effects as in the DIS case, see Ref. [9]. From this analysis we expect that the sum $\tilde{G}_M^{3,q}(x,\Delta^2,\xi) = H_q^3(x,\Delta^2,\xi) + E_q^3(x,\Delta^2,\xi)$, related to the contribution of the parton of flavor q to the dipole magnetic moment of ^{3}He, and $\tilde{H}_q^3(x,\Delta^2,\xi)$, associated to polarized targets, should be both largely dominated by the neutron contribution at low momentum transfer. Here it will be just presented the main passages of the formal calculation of the ^{3}He GPDs.

For a $\frac{1}{2}$ spin target the main quantity which describes the non perturbative hadron structure in the analyzed process is the so called light cone correlator which is parametrized, at leading twist, by the GPDs $H_q(x,\Delta^2,\xi)$ and $E_q(x,\Delta^2,\xi)$ in the unpolarized case:

$$F_{s's}^{q,A,\mu}(x,\Delta^2,\xi) = \int \frac{dz^-}{4\pi} e^{ix\bar{P}^+z^-} {}_A\langle P's'|\hat{O}_q^\mu|Ps\rangle_A|_{z^+=0,z_\perp=0} =$$

$$\frac{1}{2\bar{P}^+}\left[H_q^A(x,\Delta^2,\xi)\bar{u}(P',s')\gamma^\mu u(P,s) + E_q^A(x,\Delta^2,\xi)\bar{u}(P',s')\frac{i\sigma^{\mu\alpha}\Delta_\alpha}{2m}u(P,s)\right], \quad (1)$$

while for the polarized case one has:

$$\tilde{F}_{s's}^{q,A,\mu}(x,\Delta^2,\xi) = \int \frac{dz^-}{4\pi} e^{ix\bar{P}^+z^-} {}_A\langle P's'|\tilde{O}_q^\mu|Ps\rangle_A|_{z^+=0,z_\perp=0} =$$

$$\frac{1}{2\bar{P}^+}\left[\tilde{H}_q^A(x,\Delta^2,\xi)\bar{u}(P',s')\gamma^\mu\gamma^5 u(P,s) + \tilde{E}_q^A(x,\Delta^2,\xi)\bar{u}(P',s')\frac{i\sigma^{\mu\alpha}\gamma_5\Delta_\alpha}{2m}u(P,s)\right], \quad (2)$$

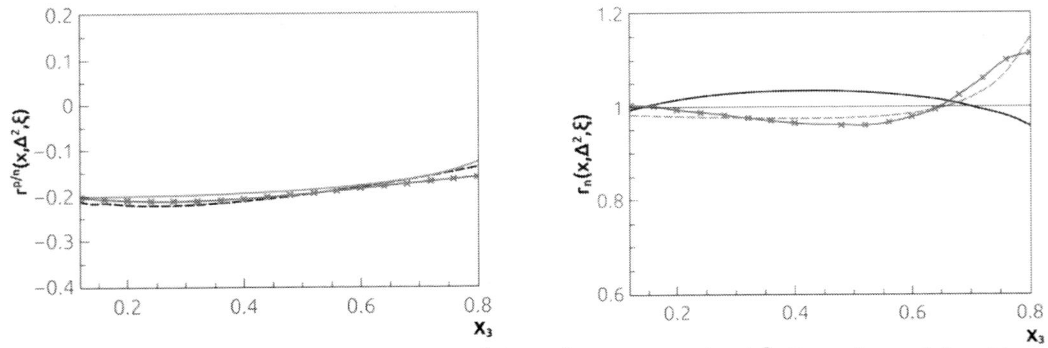

Figure 1. (Color online) The ratio $r^{p/n}(x, \Delta^2, \xi)$ and $r_n(x, \Delta^2, \xi)$ evaluated in this region: $-\Delta^2 = 0.1$ GeV^2 and $\xi = 0$, where here $x_3 = (M_3/M)x$ and using the models of Refs. [17, 19, 20].

where the initial (final) momentum and helicity are $P(P')$ and $s(s')$, respectively, and $\hat{O}^\mu_q = \bar{\psi}_q\left(-\frac{z}{2}\right)\gamma^\mu\psi_q\left(\frac{z}{2}\right)$, $\tilde{O}^\mu_q = \bar{\psi}_q\left(-\frac{z}{2}\right)\gamma^\mu\gamma_5\psi_q\left(\frac{z}{2}\right)$, $\bar{P} = (P+P')/2$, ψ_q is the quark field, m is the hadron mass.

Since we have to rely on a ^{3}He wave function, it is necessary to perform a non relativistic limit of the light cone correlator in order to have a complete consistency. Thanks to this procedure, simple relations between the GPDs and spin components of the correlator were found (see Ref. [11] for details):

$$\begin{aligned}
H^A_q(x, \Delta^2, \xi) &= \frac{\overline{P}^+}{m} F^{q,A,0}_{++}(x, \Delta^2, \xi) ,\\
\tilde{G}^{A,q}_M(x, \Delta^2, \xi) &= \frac{2\overline{P}^+}{\Delta_z} F^{q,A,1}_{+-}(x, \Delta^2, \xi) ,\\
\tilde{H}^A_q(x, \Delta^2, \xi) &= \frac{\overline{P}^+}{m}\, i\, \tilde{F}^{q,A,2}_{+-}(x, \Delta^2, \xi).
\end{aligned} \tag{3}$$

It is remarkable that one of the quantities we are analyzing is the sum $\tilde{G}^{3,q}_M(x, \Delta^2, \xi) = H^3_q(x, \Delta^2, \xi) + E^3_q(x, \Delta^2, \xi)$ which is also, in addition to the connection with the magnetic form factor (ff) of the hadron, a fundamental quantity thanks to its relation with the total angular momentum of partons inside the hadrons (Ji's sum rule [3]):

$$J^A_q = \int_{-1}^{1} dx\; x\; \tilde{G}^{A,q}_M(x, 0, 0) . \tag{4}$$

This result justifies our interest in the calculation of this particular combination of the GPDs. Starting from the non relativistic (NR) relations, one can describe the light cone correlator in terms of quantities which depend on ^{3}He wave function, by properly applying the impulse approximation (IA) (see Ref. [12] for details). Using then the same GPDs relations, equations (3), for the free nucleonic contributions, one finds:

$$\tilde{G}^{3,q}_M(x, \Delta^2, \xi) = \sum_N \int dE \int d\vec{p}\; [P^N_{+-,+-}(\vec{p}, \vec{p}\,', E) - P^N_{+-,-+}(\vec{p}, \vec{p}\,', E)] \frac{\xi'}{\xi} \tilde{G}^{N,q}_M(x', \Delta^2, \xi') \tag{5}$$

Journal of Physics: Conference Series **527** (2014) 012027 doi:10.1088/1742-6596/527/1/012027

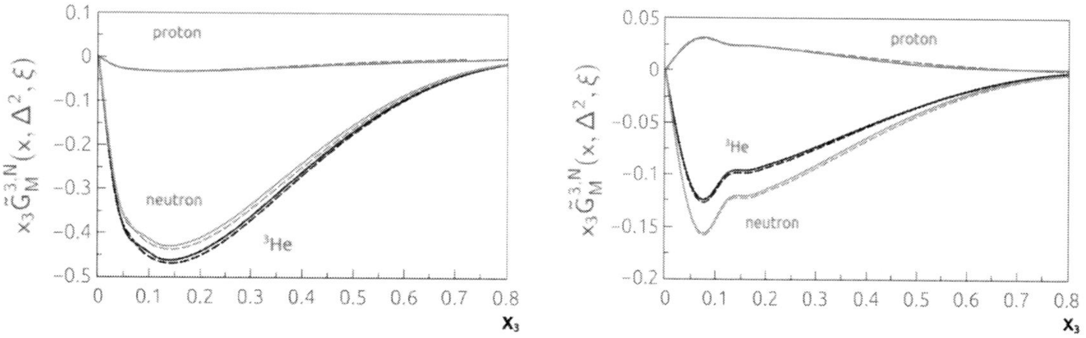

Figure 2. (Color online) The GPD $x_3\tilde{G}^3_M(x,\Delta^2,\xi)$, where $x_3 = (M_3/M)x$ and $\xi_3 = (M_3/M)\xi$, shown in the forward limit (left panel) and at $\Delta^2 = -0.1$ GeV^2 and $\xi_3 = 0.1$ (right panel), together with the neutron and the proton contribution. Solid lines represent the full IA result, equation (5), while the dashed ones correspond to the approximation equation (12).

and

$$\tilde{H}^3_q(x,\Delta^2,\xi) = \sum_N \int dE \int d\vec{p}\,[P^N_{++,++}(\vec{p},\vec{p}\,',E) - P^N_{++,--}(\vec{p},\vec{p}\,',E)]\frac{\xi'}{\xi}\tilde{H}^N_q(x',\Delta^2,\xi')\,, \tag{6}$$

where here x' and ξ', the variables for the bound nucleon GPDs, have been introduced, with $p\,(p' = p+\Delta)$ its 4-momentum in the initial (final) state and proper components appear of the spin dependent, one body off diagonal spectral function:

$$P^N_{SS',ss'}(\vec{p},\vec{p}\,',E) = \frac{1}{(2\pi)^6}\frac{M\sqrt{ME}}{2}\int d\Omega_t \sum_{s_t}\langle\vec{P}'S'|\vec{p}\,'s',\vec{t}s_t\rangle_N\langle\vec{p}s,\vec{t}s_t|\vec{P}S\rangle_N\,. \tag{7}$$

Here $E = E_{min}+E^*_R$, where E^*_R is the excitation energy of the full interacting two-body recoiling system. The main ingredient appearing in the definition equation (7) is the intrinsic overlap integral

$$\langle\vec{p}\,s,\vec{t}\,s_t|\vec{P}S\rangle_N = \int d\vec{y}\,e^{i\vec{p}\cdot\vec{y}}\langle\chi^s_N,\Psi^{s_t}_t(\vec{x})|\Psi^S_3(\vec{x},\vec{y})\rangle \tag{8}$$

between the ^{3}He wave function, Ψ^S_3, and the final state, described by two wave functions: (i) the eigenfunction $\Psi^{s_t}_t$, with eigenvalue $E = E_{min}+E^*_R$, of the state s_t of the intrinsic Hamiltonian describing the system of two *interacting* nucleons with relative momentum $\vec{t}$, which can be either a bound or a scattering state, and (ii) the plane wave representing the free nucleon N in IA. For a numerical evaluation of equations (1) and (2), the overlaps, equation (4), appearing in equation (3) and corresponding to the analysis presented in Ref. [13] in terms of AV18 [14] wave functions [15], have been used. Clearly the accuracy of this calculation, since the NR relativistic spectral function has been used, will be of the order $\mathcal{O}\left(\vec{p}^{\,2}/M^2, \vec{\Delta}^2/M^2\right)$.

Since no data for ^{3}He GPDs are actually available, the only possible checks for our calculations will be the GPDs properties, in particular the forward limit and the first moment. In Ref. [12] it has been illustrated that $H^3_q(x,\ \Delta^2,\xi)$ fulfills these constrains while, in Ref. [11], the quantity $\sum_q \int dx\,\tilde{G}^{3,q}_M(x,\Delta^2,\xi) = G^3_M(\Delta^2)$, which should give the magnetic ff of ^{3}He, has been calculated,

Journal of Physics: Conference Series **527** (2014) 012027 doi:10.1088/1742-6596/527/1/012027

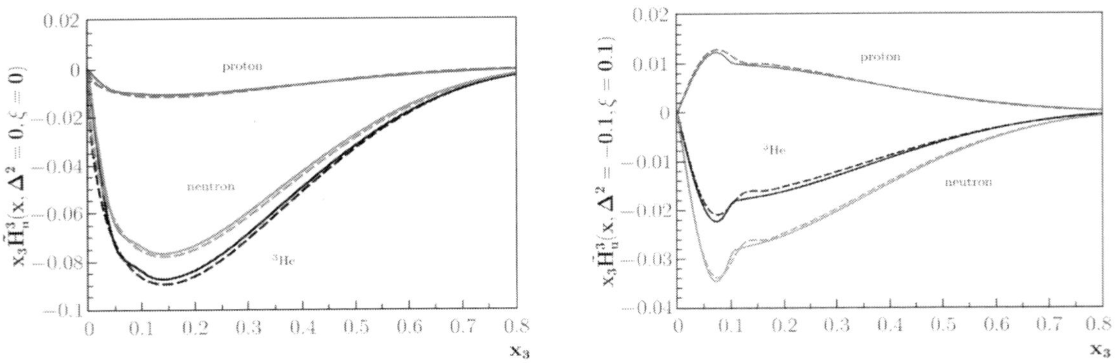

Figure 3. (Color online) The GPD $x_3\tilde{H}_q^3(x, \Delta^2, \xi)$ for the flavor $q = u$ shown in the forward limit (left panel) and at $\Delta^2 = -0.1$ GeV^2 and $\xi_3 = 0.1$ (right panel), together with the neutron and the proton contribution. Solid lines represent the full IA result, equation (6), while the dashed ones correspond to the approximation equation (11).

since no forward limit is defined for the $E_q(x, \Delta^2, \xi)$ GPD. Our result is in agreement with the one-body part of the AV18 calculation, presented in Ref. [16], and also for $-\Delta^2 \leq 0.15$ GeV^2, which is the relevant kinematics condition for the coherent DVCS at JLab, our results compare well with data. Concerning $\tilde{H}_q^3(x, \Delta^2, \xi)$, its first moment is the axial ff of ^{3}He but, since this quantity is poorly known, the only possible check for this GPD is the forward limit, which is correctly recovered.

In order to present the main results for the calculations of ^{3}He GPDs and estimating the proton and neutron contribution, describing the IA, it is necessary to explain some details on the used model for the free nucleonic GPDs, necessary to perform numerical evaluations.

3. Nucleonic models of GPDs

In this work, as pointed out in the previous section, we need to perform a numerical evaluation of ^{3}He GPDs; in particular, as suggested by the behavior of the ^{3}He wave function, these quantities should be dominated by the neutron contribution. This feature is provided by nuclear properties and should not be inferred from the nucleonic structure so that, for the calculation of $\tilde{G}_M^3(x, \Delta^2, \xi)$, we have used three completely different models:

- for our first calculation we have used a very simple model, Ref. [17], which fulfills all the properties of GPDs. In this particular case the Δ^2 dependence is factorized out from the x and ξ one by introducing the contribution of the quark of flavor q to the nucleonic ff: $F_q(\Delta^2)$. To this aim experimental values of $F_1^{n,p}(\Delta^2)$ Dirac ff have been used [18] and a parametrization has been chosen in order to have a flavor decomposition of these quantities. This model has been properly extended to describe the E_q GPD (see Refs. [11] for details).
- We have also used a description of the nucleonic structure based on a constituent quark model (along the lines of Ref. [19]), valid therefore in the valance quarks region. This model is very different from the previous one, and it does not assume any kind of factorization.
- For further analysis, a simple version of the MIT model, Ref. [20], has been used. This is a very different scenario since, here, free relativistic confined quarks are described.

Journal of Physics: Conference Series **527** (2014) 012027 doi:10.1088/1742-6596/527/1/012027

It is important to remark that our purpose is to estimate and to extract neutron information from ^{3}He GPDs so our main interest is on the nucleonic contribution to these quantities, a feature rather independent on the nucleonic model chosen, as it will be shown in the next section. Therefore, for the moment being, for the $\tilde{H}_q$ calculation, we have just extended the first model described above, where a factorization dependence on Δ^2 is assumed, by properly introducing a simple ansatz for the axial nucleonic ff. $\tilde{F}_q(\Delta^2) = \dfrac{1}{\left(1 - \frac{\Delta^2}{M_A^2}\right)^2}$ where $M_A \simeq 5.076 \ \ fm^{-1}$ (see Ref. [21] for details).

Clearly, if experiments were planned, more realistic models for nucleonic GPDs will be adopted and included into the scheme.

4. Nucleonic contributions to ^{3}He GPDs

Thanks to the nuclear and the nucleonic model described in the previous sections, and the comfort of all positive checks, the ^{3}He GPDs: $\tilde{G}^3_M(x, \Delta^2, \xi) = \sum_q \tilde{G}^{3,q}_M(x, \Delta^2, \xi)$ and $\tilde{H}^3_q(x, \Delta^2, \xi)$ have been numerically evaluated. Concerning these quantities, we found that the neutron contribution largely dominates the ^{3}He GPDs, at low momentum transfer, but, increasing Δ^2, the proton contribution grows up, in particular for u flavor. For the quantity $\tilde{G}^3_M(x, \Delta^2, \xi)$, this peculiar behavior is governed by the magnetic ff of ^{3}He, where the proton contribution is basically negligible for low Δ^2 but, increasing the momentum transfer, the size of the neutron contribution decreases (see Ref. [11]). For this reason, as already pointed out, since our interest is on the nucleonic contributions to ^{3}He GPDs, we have analyzed the ratio $r^{p/n}(x, \Delta^2, \xi) = \tilde{G}^{3,p}_M(x, \Delta^2, \xi)/\tilde{G}^{3,n}_M(x, \Delta^2, \xi)$ using the three different models described in the previous section (see figure 1 left panel). The result of this calculation shows and demonstrates that this quantity, and the relative nucleonic contributions to ^{3}He GPDs, clearly depend on the nuclear effects encoded in the ^{3}He wave function rather then the free nucleonic structure. In closing this section, let us remind that our calculations show how the neutron contribution largely dominates the ^{3}He GPDs, both in the case of $\tilde{G}^3_M$ and $\tilde{H}^3$ in the kinematics region of $-\Delta^2 \lesssim 0.1$ GeV^2, in particular for the d flavor case. Despite the limited validity of the region of IA and the neutron dominance, these studies could be a prerequisite for future experiments since the Ji's sum rule, which is the main information we are interested in, is only valid in the forward limit, so that it would be crucial to estimate GPDs at very low momentum transfer.

In spite of these promising results, a procedure to extract the neutron GPDs, from future ^{3}He data, at different values of momentum transfer, is necessary since the relations found for the nuclear GPDs, equations (5) and (6), are convolution-like equations involving a nuclear term, the off diagonal spectral function, and a nucleonic one, which is the free nucleonic GPD. In this scenario the extraction of the neutron could be not trivial.

5. Extraction procedure of neutron GPDs from ^{3}He

In this section the extraction procedure, will be presented, together with numerical results, which allows to obtain the neutron GPDs from future ^{3}He data. To this aim it is sufficient to write the ^{3}He GPDs in terms of the "light cone spin dependent off-forward momentum distributions", $h(g)^3_N(z, \Delta^2, \xi)$ which are peaked around $z = 1$ close to the forward limit.

$$\tilde{H}^3_q(x, \Delta^2, \xi) = \sum_N \int_{x_3}^{\frac{M_A}{M}} \frac{dz}{z} h^3_N(z, \Delta^2, \xi) \ \tilde{H}^N_q\left(\frac{x}{z}, \Delta^2, \frac{\xi}{z},\right) \ , \tag{9}$$

$$\tilde{G}^{3,q}_M(x, \Delta^2, \xi) = \sum_N \int_{x_3}^{\frac{M_A}{M}} \frac{dz}{z} g^3_N(z, \Delta^2, \xi) \ \tilde{G}^{N,q}_M\left(\frac{x}{z}, \Delta^2, \frac{\xi}{z},\right) \ . \tag{10}$$

Journal of Physics: Conference Series **527** (2014) 012027 doi:10.1088/1742-6596/527/1/012027

Therefore, in the region delimited by $x_3 \leq 0.7$, one can approximate the full calculation of ^{3}He GPDs with these new and simpler relations:

$$\begin{aligned}\tilde{H}_q^3(x,\Delta^2,\xi) &\simeq \sum_N \tilde{H}_q^N(x,\Delta^2,\xi)\int_0^{\frac{M_A}{M}} dz\, h_N^3(z,\Delta^2,\xi) \\ &= G_A^{3,p,point}(\Delta^2)\tilde{H}_q^p(x,\Delta^2,\xi) + G_A^{3,n,point}(\Delta^2)\tilde{H}_q^n(x,\Delta^2,\xi)\,, \end{aligned} \tag{11}$$

$$\begin{aligned}\tilde{G}_M^{3,q}(x,\Delta^2,\xi) &\simeq \sum_N \tilde{G}_M^{N,q}(x,\Delta^2,\xi)\int_0^{\frac{M_A}{M}} dz\, g_N^3(z,\Delta^2,\xi) \\ &= G_M^{3,p,point}(\Delta^2)\tilde{G}_M^{p,q}(x,\Delta^2,\xi) + G_M^{3,n,point}(\Delta^2)\tilde{G}_M^{n,q}(x,\Delta^2,\xi)\,, \end{aligned} \tag{12}$$

where here the axial (A) and magnetic (M) point like ffs have been defined: $G_{A(M)}^{3,N,point}(\Delta^2) = \int_0^{\frac{M_A}{M}} dz\, h(g)_N^3(z,\Delta^2,\xi)$.

These quantities would give the nucleon $N = n, p$ contribution to the nuclear axial (magnetic) ff if protons and neutrons were point-like particles.

In the magnetic case, in the kinematics region under scrutiny here, we found that these quantities are rather independent on nuclear potential (see Ref. [11] for details) while, concerning the axial case, we found that these ffs, in the forward limit, reproduce the so called "effective polarizations" of protons (p_p) and neutron (p_n), whose values, using the AV18 potential, are: $p_n = 0.878$ and $p_p = -0.023$. In this limit our results reproduce formally and numerically the formalism obtained in Ref. [9] for the polarized DIS off ^{3}He. Starting from equations (11) and (12) it is possible to extract the free neutron GPDs from the ^{3}He ones.

$$\tilde{H}^{n,extr}(x,\Delta^2,\xi) \simeq \frac{1}{G_A^{3,n,point}(\Delta^2)}\left\{\tilde{H}^3(x,\Delta^2,\xi) - G_A^{3,p,point}(\Delta^2)\tilde{H}^p(x,\Delta^2,\xi)\right\}\,, \tag{13}$$

$$\tilde{G}_M^{n,extr}(x,\Delta^2,\xi) \simeq \frac{1}{G_M^{3,n,point}(\Delta^2)}\left\{\tilde{G}_M^3(x,\Delta^2,\xi) - G_M^{3,p,point}(\Delta^2)\tilde{G}_M^p(x,\Delta^2,\xi)\right\}\,. \tag{14}$$

In order to test these new relations, we have compared, in figures 2 and 3, equations (11) and (12) with the full calculation of ^{3} He GPDs, equations (5) and (6) respectively. The results show that the new equations, in the kinematical region where the IA is valid, approximate very well the full calculation for $x \leq 0.7$, where the DVCS can be experimentally analyzed at the JLab. In particular it is remarkable that the only theoretical ingredients for these calculations are the axial (magnetic) ffs which are under good theoretical control. With the comfort of this successful results, in order to simulate the extraction procedure, since no data for the ^{3}He GPDs exist, we have compared the neutron GPDs extracted, from the approximated relations equations (13) and (14), with the free neutron GPDs used as input for the full calculation and using the same model for the proton GPDs. The results of this comparisons show that this procedure works perfectly for $x \leq 0.7$.

In order to have more accuracy of this analysis, as one can see in figure 1 right panel, we have studied the ratio $r_n(x,\Delta^2,\xi) = \tilde{G}_M^{n,extr}(x,\Delta^2,\xi)/\tilde{G}_M^n(x,\Delta^2,\xi)$ in a region which is beyond the forward limit and we have evaluated it using three completely different nucleonic models [17, 19, 20]. This result shows that our extraction procedure is weekly dependent even on the free nucleonic model .

6. Conclusion

In closing, we have shown the ^{3}He GPDs, which are dominated, at low momentum transfer, by neutron contribution and, due to the complicated relation between the nuclear and the nucleonic GPDs, we have proposed an extraction procedure of the neutron GPDs, which is able to take into account the nuclear effects encoded in the IA and which is rather independent on the free nucleonic model and nuclear potential. These successful results make the coherent DVCS off ^{3}He an ideal process to access the neutron GPDs. If higher momentum transfer region will be experimentally accessed, our work could be implemented by including a relativistic treatment, such as the on in Ref. [22] and/or many body currents, going beyond the IA.

Now it will be possible to study the cross section asymmetries for the DVCS processes the JLab kinematics since, for a $\frac{1}{2}$ spin target (see Ref. [23]), at leading twist, the three GPDs H, E and $\tilde{H}$ will mainly contribute. We have therefore at hand all the ingredients to perform this completely new analysis.

I thank Sergio Scopetta for a nice and fruitful collaboration on this subject in the last three years. This work was supported in part by the Research Infrastructure Integrating Activity Study of Strongly Interacting Matter (acronym HadronPhysic3, Grant Agreement n. 283286) under the Seventh Framework Programme of the European Community.

7. References

[1] Mueller D *et al.* 1994 *Fortsch. Phys.* **42** 101
[2] Radyushkin A V 1996 *Phys. Lett. B* **380** 417; Radyushkin A V 1997 *Phys. Rev. D* **56** 5524
[3] Ji X D 1997 *Phys. Rev. Lett.* **78** 610; Ji X D 1997 *Phys. Rev. D* **55** 7114
[4] Burkardt M 2000 *Phys. Rev. D* **62**, 071503 [*Erratum-ibid. D* **66**, 119903 (2002)]
[5] Mazouz M *et al.* [Jefferson Lab Hall A Collaboration] 2007 *Phys. Rev. Lett.* **99** 242501
[6] Guidal M 2010 *Phys. Lett. B* **693** 17
[7] Berger E R, Cano F, Diehl M, and Pire B 2001 *Phys. Rev. Lett.* **87** 142302
[8] Friar J L *et al.* 1990 *Phys. Rev. C* **42** 2310
[9] Ciofi degli Atti C *et al.* 1993 *Phys. Rev. C* **48** 968
[10] Liuti S and Taneja S K *Phys. Rev. C* **72** (2005) 032201; Taneja S K, Kathuria K, Liuti S, and Goldstein G R 2012 *Phys. Rev. D* **86** 036008
[11] Rinaldi M and Scopetta S 2012 *Phys. Rev. C* **85** 062201(R); Rinaldi M and Scopetta S 2013 *Phys. Rev. C* **87** 035208
[12] Scopetta S 2004 *Phys. Rev. C* **70** 015205; Scopetta S 2009 *Phys. Rev. C* **79** 025207
[13] Kievsky A, Pace E, Salmè G, and Viviani M 1997 *Phys. Rev. C* **56** 64
[14] Wiringa R B, Stoks V G J, and Schiavilla R 1995 *Phys. Rev. C* **51** 38
[15] Kievsky A, Viviani M, and Rosati S 1994 *Nucl. Phys. A* **577** 511
[16] Marcucci L E, Riska D O, and Schiavilla R 1998 *Phys. Rev. C* **58** 3069
[17] Musatov I V and Radyushkin A V 2000 *Phys. Rev. D* **61** 074027
[18] Gari M and Krumpelmann W 1986 *Phys. Lett. B* **173** 10
[19] Scopetta S and Vento V 2003 *Eur. Phys. J. A* **16** 527
[20] Ji X D, Melnitchouk W, and Song X 1997 *Phys. Rev. D* **56** 5511
[21] Ahrens L A, Aronson S H, Connolly P L, Gibbard B G, Murtagh M J, Murtagh S G, Terada S, and White D H *et al.* 1987 *Phys. Rev. D* **35** 785
[22] Pace E, Salmè G, Scopetta S, Del Dotto A, and Rinaldi M 2013 *Few-Body Syst.* **54** 1079
[23] Belitsky A V, Mueller D, Niedermeier L, and Schäfer A 2001 *Nucl. Phys. B* **593** 289

HASPECT: HAdron SPEctroscopy CenTer

E Santopinto, M Battaglieri, A Celentano, R De Vita, S Fegan
INFN Sezione di Genova, Via Dodecaneso 33, Genova 16146, Italy

V Mathieu, A Szczepaniak
Physics Department, Indiana University, Bloomington, IN 47405, USA

A Filippi
INFN,Sezione di Torino,Via Pietro Giuria, 1, 10125 Torino, Italy

A D'Angelo, A Rizzo
Università di Roma Tor Vergata and INFN, Sezione di Roma Tor Vergata, 00133 Rome, Italy

S Hughes, D Watts, L Zana
Edinburgh University, Edinburgh EH9 3JZ, United Kingdom

J Ferretti
Universidad Nacional Autónoma de México, 04510 México DF, México

S Lombardo
Cornell University, Ithaca, NY 14850, USA

D Glazier, D Ireland
University of Glasgow, Glasgow G12 8QQ, United Kingdom

E-mail: santopinto@ge.infn.it

Abstract. The main motivations for hadron spectroscopy in general and in particular for hybrid meson spectroscopy will be reviewed. The HASPECT (HAdron SPEctroscopy CenTer) project will be presented and discussed.

1. Introduction

Quantum Chromodynamics (QCD) describes the interactions of the fundamental degrees of freedom, quarks and gluons carrying the color charge, that give origin to the hadrons. Hadrons are bound states, and their properties are determined by interactions that are genuinely non-perturbative in nature. In particular, hadron masses originates mainly from gluonic self-interactions, in a way that only allows objects neutral with respect to the color charge to exist as physical, asymptotic states, and this still not completely clear phenomenon is known as confinement. As a consequence, the elementary colored degrees of freedom of the underlying theory only manifest themselves indirectly in the physical spectrum which is built from composite colorless hadrons.

On one side quark models have provided for years the main tools for studying the spectrum and the structure of the hadrons, but they lack a strict connection with QCD elementary degrees of freedom, since working with constituent degrees of freedom there are dynamical degrees of freedom not taken into account, on the other side recent developments in lattice simulations and effective-field-theory methods (considering in this category also QCD in physical gauge, chiral effective theories, etc. etc.) have opened new possible insight for investigations of hadron properties rooted deeply in non perturbative QCD. Mesons, being the simplest hadronic bound systems, are the ideal laboratory to study the non perturbative behavior of QCD, understand the role of the gluons inside hadrons in the generation of the hadron masses, and to investigate the origin of color confinement. To perform such studies, it is important to measure the meson spectrum and decays with precise determination of resonance masses and properties, looking for rare meson states and for unconventional mesons with exotic quantum numbers, i.e. that can not be obtained only starting from a quark and an antiquark. In fact, one of the most interesting part of the present and future investigation of the hadron spectrum is linked with the phenomenology of the low-energy gluons and thus regards the mapping of gluonic excitations, that may manifest themselves either as in hybrid states (states with both quarks and gluons as active or effective valence degrees of freedom) or in glue balls (states formed from gluons only). For that task the accuracy of the present and next-generation hadron spectroscopy experiments (Jlab, BESS, FAIR, CERN and BELLE) will allow for the identification also of those more elusive hadronic resonances, for which either a reliable determination of their parameters has still not been completed or their existence could not be unambiguously established before. In parallel and encoded as necessary condition for the achievement of this challenge is the development of new analysis methods incorporating into partial-waves analyses all the possible theoretical constraints. This will give the possibility of a reaction-independent determination of the pole position and residues that define a certain resonance. This is one of the goals of the HASPECT group, a strict collaboration between experimentalists and theoreticians, developing dedicated strategies and analysis methods for the Jlab12 MesonEx experiment, in strict contact with other analysis groups of other experiments.

Finally, we remind that even New Physics claims in the Standard Model and Beyond the Standard Model still need a high precision era of knowledge of the non perturbative QCD dynamics mapping program, that has still to be completed by those already cited and next coming laboratories. If there are strong interaction observables involved in an analysis claim for New Physics beyond the Standard Model, one has first to be sure to have under control all the pieces linked with non perturbative strong interaction if involved.

2. Hybrids

Hybrids, i.e states that contain both quark and gluon excitations, have been studied in various models [1, 2, 3, 4, 5, 6] but recent lattice simulations [7, 8, 9, 10, 11, 12] have generated a greater expectation, in particular also for light hybrid mesons[12]. Moreover, on the experimental side, in recent years several new states, in particular in the charmonium spectrum, have been discovered, probably including a hybrid resonance, the $Y(4260)$, discovered by Babar [13]. Conventional heavy quarkonia are well described by non-relativistic QCD, thus one can expect that hybrids containing heavy quarks could be treated in a similar way, i.e. by considering gluon excitations in presence of slow quarks. Moreover, in physical gauge, the dynamical gluons can be separated from the instantaneous Coulomb-type forces that act between color charges, thus while the non-abelian Coulomb potential is expected to be responsible for binding and confinement,

Journal of Physics: Conference Series **527** (2014) 012028 doi:10.1088/1742-6596/527/1/012028

the remaining, transverse gluon excitations could bring contribution to the spectrum. In the non-relativistic, physical gauge QCD the lowest mass charmonium hybrid multiplet has been predicted to be composed by those states $J^{PC} = 1^{--}$; $(0; 1; 2)^{-+}$ [15]. This four state hybrid multiplet identified in physical gauge calculations, has been recently identified also in lattice simulations [10, 11, 12], both in the heavy [10, 11] and light quark sectors [12], moreover it includes an exotic state (a state with exotic quantum numbers) with $J^{PC} = 1^{-+}$. In the non-relativistic, physical gauge QCD the lowest mass charmonium hybrid multiplet can be explained as due to a color-octet $c\bar{c}$ pair with ${J_q}^{P_qC_q} = 0^{-+}$ or 1^{--} corresponding to the total quark-antiquark spin with $S_q = 0$ and $S_q = 1$, respectively, coupled to a single physical, transverse gluon with predicted quantum numbers $J_g^{P_gC_g} = 1^{+-}$ [14, 15] ; the unusual positive parity of the physical gluon originates from the non-abelian nature of the Coulomb interactions as explained in Refs. [14, 15].

3. Photoproduction of hybrid and exotic mesons

Phenomenological models indicate that the photon may be more effective in producing exotic hybrids in diffractive production on a proton target than, for example, the pion. In the fact, the photon carries spin one and it can fluctuate into a qq pair with spins aligned. When a $c\bar{c}$ pair with $S = 1$ is excited into a hybrid, the production of exotic quantum numbers is expected to be favored. Phenomenological studies also indicate that exotic mesons can be produced with photon probes with cross sections comparable to ordinary mesons[17, 18]. In this respect, there will be dedicated experiments at Jlab: Gluex and MesonEx. They will use real and quasi-real photo-production on a proton target to investigate the meson spectrum in the energy range of few GeVs, looking for exotic and hybrid states, using Partial Wave Analysis to determine their mass and properties.

4. Amplitude analysis

Going back to the days of the S-matrix approaches, it was assumed that S-matrix analyticity together with requirements of crossing-symmetry and unitarity could provide constraints sufficient for computing the scattering amplitude: Regge theory determining the asymptotic behavior and dispersion techniques used to generate dynamical equations for the scattering amplitude. These general requirements from the S-matrix theory were found to be insufficient to unambiguously determine hadronic amplitudes, and $c\bar{c}$ pair later on, these ambiguities could be traced to the existence of QCD bound states. Even after 40 years of hadron physics we are still far from being able to construct hadronic amplitudes from first principles. The identification of new resonances and analysis of their properties can only be successful if there is a combined effort involving experimental analysis, theory and phenomenology. This is particularly pressing since in recent years new hadron resonances that seem to escape the established phenomenology have been found in both light and heavy quark sectors, and numerous experimental efforts in hadron spectroscopy have been started. The goal of a meson spectroscopy program is to identify resonances measuring their decay products. A resonance is formally described as a pole in the production amplitude with definite angular momentum and isospin. In practice, resonances are numerous, often broad and overlapping each other: only for a narrow and well-isolated state, the resonant structure can be identified by looking at the invariant mass spectrum of the decay products. The identification of a precise state requires the extraction of the corresponding waves from the measured distributions. This task is performed via the Partial Wave Analysis (PWA): the cross section of the process is parametrized as a coherent sum of different amplitudes with definite quantum numbers. The cross section is then fitted to the data to extract amplitudes. Fits can be performed as a function of the invariant mass of the measured decay products and other relevant kinematic variables to derive information on the dependence of the amplitudes on these variables. A resonance is uniquely characterized by its pole and residues, the position of the pole being universal, its residues depending on the decay channel in question. The problem in the precision determination of these parameters is that the experiments are limited to real, physical values of the energy. Only when a resonance is well isolated from the others and also far from thresholds, one can use simple expressions like Breit Wigner amplitudes that provide, in a limited region, a very good approximation to the result one would obtain from dispersion theory, but mathematically, in principle, are Dispersion Relations the tools to provide a rigorous way of analytically continuing amplitudes from the physical regime into the complex plane, giving us the possibility of unambiguously extracting the pole parameters of the resonance .

The cases where the distance of the resonance pole to the real axis is smaller than its distance to any other singularity, or where there is just one threshold cut nearby, have been already studied and their properties have been well established during the years. Nowadays we are trying to understand the complicated part of the spectrum, where this ideal situation does not occur and resonances are wide, with poles relatively deep in the complex plane. The effects of overlapping resonances and proximity to more than one threshold due to many possible decay channels require more elaborate techniques.

5. Conclusions

Meson Spectroscopy is a powerful tool to answer to fundamental questions in QCD, as the origin of color confinement and the role of gluons inside hadrons. Mesons are the simplest quark bound system and, therefore, the ideal laboratory to study the strong force at the non-perturbative energy scale of few GeVs. In particular, unconventional mesons would be the best experimental evidence of the active role of gluons in hadron dynamics. In this respect, from an experimental side there will be dedicated experiments at Jlab, FAIR, BESS, BELLE and CERN. In particular, at Jlab two complementary experiments will run: the Gluex and the MesonEx experiment. New high statistics and precise data need an adequate analysis and this will be task of the Haspect Center. Beyond providing a deeper understanding of the inner workings of non perturbative QCD, a theoretical control over hadronic final-state interactions is also essential for the hunt of physics beyond the Standard Model.

References

[1] Horn D and Mandula J 1978 *Phys. Rev. D* **17** 898
[2] Isgur N and Paton J E 1985 *Phys. Rev. D* **31** 2910
[3] Simonov Y A 2001 *Nucl. Phys. B* **592** 350.
[4] Szczepaniak A, Swanson E S, Ji C R, and Cotanch S R 1996 *Phys. Rev. Lett.* **76** 2011
[5] Buisseret F and Semay C 2006 *Phys. Rev. D* **74** 114018
[6] Brau F and Semay C 2004 *Phys. Rev. D* **74** 014017
[7] Morningstar C J and Peardon M J 1999 *Phys. Rev. D* **60** 034509
[8] Foster M and Michael C 1999 *Phys. Rev. D* **59** 094509
[9] Bali G S and Pineda A 2004 *Phys. Rev. D* **69** 0944001
[10] Dudek J J, Edwards R G, Guo P, and Thomas C E 2013 *Phys. Rev. D* **88** 094505
[11] Dudek J J, Edwards R G, Peardon M J, Richards D G, and Thomas C E 2010 *Phys. Rev. D* **82** 034508
[12] Dudek J J *et al.* 2011 *Phys. Rev. D* **83** 111502
[13] Aubert B *et al.* [The BABAR Collaboration] 2005 *Phys. Rev. Lett.* **95** 142001
[14] Guo P, Szczepaniak A P, Galata G, Vassallo A, and Santopinto E 2008 *Phys. Rev. D* **77** 056005
[15] Guo P, Szczepaniak A P, Galata G, Vassallo A, and Santopinto E 2008 *Phys. Rev. D* **78** 056003
[16] Eichten E and Feinberg F 1981 *Phys. Rev. D* **23** 2724
[17] Afanasev A *et al.* 1998 *Phys. Rev. D* **57** 6771
[18] Szczepaniak A *et al.* 2001 *Phys. Lett. B* **516** 72

Journal of Physics: Conference Series **527** (2014) 012029 doi:10.1088/1742-6596/527/1/012029

The SPES radioactive ion beam project of INFN

Giacomo de Angelis[1] for the SPES collaboration

G Prete[1], A Andrighetto[1], M Manzolaro[1], S Corradetti[1], D Scarpa[1], M Rossignoli[1], A Monetti[1], M Lollo[1], M Calderolla[1], J Vasquez[1], D Zafiropoulos[1], L Sarchiapone[1], D Benini[1], P Favaron[1], M Rigato[1], R Pegoraro[1], D Maniero[1], L Calabretta[2], M Comunian[1], M Maggiore[1], A Lombardi[1], L Piazza[1], A M Porcellato[1], C Roncolato[1], G Bisoffi[1], A Pisent[1], A Galatà[1], M Giacchini[1], G Bassato[1], S Canella[1], F Gramegna[1], J Valiente[1], J Bermudez[1], P F Mastinu[1], J Esposito[1], J Wyss[1], A Russo[2], S Zanella[1]

[1]INFN Laboratori Nazionali di Legnaro, I-35020 Legnaro, Italy
[2]INFN, Laboratori Nazionali del Sud, I-95123, Catania, Italy

E-mail: giacomo.deangelis@lnl.infn.it

Abstract. The SPES Radioactive Ion Beam facility at INFN-LNL is presently in the construction phase. The facility is based on the Isol (Isotope separation on-line) method with an UCx Direct Target able to sustain a power of 10 kW. The primary proton beam is provided by a high current Cyclotron accelerator with energy of 35-70 MeV and a beam current of 0.2-0.5 mA. Neutron-rich radioactive ions are produced by proton induced Uranium fission at an expected fission rate of the order of 10^{13} fissions per second. After ionization and selection the exotic isotopes are re-accelerated by the ALPI superconducting Linac at energies of 10A MeV for masses in the region $A = 130$ amu. The expected secondary beam rates are of the order of 10^7 - 10^9 pps. Aim of the SPES project is to provide a facility for high intensity radioactive ion beams for nuclear physics research as well as to develop an interdisciplinary research center based on the cyclotron proton beam.

1. Introduction

The aim of the SPES project is to provide high intensity and high-quality beams of neutron-rich nuclei to perform forefront research in nuclear structure, reaction dynamics and interdisciplinary fields like medical, biological and material sciences. SPES is a second generation Isol radioactive ion beam facility. It represents an intermediate step toward the future generation European Isol facility EURISOL [1]. The SPES project is part of the INFN Road Map for the Nuclear Physics; it is supported by the Italian national laboratories LNL (Legnaro) and LNS (Catania). It is based on the Isol method with an UCx Direct Target able to sustain a power of 10 kW. The primary proton beam is delivered by a Cyclotron accelerator with an energy of more then 40 MeV and a beam current of 200 μA. Neutron-rich radioactive ions will be produced by Uranium fission at an expected fission rate in the target of the order of 10^{13} fissions per second. The exotic isotopes will be re-accelerated by the ALPI superconducting Linac at energies of 10A MeV and higher, for masses in the region of $A = 130$ amu, with an expected rate on the secondary target of 10^7-10^9 pps.

2. Challenges of the SPES project

To understand the properties of a nucleus, apart from establishing the interaction between its components, it is necessary to determine the arrangement of the nucleons, i.e. the structure. Presently our knowledge about the structure of nuclei is mostly limited to nuclei close to the valley of stability or nuclei with a deficiency of neutrons. Only recently the availability of beams of unstable ions has given access to unexplored regions of the nuclear chart, especially on the neutron-rich side. Starting from a nucleus on the stability line and adding successively neutrons, one observes that the binding energy of the last neutron decreases steadily until it vanishes and the nucleus decays by neutron emission. The position in the nuclear chart where this happens defines the neutron drip line. It lies much farther away from the valley of stability than the corresponding drip line associated with protons, owing the absence of electrical repulsion between neutrons. The location of the neutron drip line is largely unknown as experimental data are available only for nuclei with mass up to around 30. The interest in the study of nuclei with large neutron excess is not only focused on the location of the drip line but also on the investigation of the density dependence of the effective interaction between the nucleons for exotic N/Z ratios. In fact, changes of the nuclear density and size in nuclei with increasing N/Z ratios are expected to lead to different nuclear symmetries and new excitation modes. While in the case of some very light nuclei a halo structure has been identified, for heavier nuclei the formation of a neutron skin has been predicted. The evolution of nuclear properties towards the neutron drip line depends on how the shell structure changes as a function of neutron excess. This evolution has consequences on the ground state properties of the nuclei and on the single-particle and collective excitations. In particular, studies of neutron-rich nuclei beyond the doubly magic ^{132}Sn are of key importance to investigate the single-particle structure above the N=82 shell closure and find out how the effective interaction between valence nucleons behaves far from stability. New modes of collective motion are also expected in connection with the formation of a neutron skin, namely oscillation of the skin against the core, similar to the soft dipole mode already identified in the case of very light halo nuclei. Presently, neither the thickness nor the detailed properties of the neutron skin of exotic nuclei are known. This information is needed to enable a quantitative description of compact systems like neutron stars, where exotic nuclei forming a Coulomb lattice are immersed in a sea of free neutrons, a system which is expected to display the properties of both finite and infinite (nuclear matter) objects. At the energy of SPES, it will be possible to address important questions related to the study of neutron-rich matter such as nuclear forces, level density, viscosity, barrier, neutron pairing and collective modes. With the high intensity beams delivered by SPES,a challenging and broader range of studies in nuclear spectroscopy and reaction mechanism will be performed. Interesting areas where new data will be collected are those in the very neutron rich regions, where shell evolution is an issue. Effects of how the pairing interaction is modified in the nuclear medium will receive significant inputs by measurements of multi-nucleon transfer reactions to specific nuclear states. Effects of rotational damping in the decay of high energy levels, for instance the dynamical dipole emission, will be studied by changing the N/Z of projectile and target. Neutron rich nuclei will be used to investigate the tunneling process in sub-barrier fusion reactions in presence of very positive Q-values, an issue interesting also for astrophysics.

3. The SPES radioactive ion beam facility

SPES is an INFN project involving the two national laboratories, LNL and LNS and other INFN sites in Italy, based on the Isol technique for radioactive beam production. A primary beam provided by a driver accelerator induces nuclear reactions inside a thick target. The reaction products are extracted from the target by thermal process due to the high temperature of the target-ion source system (exceeding 2000°C). Once reached the 1^+ source, the products are ionized and extracted. After an isotopic selection and a further ionization to n^+ the exotic beam is injected into a re-accelerator. SPES choice will use a proton beam to induce fission on a UCx target (Direct Target). The driver is a 70 MeV proton cyclotron with two exit ports for a total current of 750 μA. Two production Isol targets are planned to be installed and operated alternatively. As the cyclotron can supply two beams at the same time, a second independent facility can be operated in parallel. The high intensity proton beam will be used to produce radioisotopes for nuclear medicine as well as a neutron source. The expected neutron production at SPES exceeds 10^{14} n/s with energy up to 70 MeV and a peak between 0.1 and 10 MeV according to target and moderator. The proton beam of the SPES cyclotron is suited for the production of new radioisotopes for medical applications. Taking advantage of the high energy and current, it opens up the possibility to produce ^{82}Sr, ^{67}Cu, and ^{47}Sc to be used in therapy and diagnostics. A pictorial sketch of the facility is shown in figure 1.

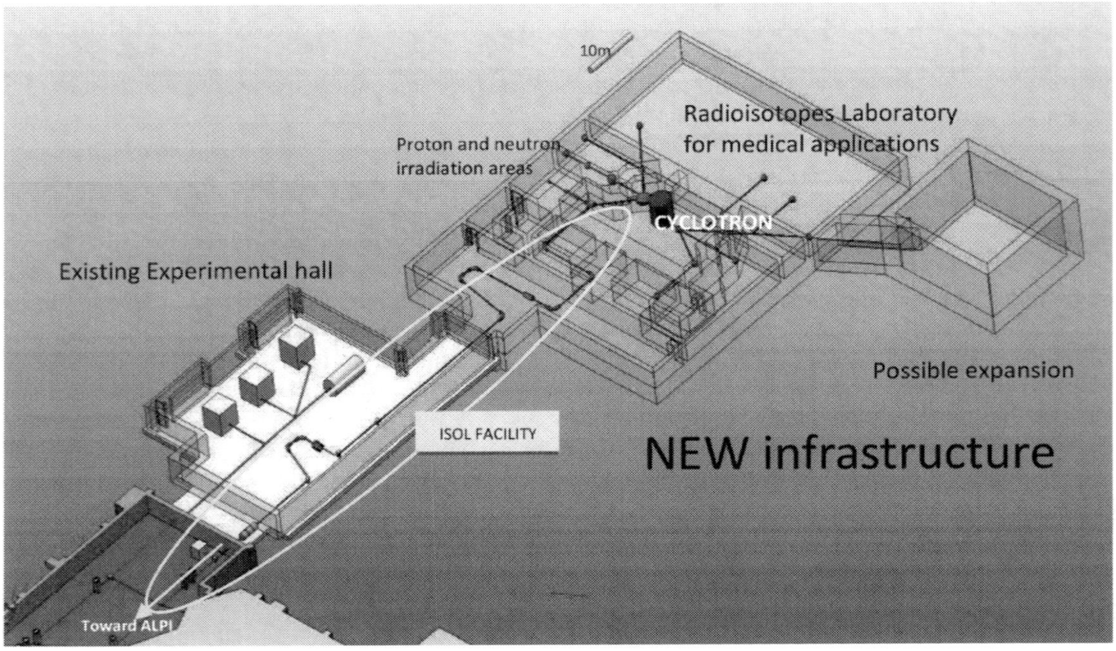

Figure 1. (Color online) Pictorial sketch of the SPES facility. The ALPI superconductive Linac (not shown) is on the left side.

3.1. The Isol front-end

The Isol front-end is the system which couples the proton beam with the UCx target, the ion source and the first part of the transport line of the exotic beam. The whole system was completed and is in operation at LNL for off-line tests.

The most critical element of the SPES Isol facility is the Direct Target. It follows the basic design of the CERN ISOLDE one but represents an innovation in term of capability to sustain the primary beam power. The design is carefully oriented to optimize the radiative cooling, taking advantage of the target system's high operating temperature, which is of the order of 2000^{0}C. The SPES target is designed in order to maximize the release efficiency and the heat dissipation. It consists of multiple thin disks housed in a cylindrical graphite box. It is composed of 7 UCx disks (diameter and thickness of 40 and 1.3 mm, respectively), appropriately spaced in the axial direction in order to dissipate, by thermal radiation, the average power of 8 kW due to the energy release of the proton beam which, passing through them, induces nuclear reactions. The graphite box is housed inside a tubular hollow tungsten ohmic heater. In fact, due to the intense heat exchange by radiation, the energy released by the proton beam is not sufficient to make the system reaching the operating temperature. This allows for a better thermal control of the target operation. An extensive simulation study of the target behaviour for thermal and release properties is at the basis of the target-ion-source design. Experimental work to benchmark the simulations was carried out in collaboration with HRIBF [2], the Oak Ridge National Laboratory Isol facility (USA). Partial results on the measured production rates (and on the expected rates of the reaccelerated SPES beam) are shown in figure 2.

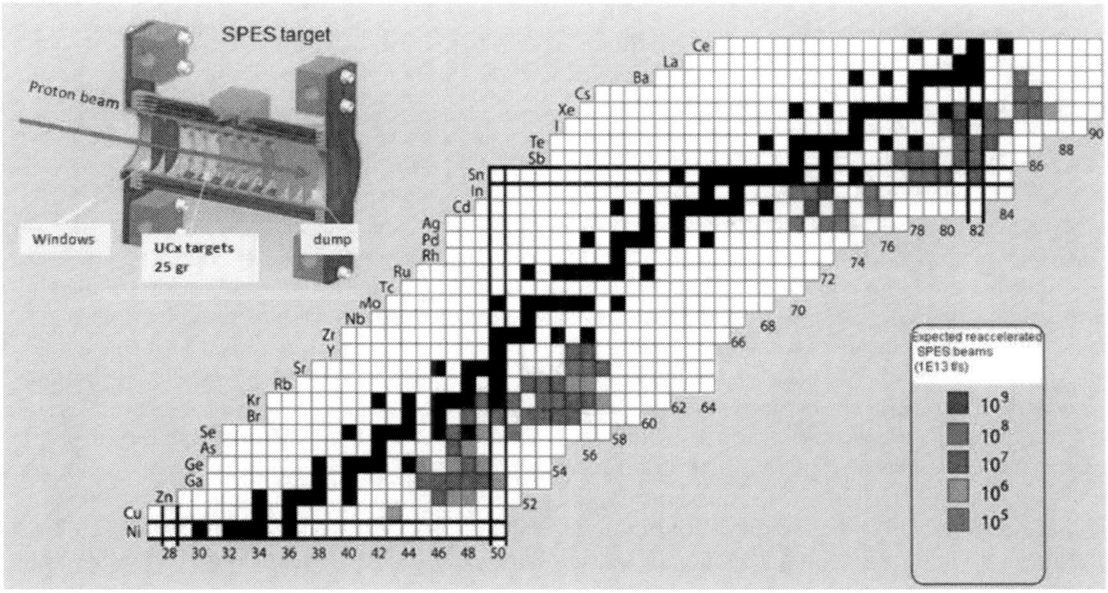

Figure 2. (Color online) Expected beam on target intensities. Experimental data scaled to: 0.2 mA proton current, 2-5% transport efficiency. Isol target operated at 2000^{+}C, ionization and extraction with 1^{+} plasma source.

3.2. The ion-source system

The interaction of the proton beam with the UCx target will produce neutron-rich fission fragments subsequently extracted by thermal motion and ionized at 1^{+} charge state by the Isol source directly connected with the production target. The hot-cavity ion source chosen for the SPES project was designed at CERN (ISOLDE) [3]. The source has the basic structure of the standard high temperature ion sources employed for on-line operation. The ionizer cavity is a thungsten tube (34 mm length, 3 mm inner diameter and 1 mm wall thickness) resistively heated to near 2000^{0}C. The isotopes produced in the target diffuse in the target material and then effuse through the transfer tube into the ionizer cavity where they undergo surface or laser ionization. The surface ionization process can occur when an atom comes into contact with a hot metal surface. In the positive surface ionization, the transfer of a valence electron from the atom to the metal surface is energetically favorable for elements with an ionization potential lower than the work function of the metal. For alkalis and some rare earth elements, high ionization efficiencies can be achieved using the surface ionization technique. Such 1^+ source has good efficiency and selectivity for the elements as Rb, Cs, Ba. For most of the other elements, the laser resonant photo-ionization, using the same hot cavity cell, is a powerful method to achieve sufficiently selected ionization. This technique, now under development, has been implemented with the aim to produce beams as pure as possible (chemical selectivity) also for metal isotopes. To ionize elements with high ionization potential, as rare gasses, the plasma source is needed. This source ionizes all the elements without any selectivity. At SPES both surface and plasma sources have been developed and are in operation at the off line test stand. The laser source is presently under development.

3.3. Beam selection and transport

The first mass selection is performed by a Wien Filter with 1/100 mass resolution, installed just after the first electrostatic quadrupole triplet inside the production bunker with the aim to confine the larger part of radioactivity inside the high shielded area. The transfer line toward ALPI is equipped with several beam handling systems to purify the beam. A Beam Cooler and a High Resolution Mass Separator (HRMS) with 1/20000 mass resolution will be installed inside the new building. Special efforts have been dedicated to design a mass spectrometer with an effective mass resolution of at least 1/20000. Such design takes advantage of a Beam Cooler, to improve the beam emittance, and of a 260 keV HV platform on which the mass separator is mounted. The mass separator is a scaled-up version of the separator designed for the CARIBU facility [4] in ANL (USA). In the SPES configuration the calculated resolution is of the order of 1/20000 constrained by an emittance of 3 mm mrad and an energy spread of 1.3 eV. Such high selectivity results in an advantage also for the safety issue, reducing the contaminations along the beam transport area, the re-accelerator and the final target location. Before the injection in the ALPI superconducting Linac it is necessary an increase of the charge state from 1^+ to n^+. This is performed by means of a Charge Breeder. The SPES Charge Breeder is based on ECR method [5] and aims to produce ions with A/q less than 7 at $A \approx 130$. The design and construction of the Charge-Breeder will be performed in the framework of the SPES-SPIRAL2 collaboration at LPCS (Grenoble, France) following an up-graded version of the Phoenix booster in operation at LPCS. After the Charge Breeder a second mass separator with 1/1000 mass resolution, will be installed to clean the beam from the contaminants introduced by the Charge Breeder itself. The same basic configuration used for the HRMS is adopted. Finally an RFQ pre-accelerator will increase the beam energy to match the ALPI acceptance. The layout of the radioactive beam line is showed in figure 3.

3.4. Exotic beam reacceleration

The secondary beams are injected into the ALPI superconducting Linac by a new RFQ that will operate in a CW mode (100% duty factor) at a resonant frequency of 80 MHz. This frequency is the same as that of the superconducting structures of the lowest energy section of ALPI. The injection energy of the ions is set to 5.7 keV/u. This choice is a compromise between the desire to reduce the ion energy to simplify the low energy injection and the RFQ bunching section design and the need to increase the beam rigidity in the 1^+ transport line to reduce space charge effects. The extraction energy of the RFQ is set to 727 keV/u to optimize the beam dynamics of the superconducting Linac ALPI. ALPI is the superconducting heavy ion Linac in operation at LNL since the early nineties. It is composed by three sections of quarter wave resonators (QWR) for a total of 80 cavities installed and operates routinely at an equivalent acceleration voltage of 50 MV. The Linac is constructed in a bended configuration: it is composed by two branches connected by an achromatic and isochronous U-bend. It uses three different kinds of cavities: Low Beta, Medium Beta and High Beta cavities, according to the different velocity along the

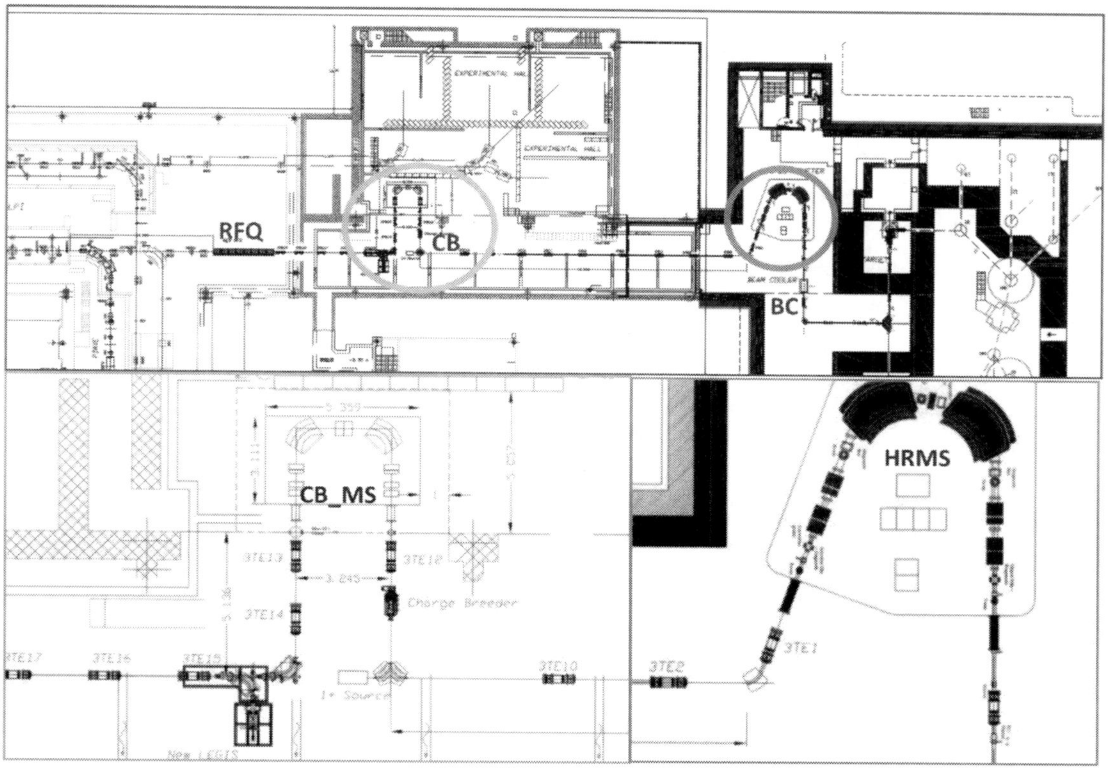

Figure 3. (Color online) Layout of radioactive beam handling and transfer line. High resolution mass separator and the Charge Breeder mass separator are shown in the lower panels.

acceleration path. The ALPI Linac has gone through a continuous upgrade in the number and performance of its accelerating cavities, and consequently in the maximum achievable beam energy. In the last years the cavities of the medium energy QWR section have been upgraded using a new Nb sputtered coating in substitution to the original Pb sputtered layer. A further upgrade program is on the way, to improve the accelerating fields of the present QWRs and adding more cavities in the Low Beta section. The final equivalent voltage will exceed 70 MV going well beyond the Coulomb barrier for nuclear reactions involving mid-A projectiles and heavy targets. In figure 4 the expected beam energies are reported.

3.5. Status of the SPES project

As already mentioned SPES is presently under construction at LNL. The building is at the excavation phase and will be completed at end of 2014. The cyclotron, provided by the BEST company in Canada, is under construction (the magnet is completed and the magnetic field was positively mapped). It is expected to be transferred to LNL, after the factory commissioning, in the second quarter of 2014. The authorizations to operate the cyclotron have been already obtained allowing the acceleration of the proton beam on non-fissile target at high power and testing the UCx target up to a current of 5 μA. A further authorization request to run the UCx target at design power will be presented to the Italian authorities at the end of this year. The physics design of the radioactive ion beam transfer line (1^+ and n^+) is completed and the reshaping of the ALPI beam line in the experimental hall is initiated to allow the

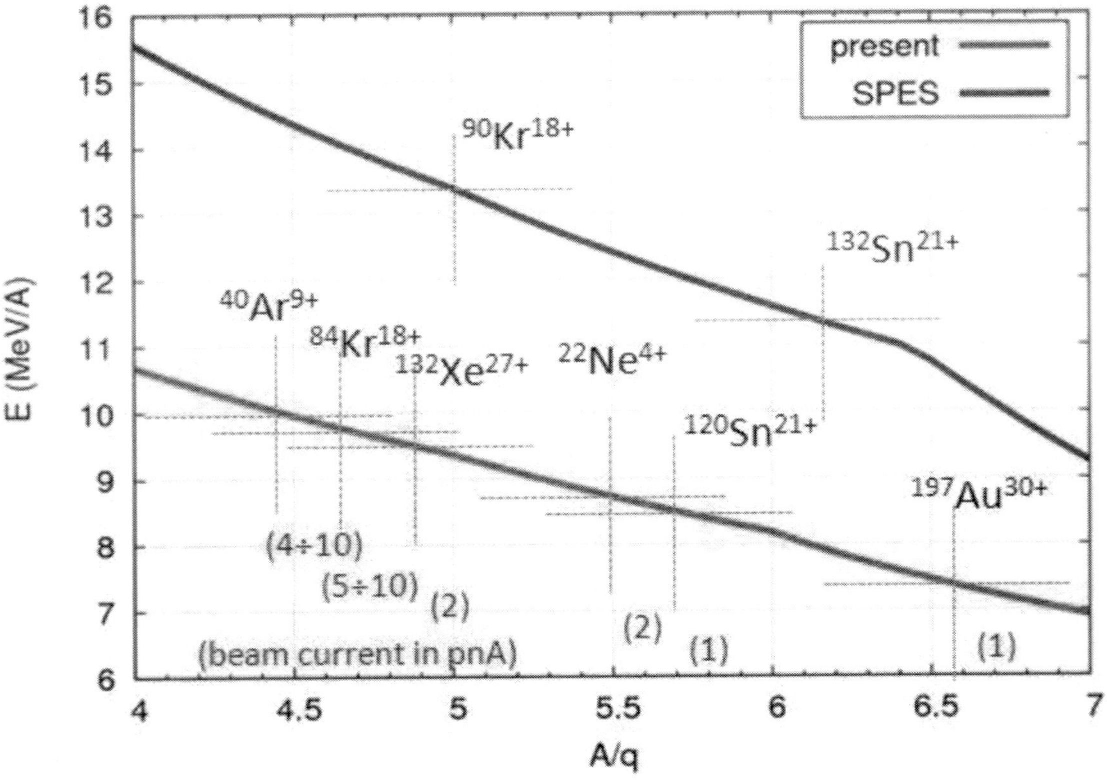

Figure 4. (Color online) Beam energy performance of ALPI superconductive Linac.

installation of the Charge Breeder and of the mass separator. The Isol front-end is completed and under operation in the off-line laboratory to test ion sources, assembly and controls. The first reaccelerated exotic beam is expected around 2018 [6].

4. References

[1] www.eurisol.org
[2] Scarpa D, Biasetto L, Corradetti S, Manzolaro M, Andrighetto A, Carturan S, Prete G, Zanonato P, and Stracener D W 2011 *Eur. Phys. J. A* **47** 32
[3] Lindroos M 2004 CERN-AB-2004-086
[4] Savard G, Pardo R, Baker S, Davids C, Levand A, Peterson D, Phillips D, Sun T, Vondrasek R, Zabransky B, and Zinkann G 2011 *Hyperfine Interactions* **199** 301
[5] Lamy T, Bouly J L, Curdy J C, Sole P, Sortais P, Thuillier T, Vieux-Rochaz J L, and Voulot D 2004 *Rev. Sci. Instrum.* **75** 1624
[6] https://web.infn.it/spes/

Journal of Physics: Conference Series **527** (2014) 012030 doi:10.1088/1742-6596/527/1/012030

Isospin physics by CHIMERA detector

S Pirrone[1], E De Filippo[1], G Politi[1,2], P Russotto[1]
for EXOCHIM Collaboration
[1] INFN-Sezione di Catania, Italy
[2] Università di Catania, Italy
E-mail: sara.pirrone@ct.infn.it

Abstract. This contribution presents some recent results relative to the studies of isospin-asymmetric nuclear matter in experiments using the 4π detector CHIMERA and the beams delivered by the Superconductive Cyclotron at INFN-Laboratori Nazionali del Sud in Catania. In particular, we report the results of the experiments named "Reverse" and "Isodec", both included in the EXOCHIM activity supported by the CNSIII of INFN. In the Isodec experiment, the two systems ^{78}Kr+^{40}Ca (neutron poor) and ^{86}Kr+^{48}Ca (neutron rich) at 10A MeV were studied, looking at the competition between the various disintegration decay paths of medium mass compound nuclei, formed by fusion processes and at the isospin influence on the decay process. The results show the presence of a relaxed component in the reaction mechanism, the evidence of staggering effects in the Z distributions, as well as different isotopic composition and neutron enrichment for the reaction products in the two systems. In the Reverse experiment, mass asymmetric projectile-target combinations ^{124}Sn+^{64}Ni and ^{112}Sn+^{58}Ni were investigated at 35A MeV. We provide evidence that the Dynamical Fission process is about two times more probable than the equilibrated (statistical) Fission one in the neutron rich ^{124}Sn+^{64}Ni system with respect to the ^{112}Sn+^{58}Ni neutron poor one. The observed difference in the strength of the dynamical effects could arise from the difference in the entrance channel Isospin (N/Z) content.

1. Introduction

In recent years, one of the principal aim in the field of heavy ion reactions is to study the influence of the entrance channel N/Z ratio on the reaction mechanism and the fragments formation and production in the exit channel. In this context, particular interest is focused on the heavy-ion reactions at low energy [1] (E/A $<$ 10-15 MeV), where the fusion process is predominant and at intermediate energy, where statistical and dynamical processes in the fragment emission can coexist in a complex way.

At low energy, indeed, one expects that the N/Z ratio, directly connected to the isospin degree of freedom, plays an important role in the compound nucleus decay and emission process, providing crucial information on fundamental nuclear parameters as the level density, the fission barrier and the viscosity. Thus the chemical composition influences the fission dynamics and a program of systematic measurements of fission cross-section for a large isotopic chain of compound nuclei, from neutron rich to neutron-poor, provides careful information on this mechanism. In this work, at section 3, we present the up to date results of the ISODEC experiment, realized to study the competition between the various disintegration modes of ^{118,134}Ba compound nuclei produced in the reactions ^{78}Kr+^{40}Ca (neutron poor system) and ^{86}Kr+^{48}Ca (neutron rich system) at 10A MeV. The investigated systems, making use of stable

Journal of Physics: Conference Series **527** (2014) 012030 doi:10.1088/1742-6596/527/1/012030

Figure 1. The CHIMERA detector at LNS.

beams, allow to produce compound nuclei in a large domain of N/Z (from 1.11 to 1.39) with similar spin distribution and excitation energy. Such a set of data will provide new constraints on sophisticated models attempting to describe statistical and/or dynamical properties [2] of excited nuclei.

At intermediate energy, and in particular in semi-peripheral reactions, the production of Intermediate Mass Fragment (IMF, $Z \geq 3$) and Light Charged Particles (LCP) in the velocity region between the PLF and TLF has been experimentally observed and the features of a dynamical origin have been clearly evidenced in many works (see [3–8] and references therein). In addition, a clear signature of an emission chronology related to the IMF size was demonstrated [3–5, 9, 10]. It was found that lighter IMF ($Z \leq 9$) in the midvelocity region are likely to be emitted during the re-separation process of the primary interacting Projectile and Target Like Fragments (PLF* and TLF* respectively) in prompt neck-rupture mechanism (time $<\sim 120$ fm/c) [3, 4, 6, 9]. Conversely, emission of heavier IMF ($Z \geq 9$) was shown to happen at the late stage of the neck expansion process (time ≥ 120 fm/c), and was associated with the "Dynamical Fission" mechanism [5, 9–12]; this is nearly a two-step sequential reaction: scattering of primary nuclei (PLF*-TLF*) followed by fast non-equilibrated fission-like process. In this work, at section 4, results of cross sections associated to Dynamical and Statistical Fission mechanisms for neutron rich $^{124}Sn+^{64}Ni$ and neutron poor $^{112}Sn+^{58}Ni$ reactions at 35A MeV were evaluated [10]. It was shown that Dynamical Fission process is about two times more probable in the neutron rich $^{124}Sn+^{64}Ni$ system than in the $^{112}Sn+^{58}Ni$ neutron poor one. This sizable difference in the cross section associated to Dynamical Fission gives indication of a strong influence of the entrance channel isospin content on the reaction mechanism.
Both the experiments were performed at the INFN Laboratori Nazionali del Sud (LNS) in Catania by using the 4π array CHIMERA with beams delivered by the Superconductive Cyclotron.

2. Experimental Setup

The presented scientific program needs the measurements of several key observables such as the cross sections, multiplicities, angular and kinetic energy distributions of the reaction products,

Journal of Physics: Conference Series **527** (2014) 012030 doi:10.1088/1742-6596/527/1/012030

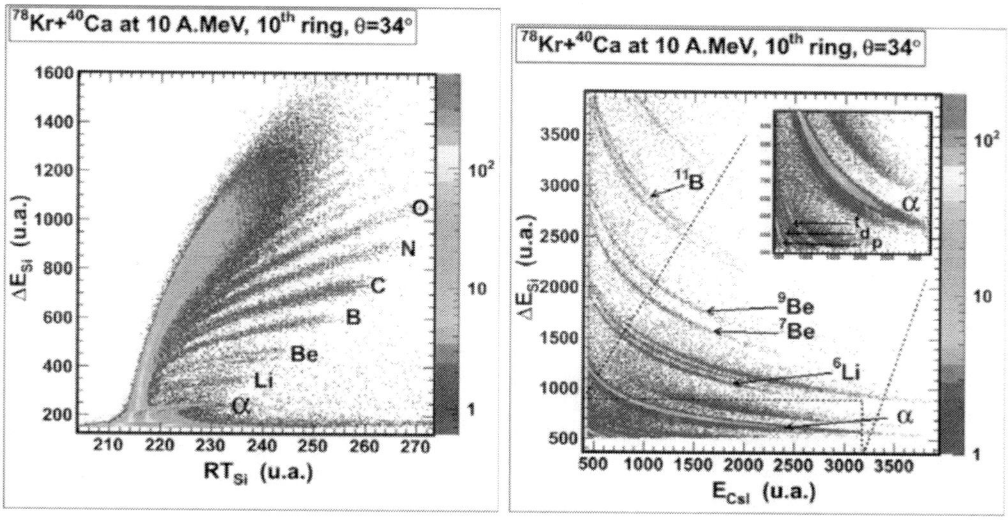

Figure 2. (Color online) Left panel: Energy-RiseTime plot for ^{78}Kr+^{40}Ca; right panel: ΔESi-ECsI(Tl) plot for ^{78}Kr+^{40}Ca.

namely Intermediate Mass Fragments (IMFs), Light Charge Particles (LCPs) and Fission Fragments (FF). The measurement of these observables requires a good isotopic resolution and a low energy thresholds for LCPs and IMFs, a high granularity and a broad angular acceptance. The presence of the second generation 4π multi detector CHIMERA [3], and the availability of beams with very good intensity and timing characteristics, make the LNS in Catania particularly favourable to perform such a program.

Briefly, CHIMERA consists of 1192 detector telescopes, arranged on 9 rings in the forward part and 17 rings, in spherical configuration, in the backward part; the covered polar angle goes from 1° to 176° and the efficiency is the 94% of the total solid angle. In figure 1, a picture of the detector in its experimental scattering chamber at LNS is presented. The single detection cell consists of a silicon detector (Si, thickness about 300 μm) followed by a Caesium Iodine Thallium doped crystal, (CsI(Tl), thickness from 3 cm to 12 cm), coupled to a photodiode. The identification methods employed are: ΔE-E for charge identification of particles punching through the Si detector and stopped in the CsI(Tl) with also mass identification for particles with $Z<10$; E-TOF (Time Of Flight) for mass identification of particle stopped in the Si detector and velocity measurement of all the particles; PSD (Pulse Shape Discrimination) in CsI(Tl), for isotopic identification of more energetic light charge particles; PSD (Pulse Shape Discrimination) in Si detector, for charge identification of the particles stopped in the Silicon detector. This latter technique [13, 14] was the last implementation made on the CHIMERA array and allows the use of this device to study reaction mechanism also in the low energy domain, extending the range of reaction mechanism investigation from low energy fusion to multifragmentation reactions. In figure 2, left panel, is reported as an example the Energy (E_{Si}) vs Rise Time (RT_{Si}) plot obtained by the PSD methods in silicon detector, for the n-poor system ^{78}Kr+^{40}Ca at 10A MeV at ϑ_{lab}=34°.

The CHIMERA multidetector is characterized by a low energetic detection threshold, that is less of 0.5 MeV/A for heavy ions and 1 MeV/A for light particles. These characteristics allowed the complete identification of LCP in a wide energy range, the complete identification in charge and mass of the IMF ($3<Z<8$) products, the charge identification up to Z =14-16 for products

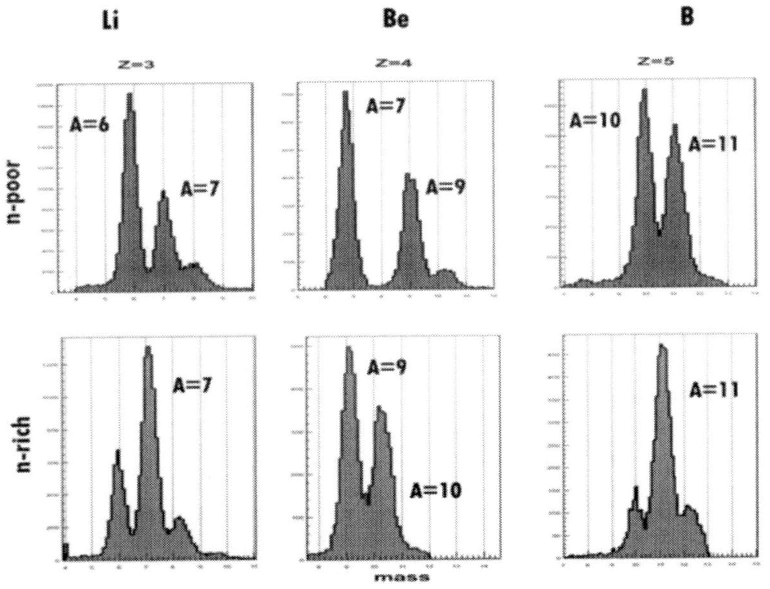

Figure 3. (Color online) Mass distributions for Z=3,4,5, for the n-poor system and n-rich system.

stopped in the Silicon detectors, and up to about Z = 30 for the most energetic particles stopped in CsI. In figure 2, right panel, is shown an example of the typical matrix ΔE-E illustrating the good charge resolution discrimination of the array.

3. Isodec, experimental results

In order to study the influence of the isospin degree of freedom it is essential to get the isotopic identification at least of part of the detected fragments. The capability of the CHIMERA detector allows us to get the mass distributions for the IMF with charge $2 \leq Z \leq 8$. In figure 3 the mass distribution for Z = 3, 4, 5 and for the two studied systems are reported [15–17]. It is clearly seen that, for the same charge Z, in the two systems the isotopic composition and relative neutron enrichment is extremely different. In particular this effect is evident for the Be element where the isotopic composition spans from $A = 7, 9$ for the n-poor system to $A = 9, 10$ for the n-rich one. These differences in the two systems show that a memory of the entrance channel is still present in this class of reaction products; it seems evident the influence of the isospin degree of freedom of the entrance channel on reaction mechanism or in the structure properties of the emitted fragments.

Relevant information on the reaction mechanism can be obtained from the kinetic energy spectra of the detected fragments. In figure 4 the centre of mass energy spectra for Z=12, at laboratory angle $\vartheta = 10.75°, 12.25°, 13.75°, 15.25°$ and for the neutron poor system is shown. One notes the Gaussian-like shape of the spectra and that they are centred at the same energy value, independently from the detection angle. The same result is obtained for the energy spectra of fragments with charge $11 \leq Z \leq 17$ for both the studied systems.

As said before, one of the principal aim of the experiment is the study of the influence of the isospin and the related observables, on the reaction mechanism and on the productions and de-excitation properties of the emitted fragments, as already observed in the Ganil [1] experiment. A

Journal of Physics: Conference Series **527** (2014) 012030 doi:10.1088/1742-6596/527/1/012030

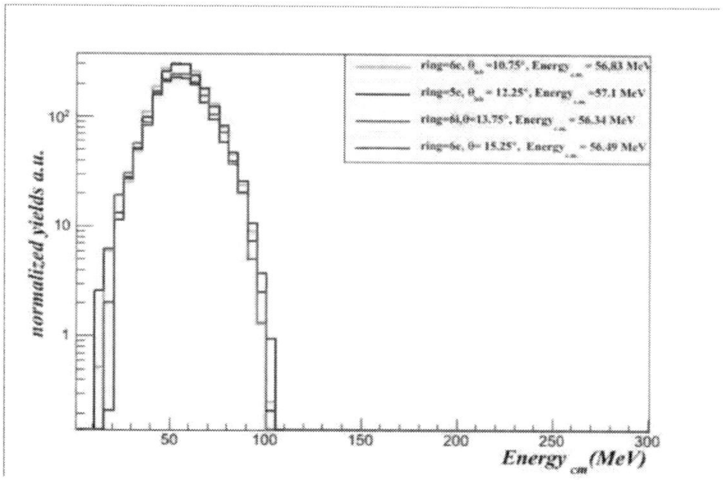

Figure 4. (Color online) Centre of mass Energy spectra for Z=12 at different angles, for the neutron poor system.

preliminary comparative study between the fragments yields of the studied systems is presented in figure 5. The results are obtained by integrating the yields of particles with charge $3 \leq Z \leq 15$, identified by ΔE-E and PSD method on Si detector, in the angular range $10° \leq \vartheta \leq 16°$.

Figure 5 shows a strong odd-even staggering of the Z yield for $Z<10$ detected fragments, and this effect persists for higher Z with a smaller amplitude. Indeed, by comparing the two systems, the yields of the IMFs exhibit an even-odd staggering that is more pronounced for the neutron-poor system. The same effect, by comparing systems with different N/Z neutron to proton ratio, appears also for other reactions in the literature [18]

It seems that the neutron excess in the entrance channel affects the yields of the light fragments; the extraction of the absolute cross section of the reaction products, which is in progress, will confirm this result. Structure effects linked to nuclear pairing forces certainly play a role [19] and their understanding is relevant, for example, for constraining the density dependence of the symmetry energy in asymmetric nuclear matter.

In figure 5 the line show results from preliminary calculations using the dynamical DNS (Di

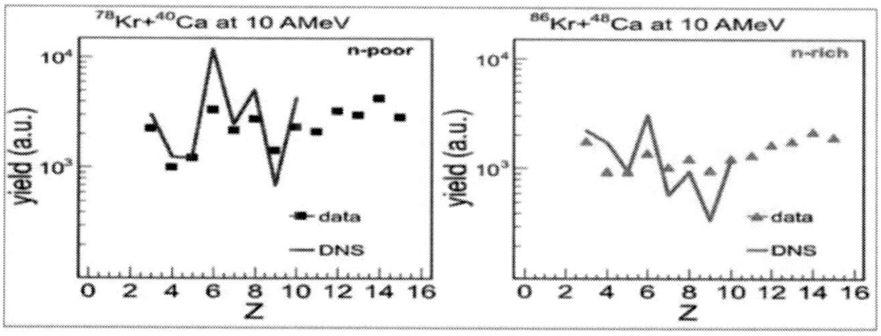

Figure 5. (Color online) Charge distribution for the two studied systems and comparison with preliminary results from DNS-model.

Nuclear System) model, for Z<10. In this model a dinuclear system is formed in the beginning of the reaction, that can go towards a not fully equilibrated fast-fission or towards an equilibrated fissioning compound nucleus. A detailed description of the model can be found in [20]. The calculation, normalized at Z=6, seems to be in a better agreement with the neutron poor system data, where the oscillation of the charge distribution yields staggering is decreasing as the atomic number increases in agreement with the experimental findings.

4. Dynamical Fission, experimental results

For the two reactions ^{124}Sn+^{64}Ni and ^{112}Sn+^{58}Ni studied at 35A MeV, in order to select peripheral collisions, the method of Cavata [21] was used to estimate the impact parameter from the total charged-particle multiplicity. This method assumes a purely geometrical monotonic correlation between the total charged-particle multiplicity of the collision process and the impact parameter. Figure 6 shows the correlation between the experimental total charged particles multiplicity and the estimated reduced impact parameter $b_{red} = b/b_{Max}$ where b_{Max} corresponds to the total geometrical cross section. Note that at a fixed b_{red} the neutron poor system produces a higher multiplicity. In fact, as predicted by calculations [22, 23], in the neutron poor system the dynamical and pre-equilibrium phases lead to more light charged particles (mainly protons) emission than in neutron rich one where, in contrast, neutron emission is favored.

After selecting, in both systems, peripheral collisions with condition $b_{red} > 0.7$, the two heaviest fragments of the chosen subset of events, named as Heavy (H) and Light (L) according to their atomic number Z, were analyzed in those events that satisfy the following conditions: (i) the combined charge of the two selected fragments $Z_{2F} = Z_H + Z_L$ is close to the charge of the projectile ($Z_{proj} = 50$), that is, $37 < Z_{2F} < 57$, and (ii) the Heavy-to-Light-fragment mass ratio is $A_H/A_L < 4.6$, so that the Light fragment has charge $Z_L >\sim 9$. Applying such conditions, the Heavy fragments have the component of the velocity parallel to the beam axis (V^H_{par}) very close to the value of $\sim$ 7.5 cm/ns, that is, slightly below the beam velocity of $\sim$ 8 cm/ns, while the Light fragments have a wider distribution of the parallel velocity, ranging from the velocity of the TLF ($\sim$ 1 cm/ns) up to velocities exceeding the PLF ones. This is shown in figure 7, where, using a logarithmic intensity scale, the V_{par} versus V_{per} Galilean-invariant plots for Light fragments produced in the ^{124}Sn+^{64}Ni reaction are presented, for three ranges of mass asymmetry A_H/A_L (columns), and for three ranges of the total kinetic energy of the two selected

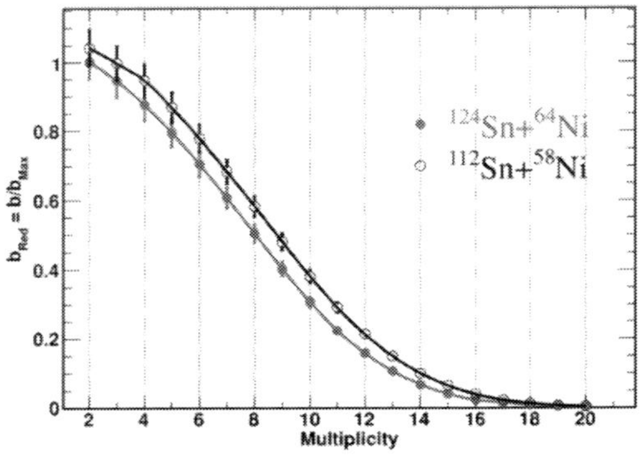

Figure 6. (Color online) Estimated reduced impact parameter bred plotted as function of the experimental total charged particles multiplicity for the ^{124}Sn+^{64}Ni and ^{112}Sn+^{58}Ni reactions at 35A MeV, as obtained by applying the method of Cavata.

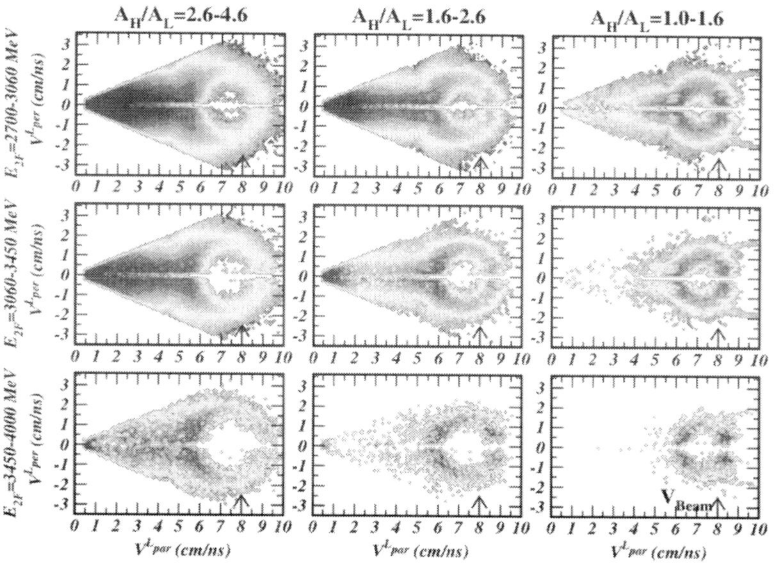

Figure 7. (Color online) Invariant V_{par} versus V_{per} plots for the Light fragments in the ^{124}Sn+^{64}Ni reaction at 35A MeV. Different panels correspond to different values of mass asymmetry A_H/A_L and total kinetic energy $E_{2F} = E_H + E_L$. The distributions are shown in logarithmic scale. The red color corresponds to the largest cross sections and the arrows indicate the beam velocity.

fragments $E_{2F} = E_H + E_L$ (rows). The quantity E_{2F} is a measure of the collision violence: larger values of E_{2F} are associated with more gentle (peripheral) collisions, while more violent collisions are associated with smaller E_{2F} values. In all panels of figure 7 it is possible to observe the characteristic Coulomb rings centered slightly below the beam velocity; the presence of such rings points to PLF* as a well-defined decay source and proves the scenario of two separate reaction steps: first the scattering of the PLF*, followed by its splitting into two fragments (H and L). In almost symmetric divisions after less dissipative collisions [E_{2F}=3450-4000 MeV and A_H/A_L =1.0-1.6], the Light fragments distribution is forward-backward symmetric, that is, the Light fragments have equal probability to be emitted forward or backward in the reference frame of the PLF* source. This result is characteristic of an equilibrated fission, where the nucleus is supposed to be completely equilibrated in all its degrees of freedom; there, the PLF* splitting is expected to occur a long time (1000 fm/c or more) after at least one complete rotation.

In more dissipative collisions and/or more asymmetric splits, the population of the Coulomb ring is no longer forward-backward symmetric. In fact, Light fragments have the tendency to populate preferentially the low-velocity side of the Coulomb ring, which means that they are backward emitted in the PLF* reference frame, that is, toward the TLF*. Nonetheless, a forward-backward symmetric component is still present. Therefore, the observed distributions can be interpreted as a superposition of a forward-backward symmetric component and an asymmetric one. The observed forward-backward asymmetry is the main signature of the Dynamical Fission; that indicates that PLF* fission-like splitting has to be a fast process. After selecting Light fragments populating the Coulomb ring by gating on V^L_{par} (see Ref. [5] for details), in order to disentangle and estimate isotropic and anisotropic fission-like splitting in the two Sn+Ni sytems, the differential cross sections $d\sigma/cos(\theta_{prox})$ have been evaluated. That

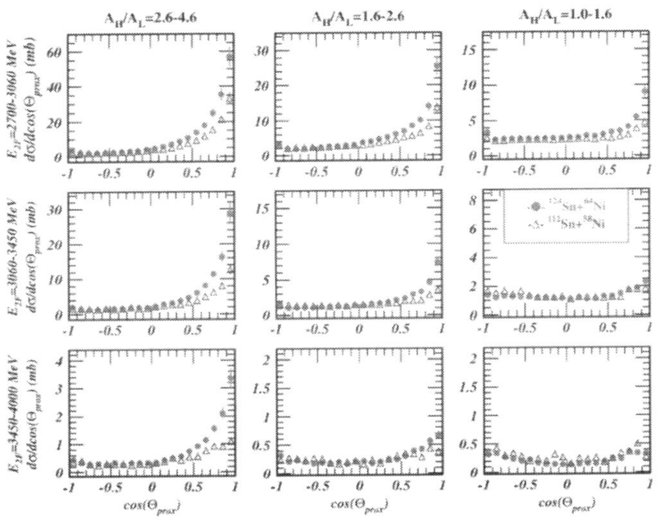

Figure 8. (Color online) $cos(\theta_{prox})$ angular distributions of the PLF break-up fragments for the $^{124}Sn+^{64}Ni$ (circles) and $^{112}Sn+^{58}Ni$ (triangles) at 35A MeV, for three different bins of the total kinetic energy (E_{2F}) and mass asymmetry A_H/A_L.

is shown in figure 8 for the three bins of the mass asymmetry A_H/A_L and three bins of the E_{2F}. The θ_{prox}, defined in left panel of figure 9, is the angle betwen the PLF* flight direction (in the CM reference frame) before fission-like splitting , and the breakup axis, defined by the relative velocity of the two fission-like fragments $\mathbf{V}^H - \mathbf{V}^L$. The value of $cos(\theta_{prox}) = 1$ corresponds to the Heavy fragment moving forward along the PLF* flight direction, while the Light fragment is emitted backward along the PLF*-TLF* direction. In the case of equlibrated

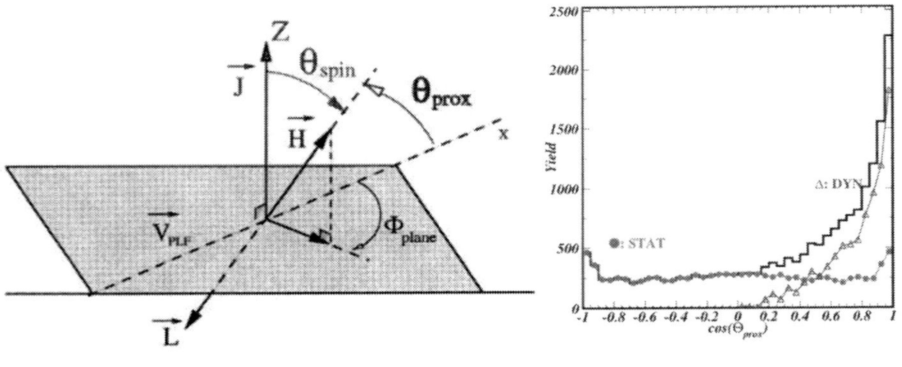

Figure 9. (Color online) Left panel: θ_{prox} definition: angle between the breakup axis, oriented from the light L to the heavy fragment H, and the recoil velocity in the center of mass of the projectile-like fragment (noted VPLF) reconstructed with the two fission fragments, from [12]; right panel: example of extrapolation of Dynamical and Statistical Fission from $cos(\theta_{prox})$ distribution.

splitting, a symmetric ditribution around the value of $cos(\theta_{prox}) = 0$ is expected, as observed in figure 8 for almost symmetric splitting in more peripheral collisions. Instead, by progressively increasing the mass asymmetry and inelasticity (lower E_{2F}), an increase of a forward peaked distribution, superimposed to a symmetric one around the value of $cos(\theta_{prox}) = 0$, is clearly seen: that is associated to the Dynamical Fission. In order to isolate the equilibrated component, a symmetrization around $cos(\theta_{prox}) = 0$ of the backward part, $cos(\theta_{prox}) < 0$, of the distribution has been done, by assuming that the latter is not influenced by Dynamical Fission. The Dynamical contribution is then obtained by subtracting the extrapolated Statistical Fission distribution from the total experimental one. A sketch of this procedure is shown in the right panel of figure 9. It results that the Statistical Fission cross section is approximately the same in the two systems. In contrast the Dynamical component is larger for the neutron rich system by a factor of about 2 as compared with the neutron poor one, as can be found in tables published in Ref. [10]. This sizable difference in the cross section associated to Dynamical Fission indicates a strong influence of the entrance channel on the reaction mechanism. Since the two systems have a large difference in the Isospin content it appears quite natural to link this to an Isospin effect; however, a possible influence of the system size can not be a priori excluded.

In order to disentangle entrance channel Isospin effects from the possible dependence of these results upon the initial different mass of the two systems, in April 2013 we carried out a new experiment, named InKiIsSy (Inverse Kinematic Isobaric Systems) [24], using a projectile/target combination having the same mass as the neutron rich ^{124}Sn+^{64}Ni system and a N/Z near to value of the neutron poor ^{112}Sn+^{58}Ni one, that is ^{124}Xe+^{64}Zn at the same bombarding energy of 35A MeV; the analysy is in progress.

From theoretical side, the Dynamical Fission phenomenon has been studied by transport model simulations. The Stochastic Mean Field simulation (SMF) [25] is a good approach in order to describe the gross features of Dynamical Fission. However, the simulation can hardly follow the complete time evolution of the deformed PLF up to the scission point. Many features of the Dynamical Fission processes could be reproduced in quantum molecular dynamics models like the Constrained Molecular Dynamics (CoMD-II) [9]. The main feature of this last model is a self-consistent N-body approach that solves the equations of motion using procedures to satisfy event-by-event the Pauli principle and the total angular momentum conservation law. Then, we plan to perform future studies using both SMF and CoMD-II models in order to shed light on the observations here reported.

Acknowledgments

The authors wish to acknowledge the colleagues of the EXOCHIM collaboration, with which were realized the experiments and the data analysys presented in this work.

References

[1] Wieleczko J P *et al.* 2008 *AIP Conf. Proc.* **64** 1098; Ademard G *et al.* 2011 *Phys. Rev. C* **83** 054619
[2] Nadtochy PN *et al.* 2007 *Phys. Rev. C* **75** 064614
[3] Pagano A *et al.* 2004 *Nucl. Phys. A* **734** 504
[4] De Filippo E *et al.* 2005 *Phys. Rev. C* **71** 044602
[5] De Filippo E *et al.* 2005 *Phys. Rev. C* **71** 064604
[6] Di Toro M *et al.* 2006 *Eur. Phys. J. A* **30** 65
[7] Piantelli S *et al.* 2007 *Phys. Rev. C* **76** 061601
[8] McIntosh A B *et al.* 2010 *Phys. Rev. C* **81** 034603
[9] Papa A *et al.* 2007 *Phys. Rev. C* **75**, 054616
[10] Russotto P *et al.* 2010 *Phys. Rev. C* **81** 064605
[11] Stefanini A A *et al.* 1995, *Z. Phys. A* **351** 167
[12] Bocage F *et al.* 2000 *Nucl. Phys. A* **676** 391 067604 (2010)
[13] Politi G *et al.* 2006 *IEEE Nuclear Science Symposium Conference Record* **N28** 1140; Alderighi M *et al.* 2005 *IEEE Trans. Nucl. Sci.* **52** 1624

[14] Bassini R *et al.* 2006 *IEEE Nuclear Science Symposium Conference Record* **N14** 173
[15] Pirrone S *et al.* 2011 *EPJ Web of Conferences* **17** 16010
[16] Politi G *et al.* 2012 *EPJ Web of Conferences* **21** 02003
[17] Pirrone S *et al.* 2013 *AIP Conf. Proc.* **7** 1524
[18] Lombardo I *et al.* 2011 *Phys. Rev. C* **84** 024613; Casini G 2012 *Phys.Rev. C* **86** 011602(R)
[19] D'Agostino M *et al.* 2011 *Nucl. Phys. A* **861** 47
[20] Kalandarov Sh A, Adamian G G, Antonenko N V, and Scheid W 2010 *Phys. Rev. C* **82** 044603; Zubov A S, Adamian G G, Antonenko N V, Ivanova S P, and Scheid W 2003 *Phys. Rev. C* **68** 014616
[21] Cavata C *et al.* 1990 *Phys. Rev. C* **42** 1760
[22] Baran V, Colonna M, and Di Toro M 2004 *Nucl. Phys. A* **730** 329
[23] Płaneta R *et al.* 2008 *Phys. Rev. C* **77** 014610
[24] Russotto P, De Filippo E, Pagano A and the EXOCHIM collaboration, InKiIsSy Experiment approved by the INFN-LNS PAC 2012
[25] Baran V *et al.* 2005 *Phys. Rep.* **410** 335

AUTHOR INDEX

AUTHOR INDEX

Institute of Physics
Dirac House, Temple Back
Bristol BS1 6BE UK

CURRAN ASSOCIATES INC
proceedings.com

International Workshop on Beamed and Unbeamed Gamma-Rays from Galaxies 2011

Journal of Physics: Conference Series Volume 355

Muonio, Finland
11-15 April 2011

ISBN: 978-1-62276-772-4
ISSN: 1742-6588

IOP Institute of Physics